The Palgrave Macmillan Animal Ethics Series

Aim of the series
In recent years, there has been a growing interest in the ethics of our treatment of animals. Philosophers have led the way, and now a range of other scholars have followed from historians to social scientists. From being a marginal issue, animals have become an emerging issue in ethics and in multidisciplinary inquiry. This series will explore the challenges that Animal Ethics poses, both conceptually and practically, to traditional understandings of human-animal relations. Specifically, the Series will:

- provide a range of key introductory and advanced texts that map out ethical positions on animals
- publish pioneering work written by new, as well as accomplished, scholars;
- produce texts from a variety of disciplines that are multidisciplinary in character or have multidisciplinary relevance.

More information about this series at
http://www.springer.com/series/14421

Jodey Castricano • Rasmus R. Simonsen
Editors

Critical Perspectives on Veganism

palgrave
macmillan

Editors
Jodey Castricano
The University of British Columbia
Kelowna, British Columbia, Canada

Rasmus R. Simonsen
Copenhagen School of Design and Technology
Copenhagen, Denmark

The Palgrave Macmillan Animal Ethics Series
ISBN 978-3-319-33418-9 ISBN 978-3-319-33419-6 (eBook)
DOI 10.1007/978-3-319-33419-6

Library of Congress Control Number: 2016950059

Printed on acid-free paper

This Palgrave Macmillan imprint is published by Springer Nature
The registered company is Springer International Publishing AG Switzerland

Foreword

The Rise of Veganism

Veganism is an ideology whose time has come. The vegan movement is arguably one of the fastest-growing social justice movements in the world today, and it is likely that this trajectory of growth will even accelerate as the movement gains greater traction.

But why now? Why is it that centuries after the inception of ethical vegetarianism, in just the past decade—and in some regions in the past two to five years—veganism has gone from an unknown, fringe vegetarian submovement to a way of life embraced by some of the world's top celebrities, businesspeople, politicians, and thought leaders? The reasons are, of course, diverse, including everything from shifts in agricultural practices (e.g., the corporatization of agribusiness and the subsequent consumer demand for democratization of, and transparency in, food production) to the obesity epidemic in the USA that has led to a radical reenvisioning of the role of food and nutrition in health and to the realization that animal foods are key causal factors of not only obesity, but of a range of preventable and treatable diseases. So, veganism as an alternative to some of the practical problems caused by animal agriculture has begun to become a realistic, and even commendable, option.

However, it is not only the practical dimensions of veganism that are to account for its current expansion. Perhaps even more important are the psychological, and thus, ethical dimensions of the ideology. And there are two key factors that have led to the shift in attitudes toward veganism: the *visibility* of farmed animal suffering and the *viability* of veganism as a personal and thus moral choice. Thanks to the advent of the Internet and to the efforts of vegan advocates, many people today are aware of the intensive and extensive suffering of farmed animals. And, due to the modernization of food production, unless one is geographically or economically unable to make her or his food choices freely, eating animals is no longer a necessity and is therefore a choice. When a behavior becomes a choice, it takes on a much more significant ethical dimension. Thus, when consumers become aware of the fact that they have a choice of what—or whom—they eat, they must grapple with an ethical dilemma that they didn't have previously.

One can see similar patterns in historical shifts from oppressive to liberatory attitudes and behaviors. It becomes difficult if not impossible to continue justifying the oppression of others (enslaved people of color, women, etc.) when doing so is no longer believed to be a matter of self-preservation. Indeed, virtually all cases for mass oppression rest on the argument that doing so is necessary for the preservation of the dominant group, the social order, and sometimes even the species. And as this argument is increasingly disabled, the oppressive ideology it upholds becomes increasingly challenged by the social justice movement that seeks to replace it.

Although oppressive ideologies are still an unfortunate part of social reality, we can see tremendous shifts of consciousness and genuine progress in transforming them. While sexism is still globally pervasive and deeply problematic, in many parts of the world women enjoy freedoms not even imaginable a 100 years ago. Though racism is no doubt still woven through the fabric of social life in virtually every region of the globe, race relations have been transformed in myriad ways, and there is a sustained, powerful, and highly successful global effort to abolish racial discrimination. When people become aware of oppression and feel empowered to act against it (e.g., they do not feel that their survival is threatened), history shows us that they rise up and say no to injustice and yes to compassion.

What, then, is the oppressive system, or ideology, that veganism challenges? *Carnism* is the ideology that conditions people to eat animals.

It is the counterpoint to veganism, just as patriarchy, for example, is the counterpoint to feminism. And, as with all ideologies, carnism is social and psychological in nature. Understanding carnism helps one understand not only why veganism—the ideology that seeks to, and no doubt one day will, replace carnism—is on the rise. It also helps one understand how to maintain and even bolster the growth of veganism. Exposing carnism for what it is, and demonstrating how veganism is an ethical alternative and imperative, helps ensure that carnism continues to follow the trajectory of other oppressive *isms* and is, eventually, abolished.

Understanding Carnism

Most people born into a prevailing animal-eating culture have inherited a certain paradoxical mentality. They know that the animals they eat are individuals, yet they'd rather not know it. They cringe when confronted with images of animal suffering, yet they dine on animals' bodies several times a day. They refuse to ingest certain animals, yet they thoroughly enjoy eating others. And in so doing, they experience no noticeable inconsistency.

The presence of ambivalent and contradictory attitudes toward eating animals is indicative of carnism. Oppressive ideologies such as carnism require, and enable, rational, humane people to partake in irrational, inhumane practices while failing to notice the inherent contradictions involved. Thus, eating animals is not simply a matter of "personal ethics." Rather, it is the unavoidable consequence of a deeply entrenched, all-encompassing oppressive *ism*. Eating animals is an issue of social justice.

Most people, however, practice carnism unwittingly, as they are unaware that they have a choice when it comes to eating—or not eating—animals. This lack of awareness is the result of being socialized in an environment, in which the practice of eating animals is omnipresent and virtually always uncontested.

And carnism is not only widespread; it is also violent. It is organized around excessive and unnecessary violence toward sentient beings. Even the production of so-called "humane" or "happy" animal products, which form only a tiny percentage of animal foods today, involves brutality in various forms.

Carnistic Defense Mechanisms

Like other oppressive ideologies, carnism runs counter to core human values. Therefore, it needs to disable people's natural empathy toward, and thus compassion for, animals so as to make it possible for them to support unnecessary violence toward nonhuman others without experiencing any moral discomfort in the process. To this end, carnism employs a set of social and psychological defense mechanisms that distort reality and dissociate people, psychologically and emotionally, from their actual experiences. Only then can most people partake in a violent system they most likely would otherwise oppose.

Denial

The main defense of carnism is denial. Denying the existence of an oppressive system implies denying there is a problem in the first place; denying the existence of a problem absolves one from addressing it. Denial finds its expression in invisibility: the ideology itself remains invisible by remaining unnamed, and the victims of carnism are kept out of sight, and thus, conveniently out of mind. Although the body parts of slaughtered animals are essentially everywhere one turns, one hardly ever sees any of these animals alive. However, owing to the excellent work of vegan advocates, as well as the advent of the Internet, denial has been largely destabilized. Denying the existence of at least the most horrendous practices of animal agriculture therefore no longer seems to be a viable option. So, justification—another carnistic defense—has taken on a more central role in sustaining carnism.

Justification

People learn to justify eating animals by learning to believe that the myths of meat, eggs, and dairy are the facts of meat, eggs, and dairy. In one way or another, these myths fall under the *Three Ns of Justification*: eating animals is *normal*, *natural*, and *necessary*. And of course these justifications are anything but new. Throughout human history, they have been used to justify violent behaviors and beliefs (including war, slavery, misogyny, homophobia, etc.) in order to exploit disempowered groups of others. And these myths have been used to discredit progressive movements by depicting progressive ideologies as *ab*normal, *un*natural, and *un*necessary.

Eating animals is normal: The problem with this justification is that what is called "normal" simply reflects the beliefs and behaviors of the dominant culture, the *carnistic norm*. The mere existence of a dominant norm, however, does not justify it.

Eating animals is natural: The problem with this justification is that what is called "natural" simply reflects the dominant culture's interpretation of history. This justification refers not to *human* history, but to *carnistic* history. The reference used is not our early fruit-eating ancestors, but their later flesh-eating descendants. Indeed, many practices that are today considered morally unacceptable, such as infanticide, murder, and rape, are probably as long-standing—and therefore arguably as natural—as eating animals. Yet, no one seriously invokes the longevity of these practices in order to justify them.

Eating animals is necessary: The problem with this justification is that what is called "necessary" simply serves to sustain the dominant culture, the carnistic status quo. Depicting the practice of eating animals as a biological or nutritional necessity demoralizes a fundamentally moral issue. If a diet without animal products were nutritionally deficient, eating animals would pose a much smaller ethical challenge indeed. However, since there is now overwhelming evidence that a vegan diet is not only nutritionally sound but likely even healthier than a carnistic one, people who are economically and geographically able to make their food choices freely cannot logically defend their eating of animals based on the argument that eating animals is necessary.

Still, most people, including social reformers, have not (yet) rejected the Three Ns of carnism. The reason is that carnism is structural, that is to say that it is subtly integrated into the very structure of society and thus represents an institutionalized form of oppression. And institutionalization begets internalization; when an ideology is embraced and maintained by all major social institutions, it becomes internalized, forming an internal psychological system that reflects the external social system. So, even many of those who work toward progressive social change have learned to look at the world through the carnistic lens, and they therefore fail to recognize carnism and its Three Ns of Justification for what they are.

Neocarnism

As denial, the main defense of carnism, has become increasingly destabilized and justification has come to play a more prominent role in maintaining the system, each of the Three Ns has morphed into a new ideology. These new ideologies constitute *neocarnism*. Unlike traditional carnism, in which the consumption of animals is virtually entirely unexamined, neocarnisms incorporate ethical considerations of eating animals into their analyses. So, neocarnisms appeal in particular to consumers who have begun to critically reflect on the validity of eating animals. Unlike veganism, however, neocarnisms do not arrive at the conclusion that the solution is to *stop* eating animals; instead, they recommend changing *the way* one eats animals.

Neocarnism belongs to the category of "secondary carnistic defenses." While "primary carnistic defenses" are pro-carnist, in that they aim to validate carnism, secondary defenses are anti-vegan, in that they aim to *in*validate veganism. Neocarnisms seek to provide carnistic justifications to invalidate veganism primarily by invalidating three central elements of the vegan argument: animal welfare, environmental protection, and human health.

Compassionate Carnism: Invalidating the Animal Welfare Argument

Compassionate carnism developed out of the idea that eating animals is normal; it supposedly addresses animal welfare concerns. While recognizing animal welfare as an issue, compassionate carnism rejects veganism as "extreme" and thus impractical. Instead, compassionate carnism suggests a more practical alternative: eating "humane" or "happy" animal products. So, to solve the dilemma between caring about animals and eating them, compassionate carnism recommends moderation—that one should not stray too far outside the carnistic norm. The problem here is that although compassionate carnism might imply a step toward veganism, often the opposite is true, as eating "humane" animal products tends to soothe one's conscience so that veganism is no longer a necessary alternative. Moreover, in reality, it would seem more difficult to avoid "inhumane" animal products with any consistency than to stop eating animals altogether.

Ecocarnism: Invalidating the Environmental Protection Argument

Ecocarnism developed out of the idea that eating animals is natural; it supposedly addresses environmental concerns. Ecocarnism holds that the problem is not *animal* agriculture, but *industrial* agriculture. The ecocarnist solution is not to stop eating animals, but to only eat "sustainably" produced animal products. Ecocarnism tries to invalidate veganism in several ways. First, it portrays veganism as unnatural and unsustainable, focusing only on those processed vegan specialty foods whose production methods are environmentally problematic. Second, it denounces people's aversion to killing animals as a modern aberration, portraying veganism as a movement of middle-class city-dwellers who are "soft" and "disconnected" from nature.

Both ecocarnist arguments are problematic. First, the fact is overlooked that many vegans do consciously support a sustainable whole foods diet. Besides, a vegan diet is more likely to be sustainable than one that includes animal products, even when one takes manufacturing and transportation of produce into account. Veganism is in fact the solution to plenty of sustainability issues. Second, the question arises as to why modern human sensitivity to killing should be seen as weakness rather than as a sign of moral progress and integrity.

Biocarnism: Invalidating the Human Health Argument

Biocarnism developed out of the idea that eating animals is necessary; it supposedly addresses human health concerns. Biocarnism rejects veganism by assuming that since eating animals is necessary for human health, this practice is exempt from ethical reflection. In so doing, biocarnism refers to medical claims that allegedly demonstrate the unhealthy nature of veganism. It bases the case against veganism on the assumption that "man, the omnivorous hunter," serves as the prototype for the human food consumer. The problem here is that biocarnism invokes *carnistic* history rather than *human* history for evidence of human physiology and nutritional needs; biocarnism looks not to our early fruit-eating ancestors, but to their later flesh-eating descendants, for confirmation of what is necessary for an optimal human diet. Moreover, biocarnism apparently ignores the official positions of various notable institutions, such as the Academy of Nutrition and Dietetics, according to which plant-based diets are nutritionally complete and may even be healthier than animal-based ones.

Cognitive Distortions

Cognitive distortions comprise another set of carnistic defenses. Like other violent ideologies, carnism employs a set of cognitive defenses that aim to distort perceptions. These defenses work as psychological and emotional distancing mechanisms. Accordingly, carnism causes people to see farmed animals as objects, as some*thing* rather than some*one*. It also causes people to see animals as abstractions, as representatives of an abstract group without any individuality or personality and who have often been given numbers rather than names. And, finally, carnism places animals in rigid mental categories in order to enable people to harbor different feelings and behave in different ways toward different species: dogs and cats are friends and family; pigs and cows are food.

Carnism and Intersection

Carnism is just one of the many violent ideologies that are an unfortunate part of the human legacy. And while the experiences of each group of victims, and every individual victim, are always different and somewhat unique, the respective ideologies that cause victimization are structurally similar. Basically, the same mentality grounds these ideologies and enables all forms of violence: the mentality of domination and subjugation, of privilege and oppression; the mentality of might-makes-right; the mentality that justifies oppressing and exploiting vulnerable others simply because they are, after all, "only" savages, women, animals.

Many socially conscious individuals acknowledge the fact that the various, superficially different forms of oppression are in fact intersecting and are thus mutually reinforcing. This insight has important implications: bringing about social reform and transformation requires not merely liberating specific groups of victims, but challenging the very foundations of oppression; it requires getting to the root of the problem. Failing to address the foundation of the problem will inevitably allow for further atrocities in new and different forms, and the abolition of one form of oppression might even reinforce others: yesterday's oppressed can easily become tomorrow's oppressors—a mechanism plainly visible, for instance, when oppressed people demand not to be treated "like animals." Thus, to realize a truly humane and just society, carnism must be included in the analysis of oppressive ideologies. This, however, requires a paradigm shift: the systemic and ideological nature of eating animals needs to be appreciated. Challenging carnism is not simply a matter of encouraging a shift of personal food choices, but it is an integral part of working toward genuine social justice.

Solidarity among those working against oppression is essential. The beneficiaries of oppression often employ a divide-and-conquer approach, pitting oppressed groups against one another to divert attention from the true matter at hand. When possible, those on the receiving end of oppression must attempt to thwart this strategy. All victims of oppression and exploitation ought to appreciate that they are united—not because their respective suffering is identical or somehow comparable, but because their suffering is owed to the very same systemic and institutionalized mentality. The same goes for those actively working to end the various forms of oppression and exploitation. Although one cannot take on all causes, it is crucial to value any cause that is dedicated to creating a more just and compassionate society. Only then may we create a better world for all creatures—human and animal alike.

Food for Thought

If veganism is an ideology whose time has come, then Vegan Studies is a field of research whose time has come, too. As veganism is growing, so, too, is the need for critical reflection on theoretical and practical aspects of this ideology. New possibilities inevitably raise new questions and pose new challenges. Vegan Studies will no doubt play a central role in dealing with this development: it may help to better establish the vegan issue in academia, spread the professional discourse further, and attract new researchers to the field. Beyond academia, Vegan Studies can help to clarify and deepen the understanding of veganism and its theoretical underpinnings and practical implications; and it can help to establish veganism not merely as a fashionable lifestyle, but as an ideology and practice with fundamental ethical, political, and cultural ramifications.

We therefore welcome the publication of this collection of essays, as it promises to make a significant and important contribution to the field of Vegan Studies by critically examining ethical, political, and cultural aspects of veganism in various contexts. This volume includes an array of perspectives and recommendations that help readers see the problem more clearly and approach the solution more dynamically and effectively. The contributors to this volume offer unique, relevant, and important insights as to how to not only address carnism, but also to move beyond carnism toward an ethical vegan practice and psychology. We are thrilled to see how the concept of carnism serves as a starting point for very different fascinating and original approaches. This collection proves to be a helpful tool of analysis in this field of research, helping to illuminate, understand, and explicate various issues related to veganism.

We also appreciate that this anthology draws on insights from various academic fields, thus enabling interdisciplinary exchange, which is indispensable when approaching such a complex topic. And this collection of course provides an excellent overview of the various aspects and debates, highlighting some of the most important and pressing issues such as the broad spectrum of ethical and political positions in vegan discourse, ranging from more principled and ideological ones to more pragmatic and strategic ones; the multiplicity of cultural approaches that can help to illuminate carnism and veganism in various aspects of life; the possible tension between total liberation and single-issue approaches; veganism and commercialization; veganism and its engagement with technological progress (such as in vitro animal products); veganism and intersectionality (interlocking issues of race, gender etc.); veganism and its engagement with religious and traditional practices; vegan identity; and the fundamental question which also underlies this anthology, namely, whether the trend toward normalization strengthens or detracts from the radical impetus of veganism as a politics.

By pulling together the growing body of research being done on veganism and its antithesis, carnism, this collection furthers critical debate and encourages rethinking on how one understands and practices veganism in the twenty-first century. Therefore, this volume can be an invaluable asset for those working in Vegan Studies and for everyone interested in the subject.

Moving beyond carnism and toward veganism will one day, we believe, be looked back upon as one of the greatest transformations in human history. It will be an expression of unparalleled moral, political, and cultural progress.

This anthology makes an important contribution to this end as it reflects an increasing worldwide sensitivity to the devastating effects of our established, dominant patterns of food consumption. And it adds a strong voice to the growing chorus of those calling for fundamental change. We hope it will persuade many people to critically assess and reconsider their attitudes and behaviors in terms of their impact on animals, themselves, other humans, and the planet. So, quite obviously, *Critical Perspectives on Veganism* is an anthology whose time has come.

Melanie Joy and Jens Tuider

Series Editors' Preface

This is a new book series for a new field of inquiry: Animal Ethics.

In recent years, there has been a growing interest in the ethics of our treatment of animals. Philosophers have led the way, and now a range of other scholars have followed from historians to social scientists. From being a marginal issue, animals have become an emerging issue in ethics and in multidisciplinary inquiry.

In addition, a rethink of the status of animals has been fuelled by a range of scientific investigations which have revealed the complexity of animal sentiency, cognition, and awareness. The ethical implications of this new knowledge have yet to be properly evaluated, but it is becoming clear that the old view that animals are mere things, tools, machines, or commodities cannot be sustained ethically.

But it is not only philosophy and science that are putting animals on the agenda. Increasingly, in Europe and the USA, animals are becoming a political issue as political parties vie for the "green" and "animal" vote. In turn, political scientists are beginning to look again at the history of political thought in relation to animals, and historians are beginning to revisit the political history of animal protection.

As animals grow as an issue of importance, so there have been more collaborative academic ventures leading to conference volumes, special journal issues, indeed new academic animal journals as well. Moreover, we have witnessed the growth of academic courses, as well as university posts, in Animal Ethics, Animal Welfare, Animal Rights, Animal Law, Animals and Philosophy, Human–Animal Studies, Critical Animal Studies, Animals and Society, Animals in Literature, Animals and Religion—tangible signs that a new academic discipline is emerging.

"Animal Ethics" is the new term for the academic exploration of the moral status of the nonhuman: an exploration that explicitly involves a focus on what we owe animals morally, and which also helps us to understand the influences—social, legal, cultural, religious, and political—that legitimate animal abuse. This series explores the challenges that Animal Ethics poses, both conceptually and practically, to traditional understandings of human–animal relations.

The series is needed for three reasons: (i) to provide the texts that will service the new university courses on animals; (ii) to support the increasing number of students studying and academics researching in animal-related fields, and (iii) because there is currently no book series that is a focus for multidisciplinary research in the field.

Specifically, the series will:

- provide a range of key introductory and advanced texts that map out ethical positions on animals;
- publish pioneering work written by new, as well as accomplished, scholars, and
- produce texts from a variety of disciplines that are multidisciplinary in character or have multidisciplinary relevance.

The new Palgrave Macmillan Series on Animal Ethics is the result of a unique partnership between Palgrave Macmillan and the Ferrater Mora Oxford Centre for Animal Ethics. The series is an integral part of the mission of the Centre to put animals on the intellectual agenda by facilitating academic research and publication. The series is also a natural complement to one of the Centre's other major projects, the *Journal of Animal Ethics*. The Centre is an independent "think tank" for the advancement of progressive thought about animals, and is the first Centre of its kind in the world. It aims to demonstrate rigorous intellectual enquiry and the highest standards of scholarship. It strives to be a world-class centre of academic excellence in its field.

We invite academics to visit the Centre's website www.oxfordanimalethics.com and to contact us with new book proposals for the series.

Andrew Linzey and Priscilla N. Cohn
General Editors

Contents

Notes on Contributors

Julie Andreyev is an artist, vegan, researcher, and educator. Andreyev's art practice, called Animal Lover, examines human relations with other beings, and explores animal agency and creativity using modes of interspecies collaboration. Furthermore, Andreyev is Associate Professor at Emily Carr University of Art + Design in Vancouver, and Artistic Director of a biannual symposium/exhibition called Interactive Futures. She is co-founder of the relational art group Vegan Congress that holds events intended to develop awareness and compassion for nonhuman beings. Finally, Andreyev is a Joseph Armand Bombardier Scholar completing her Ph.D. research at Simon Fraser University in Vancouver.

Francesco Buscemi is a lecturer in the Faculty of Media and Communication at Bournemouth University and also teaches at the Catholic University of Milan. His Ph.D., obtained at Queen Margaret University, looks at how representations of food in the media support national ideologies in Italy and Britain. Another strand of his research involves meat, cultured meat, and their links to the living animal, death, religion, blood, gender, and the relationships between Nature and Culture. In 2012, Francesco was awarded the Santander Grant Fund for the research *Edible Lies: How Nazi Propaganda Represented Meat to Demonize the Jews*. From a historical perspective, he has also investigated meat representations in the propaganda of the Italian regency of Fiume, Italian Fascism, and the East Germany regime. He has published a book on the Italian film director Liliana Cavani as well as various articles and book chapters on food and media studies. He has also reviewed articles for various refereed and indexed international academic journals, and has presented his studies in many European and American universities. He is currently a member of the Semiotic Society of America.

Jessica Carey is a Professor of Literary and Cultural Studies at Sheridan College in Ontario, Canada. She received her doctorate in 2011 from McMaster University, and her dissertation analyzed cultural echoes of factory-farm practices in contemporary biopolitical discourse. Her ongoing research focuses on the biopolitics of human–animal relationships, food politics, and environmental discourse. She has published in various journals and anthologies on topics that include scientist Temple Grandin's animal-oriented rhetoric, discourses of care in animal cloning, the cultural politics of popular food movements such as "nose-to-tail" eating, and human–dog ecologies in the Canadian novel *Wild Dogs*. Her chapter for this volume is her first foray into her new research on cultural memory and speciesism.

Jodey Castricano is an associate professor in the Faculty of Creative and Critical Studies at the University of British Columbia, Okanagan, and a research fellow in the Oxford Centre for Animal Ethics. Her primary areas of teaching/research are in posthumanist philosophy and critical animal studies with extended work in ecocriticism, ecofeminism, and ecotheory. She is the editor of *Animal Subjects: An Ethical Reader in a Posthuman World* and the contributing co-editor of the forthcoming *Animal Subjects 2.0* (Wilfrid Laurier University Press 2016).

David L. Clark is Professor in the Department of English and Cultural Studies and Associate Member of the Department of Health, Aging and Society at McMaster University, where he teaches Critical Theory, Critical Animal Studies, and Romantic Literature. He has published research on a wide range of subjects, from the question of the animal to the work of Jacques Derrida, and from Kant's late writings to HIV/AIDS. He was George Whalley Visiting Professor in Romanticism at Queen's University in 2012 and Lansdowne Visiting Scholar at the University of Victoria in 2013.

Parag Kumar Deka is a doctoral fellow in the Department of English at Gauhati University, India. He did his Masters in English Literature from Tezpur University. He was awarded the MPhil degree for his work titled *The Body of the Protagonist in J.M. Coetzee's Life & Times of Michael K.* His areas of interest include animal studies, cognitive linguistics, and Assamese literature.

Juawana Grant is a Master's candidate at the University of British Columbia, Okanagan, where she works at the intersections of feminism and critical animal studies. Her research interests include representations of social movements in popular culture, alternative activist media, and radical pedagogy.

A.G. Holdier is currently the program director for Idaho's Minidoka Christian Education Association, as well as an instructor for Colorado Technical University. His research interests lie at the intersection of philosophy, theology, and aesthetics with a particular focus on the ontology of creativity and the function of stories as cultural artifacts. He has published in *The Journal for Cultural and Religious Theory* and contributed chapters to several volumes of Open Court's *Pop Culture and Philosophy* series. Additionally, he has presented at conferences like the Northwest Philosophy Conference and the Rocky Mountain Ethics Congress, among others. He holds an M.A. in the philosophy of religion from Denver Seminary.

Robert C. Jones is currently Associate Professor of Philosophy at California State University. He is also a member of the Advisory Council of the National Museum of Animals and Society, and a speaker with the Northern California Animal Advocacy Coalition. Prof. Jones has published numerous articles and book chapters on animal ethics, animal cognition, food ethics, and research ethics, and has given nearly 40 talks on animal ethics at universities and conferences across the globe. A 2012 Summer Fellow with the Animals & Society Institute, Prof. Jones lives in Chico, a small agriculture community in Northern California, where he spends time arguing about animal rights with local ranchers.

Melanie Joy is a Harvard-educated psychologist, professor of psychology and sociology at the University of Massachusetts, Boston; celebrated speaker; and author of the award-winning book *Why We Love Dogs, Eat Pigs, and Wear Cows,* soon to be published in nine languages and a top book pick by television host Ellen DeGeneres. Dr. Joy was the eighth recipient of the Institute of Jainology's Ahimsa Award (past recipients include the Dalai Lama and Nelson Mandela), which she was presented with in the House of Commons in London. She also received the Empty Cages Prize, presented by Milan city councilors in Italy. Dr. Joy's work has been featured on stations and programs including National Public Radio, PBS, the BBC, Radio Canada, Germany's ARD (the world's second-largest public broadcaster), Luxembourg's RTL (Europe's second-largest media production company), ABC Australia, and Good Morning Croatia. Her work has also been highlighted in publications including *The New York Times*, Canada's *Le Soleil* and *The Huffington Post Quebec*, *Süddheutsche Zeitung* (Germany's largest national subscription daily newspaper), *Spiegel Online*, Luxembourg's *Tageblatt*, Italy's *Di la Repubblica* and *Le Scienze*, Austria's renowned *Der Standard*, Belgium's *De Standaard*, Sweden's *Svenska Dagbladet*, and *Jana*

(Slovenia's leading women's magazine). Dr. Joy has given her critically acclaimed carnism presentation across the USA and in 16 other countries. She is also the author of *Strategic Action for Animals*, and she has written a number of articles on psychology, animal protection, and social justice. Dr. Joy is the founder and president of the Carnism Awareness & Action Network.

Sarah Lingo is a master's student in English at Virginia Tech and studies rhetorical humor and how it can contribute to productive conversations among vegans and non-vegans.

Brittni MacKenzie-Dale earned her B.A. from the University of British Columbia in Creative Writing in 2015. In addition to fiction writing and previous publications in philosophy/religion journalism, she seeks to aggregate her scholastic interests of the nonhuman with creative mediums in hopes of raising timely psychosocial questions.

Jennifer Polish teaches writing at CUNY Queens College and is a Ph.D. student in English at the CUNY Graduate Center, from where she received her Master's degree in Liberal Studies. Her research interests include the intersections of dis/ability, race, and animality in children's literature and media. She has published an article on queerness and dis/ability in group homes for people with intellectual disabilities in *Zeteo: The Journal of Interdisciplinary Writing*. She is currently pursuing the relationship between affective whiteness and dis/ability in composition classrooms. She has taught and written extensively about trauma and dis/ability in *The Hunger Games* and other young adult media, and is currently working on her first novel, a queer young adult fantasy.

Alexis Priestley is a Ph.D. student in Rhetoric and Writing at Virginia Tech; she researches the relationship between food rhetorics, intersectionality, and the ethos of people who speak about food practices in public spaces.

Margaret Robinson is a vegan Mi'kmaq scholar, and a member of the Lennox Island First Nation. Margaret grew up on the Eastern Shore of Nova Scotia, and holds a Ph.D. in Theology from St. Michael's College in Toronto. Her work examines issues of food justice, Indigenous health, two-spirit identity, and cultural continuity. She is a Researcher in Residence in Indigenous Health at the Ontario HIV Treatment Network, and an Affiliate Scientist with the Centre for Addiction & Mental Health in Toronto, Ontario. She currently lives at the corner of Chinatown and Kensington market with her partner of 20 years and four cats.

Jeanette Rowley is in the final year of a Ph.D. research project that examines law in relation to veganism. She has given presentations on veganism in law in the UK; Europe; and, recently, in Australia. She is the UK representative for the International Vegan Rights Alliance and sits on the Academic Research Committee of the United Kingdom Vegan Society. Jeanette is also an Academic Tutor and a Fellow of the United Kingdom Higher Education Academy. Jeanette comes from a three-generation vegan family and is a long-standing vegan animal rights activist.

Peter Royal is a Master's student in English at Virginia Tech who examines scientific rhetoric, especially the use of visuals to represent science to the public. He is also interested in how conversations about health and food in online spaces draw on scientific rhetoric to promote particular lifestyles.

Joshua Schuster is Associate Professor of English at Western University, London, Ontario. His first book is *The Ecology of Modernism: American Environments and Avant-Garde Ethics* (U of Alabama P, 2015). Recent essays on animals and ecology have been published in *Humanimalia*, *Minnesota Review*, and *Photography & Culture*. He is currently working on a new book, *What Is Extinction? A Cultural and Natural History of Last Animals*.

Adam D. Shprintzen is a historian of nineteenth-century and early America at Marywood University in Scranton, Pennsylvania. Dr. Shprintzen's research and pedagogy focus on topics including American reform movements, cultural history, public history, and social history. Dr. Shprintzen's first book, *The Vegetarian Crusade: The Rise of American Reform Movement 1817–1921*, was published by the University of North Carolina Press in 2013, and he is currently editing a collection of primary sources related to interactions between humans and nonhuman animals that will be released in 2017.

Rasmus R. Simonsen is Assistant Lecturer at the Copenhagen School of Design and Technology, where he teaches communication and media courses, drawing on the interplay between semiotic analysis and practical examples from the contemporary media and design landscapes. He is the author of "A Queer Vegan Manifesto," which was translated into Italian and published as a small volume by Ortica Editrice in 2015. Additionally, he has published articles in *ESQ: A Journal of the American Renaissance*, *Leviathan: A Journal of Melville Studies*, *Children's Literature*, *Journal for Critical Animal Studies*, and *American Studies in Scandinavia*.

Michael D. Sloane is a professor at Fanshawe College, London, Ontario. His work focuses on ecological objects in American modernist poetry. Having published on the poetics of contemporary waste, he is now working on an edited book chapter for *Modernism and the Anthropocene* and one for *Modernism in the Green*. His poetry has appeared in *The Goose*.

Jens Tuider is pursuing a Ph.D. in philosophy at the University of Mannheim, Germany, with a focus on animal ethics, and has presented at a number of international events such as the Minding Animals Conference 2012 in Utrecht, the Netherlands, and the International Animal Rights Conference 2013 and 2014 in Luxembourg, and the Tierrechtskongress (Animal Rights Congress) in Vienna, Austria. He has published various academic articles, and in 2012 he won an award from Germany's Gesellschaft für analytische Philosophie (Society for Analytical Philosophy) for his essay on the ethics of eating animals. Tuider is part of an interdisciplinary working group that organizes lectures on the subject of animal ethics, and he runs an online database providing a comprehensive and systematically structured collection of responses to frequently asked questions about vegetarianism, veganism, and animal concerns.

Ophélie Véron a former student at the Ecole Normale Supérieure of Paris and at the University of Oxford, and lecturer at Sciences-Po Paris, completed her Ph.D. in Political Geography at University College London in 2014. Her work focuses on social movements, vegan and environmental activism, and issues of domination and resistance in public space.

Introduction: Food for Thought

Jodey Castricano and Rasmus Rahbek Simonsen

This edited collection examines the ethics, politics, and aesthetics of veganism in contemporary culture and thought. Traditionally located on the margins of western culture, veganism has now been propelled into the mainstream. A Google search for "new vegan restaurants," for example, results in 16,800,000 hits. In recent years, a burgeoning array of vegan cookbooks—*The Joy of Vegan Baking, How it all Vegan!, Sinfully Vegan*—has departed from earlier stereotypes of healthy but ostensibly unappealing vegetarian food to emphasize pleasure and taste as key components

J. Castricano (✉)
The Faculty of Creative and Critical Studies, University of British Columbia, Okanagan, Kelowna, BC, Canada
e-mail: jodey.castricano@ubc.ca

R.R. Simonsen
The Faculty of Design and Business, Copenhagen School of Design and Technology, Copenhagen, Denmark
e-mail: rasi@kea.dk

J. Castricano, R.R. Simonsen (eds.), *Critical Perspectives on Veganism*, The Palgrave Macmillan Animal Ethics Series,
DOI 10.1007/978-3-319-33419-6_1

of a diet free from animal products. Contemporary popular culture offers up humorous representations of vegans such as those in *Scott Pilgrim vs. the World* and David Agranoff's *The Vegan Revolution…with Zombies.* At the same time, concern over the environmental impact of agribusiness and meat consumption, sometimes referred to as "carnism," is leading many omnivores to also consume meat-free meals. A wider range and increased availability of commercial meat and dairy substitutes makes a vegan diet easier to adopt; however, this shift is also consolidating vegans as a new market category. And although People for the Ethical Treatment of Animals (PETA), of course, is responsible for making veganism as "sexy" as any other product that corporate America wants us to buy, ethical veganism is on the rise to the extent that in Ontario, Canada, vegans are seeking legislation to have veganism become a protected human right, and an online feminist nonprofit organization, "Our Hen House," promotes veganism as germane to creating change for animals in a carnist society. In following the focus on veganism in mainstream culture, a central question of this volume is whether the trend toward normalization strengthens or detracts from the radical impetus of veganism as a politics. Despite the potential challenges to veganism as a radical category, animal-free diets and lifestyles proliferate. This collection addresses these cultural shifts and asks how veganism might be rethought and repracticed in the twenty-first century.

Many of the contributors take as their starting point psychologist Melanie Joy's concept of "carnism" to analyze the mounting prevalence of veganism as it appears in different spheres of the culture. Carnism defines the gap in consciousness that draws attention to the culturally specific categorization of some animal species as either "edible" or "inedible" as food objects, while others are allowed to live as pets, for example. As Joy, with Jens Tuider, explains in the foreword to this book, carnism "is the counterpoint to veganism, just as patriarchy, for example, is the counterpoint to feminism". Veganism—and thus, this collection—engages with the intellectual and philosophical gap inherent to carnism to interrogate yet other processes of exclusions that are nevertheless tied to consumption. Contributors to this collection connect veganism to a range of topics—from gender and sexuality and race to legal discourse, and from popular TV shows to the foodie blogosphere—with the shared intention to show how something as basic as food choice continues to impact and matter for cultural, political, and philosophical conversations in the new millennium where *the question of the animal* is inextricable from ethics and from the environmental devastation linked to agribusiness.

As the collection indicates, veganism involves an ethical turn and provides a powerful analytic that opens up the field of food studies to often surprising connections.

Common to the essays collected here is a desire to interrogate and seek out intellectually demanding as well as creative responses to the questions that veganism demands we ask of the different food cultures we inhabit in the western world and beyond. As vegan studies is claiming a place as a field of study in the academy, it is imperative to examine its scope—geographically, aesthetically, politically, philosophically, and culturally. In 2015, the first vegan studies monograph to be published by a university press came to light. Laura Wright's important book *The Vegan Studies Project* discusses the current discourse of veganism specifically in relation to the 9/11 attacks and the subsequent "war on terror." The book's primary focus is "the construction and depiction of the U.S. vegan body" (19). Her study is ubiquitously American, and specific to a personal experience of the 9/11 attacks, which, for the author, are inextricably linked to her decision to become vegan. Wright presents a powerful account of veganism that blends personal reflection with cultural critique.

Food for Thought adds to the ongoing discourse of veganism in the academy, and, at the same time, attempts to present a more encompassing view of vegan studies by considering veganism in a global context. The beauty of an edited collection is that many different voices can be put into conversation (and sometimes dissent) with one another. Our authors come from six countries, spread out over three continents, and an array of critical and cultural backgrounds. It is our hope, then, that the insights collected here about the current state of veganism in society and academia contribute to an increased interest in and probing of veganism as much more than a specialized diet.

To emphasize the inherently multidisciplinary and intersectional scope of vegan studies, we have divided the chapters into the following four sections: (1) Ethics, Politics, and Philosophy; (2) Aesthetics and Representation; (3) Food, Memory, Histories; and (4) New Media Is the Message. The chapters in the first section take a critical look not only at the tradition of thought that subtends the exploitation and consumption of animals but also at the intellectual and ethical strength of veganism itself.

Ethics, Politics, and Philosophy begins with the chapter, "Veganisms," in which Robert Jones criticizes the kind of vegans that tend to focus solely on "the practitioner's self-image, state of mind, and attitude (particularly regarding themselves vis-a-vis non-vegans)" or on the conspicuous consumption of meat alternatives as the basis of activism. In his analysis, he makes the provocative point that "it's probable that a Michael Pollanesque omnivore who has no children, doesn't own a car, rides her bike everywhere, and doesn't travel by plane nor shop at Walmart can have a less-damaging welfare/environmental footprint/hoofprint than a conscientious boycott vegan who produces two children, drives a Prius, often travels by plane, and purchases vegan products at Walmart" (p). To address this, Jones calls for what he names "revisionary political veganism," which is a kind of veganism that takes into account the intersectional violence and oppression of western consumer culture—from the economic and social exploitation of migrant workers to the environmental precariousness of soy burger packaging.

A.G. Holdier takes a different and unusual approach to vegan advocacy, as his chapter, provocatively, makes an *anthropocentric* argument for veganism. Going through the harmful effects of concentrated animal feeding operations (CAFOs), he asserts that "even the most devout speciesist could still conclude, on the sole basis of his or her concern for *homo sapiens*, that a de facto vegan diet is morally obligatory in most Western contexts." Because a key concern for vegan movements is the protection of endangered animals, this chapter argues that the most expedient route to the preservation of would-be slaughter victims lies in an argument based on a premise most carnists already affirm: Human flourishing should generally be promoted. Holdier first presents four areas of human suffering promoted by a carnistic diet before arguing that the simplest course of action to simultaneously undermine each is to adopt a vegan diet—simultaneously, if coincidentally, defending the lives of nonhuman creatures in the process. Ultimately, if the speciesistic attitude of this stopgap measure contributes to the expedited prevention of creaturely slaughter, then vegans could temporarily rest somewhat more comfortably on this pragmatic beachhead, the author concludes.

Jeanette Rowley's chapter similarly works with what we could call a productive anthropocentrism, as it introduces ethical vegan identity to the narrative of suffering in critical human rights scholarship. Arguing that human rights have failed to ameliorate suffering, she shows how the egoism of Kantian autonomy, entrenched in the foundational architecture of human rights, has helped to construct a humanity devoid of care and compassion for others. To challenge this, critical scholars call for a reorientation of the grounding principles of human rights, from Kantian autonomy to Levinasian otherness. The chapter argues that this narrative of suffering implicitly concerns nonhumans and that ethical veganism, already operational in human rights practice, illustrates the sought after transcendental principle of otherness. Modifying the inherent anthropocentrism of human rights discourse, Rowley concludes that ethical veganism brings human rights to a transformational threshold at which otherness and the moral standing of nonhumans are visible.

In chapter 4, "Hegel, Eating: Schelling and the Carnivorous Virility of Philosophy," David L. Clark explores the philosopher Friedrich Schelling's renunciation of G. W. F. Hegel's ideas regarding the voracious nature of *power* that treats everything in its path "merely as 'food' for thought." In this chapter, Clark argues that Schelling places Hegel at a "grim" ontological meal, "a scene of incomparable appetite and predation" that, as Clark shows, "culminates in nothing less than the ingestion of everything that is." Clark draws attention to the question uppermost for Schelling and relevant to veganism and philosophy: "What is 'eating' and what must 'being' be if Hegel's mature philosophy claims to absorb it without remainder."

In the second section of this collection, *Aesthetics and Representation*, the chapters consider the interconnectedness of animal exploitation, consumption, and the construction of art, broadly speaking. Several artists have in recent years dealt with the arbitrary division of nonhuman species that supports an anthropocentric and carnistic worldview. In his contribution to this volume, Michael Sloane examines Matthew Herbert's uncomfortable audio work, *One Pig*, which features tracks made from field recordings of a pig's life from birth to consumption. In telling the story of Herbert's controversial record, Sloane's chapter offers the idea,

"I explore what it means to veganism. Asking what is dangerous about *One Pig*..." In this regard, Sloane's chapter offers the idea of dark veganism, which is a nuanced part of and perspective on veganism that helps one to unpack and understand paradoxical cultural forms of animal activism. Sloane thus presents an in-depth discussion of instrumental intimacies—namely, the fleshy specters, indirect complicity, and the pain–pleasure problematic of *One Pig*. Drawing on Derrida's concept of "hauntology," Sloane's reading of and listening to Herbert's album yields an as of yet unexplored dimension of vegan studies—dark veganism.

Julie Andreyev's chapter, "The Compassion Manifesto: An Ethics for Art + Design and Animals," appears as a kind of companion piece to Sloane's chapter, as she engages directly with ethical questions of reflexivity, intentionality concerning the appropriation of animal plight to create artworks. The manifesto was written for a recent event held by the Vegan Congress, a relational art group that practices mindful action toward eliminating suffering of nonhuman animals, founded by the author and other faculty, staff, and students at Emily Carr University of Art + Design, Simon Fraser University, and the University of British Columbia, Okanagan. The original text was delivered as a public reading on 19 March 2015. Its form is influenced by historical art groups, such as the Situationist International and the surrealists who issued public manifestos as impassioned resistance to oppressive social structures. Challenging the practices of artists like Herbert, the text asks: Should there be limits on artistic expression when other beings are involved? Are there positive modes of depiction or collaboration with nonhuman animals that contribute to compassionate relations with nonhuman worlds?

The two final chapters of this section speak of an attempt to answer these questions and others, as they engage with the fictional writings of J. M. Coetzee and Phillip K. Dick. Parag Kumar Deka's contribution examines how Coetzee's works of fiction are marked by an emphasis on the intrinsic value of life in any form and on the joyfulness of embodied existence. For Coetzee, Deka argues, this joyfulness is to be achieved by foregoing the Cartesian duality of the material body and the disembodied mind. Deka further indicates that, as animals lack this mind–body duality, they live more intensely joyful lives than human beings. This, according to the author, invalidates the contemporary ecological philosophy that places the abstract idea of species above individual animals. Most significantly, as Deka points out, Coetzee stresses the potential role of human sympathy in veganism and animal ethics. He identifies this capacity of sympathy as the primary source of all meat taboos and rituals related to meat eating.

Joshua Schuster's essay provides a discussion of veganism in Philip K. Dick's novel *Do Androids Dream of Electric Sheep?*, placing it in the context of recent debates on how veganism engages with a number of contemporary philosophical, ethical, and ecological crises. Schuster argues that veganism embraces both rational commitment to principled living and openness to unknown and unknowable others. He takes inspiration from the world of the protagonist, Deckard, as well as from its troubled vegans to look closer at some of veganism's own troubles today. By relating Dick's treatment of the ambiguities of animal care, Schuster traces his argument through Derrida's ambiguous statements that connect the deconstruction of animality and sovereignty together. Finally, the chapter takes a brief look at how vegan abolitionism today might connect to a wider consideration of what abolition could entail when ecology and human sovereignty are in question.

The third section, *Food, Memory, Histories*, begins with a historical account of meatless Tuesdays in the USA. Adam Shprintzen's essay shows how, in October 1917, the recently formed United States Food Administration began encouraging American citizens to take part in its Meatless Tuesdays program, aimed at encouraging individuals to pledge to have one meatless meal a week. Ironically, this program sought to save meat that could then be shipped to allied troops abroad. Despite this fact, vegetarians embraced the diet's association with the program. This chapter explores intersections between vegetarianism and meatless Tuesdays, illustrating how a plant-based diet influenced culinary trends associated with the program while simultaneously embracing the government's positioning of the diet as a patriotic act. By analyzing cookbooks, restaurant menus, and newspaper reports, this chapter illustrates indelible links between the vegetarian movement and the government's attempts at patriotic meat abstention, while considering the longer-term implications for vegetarians in America.

Jessica Carey's chapter, "Veganism and the Politics of Nostalgia," also takes on the cookbook genre. Her context is contemporary, however, as she compares mainstream food advertising with selected vegan cookbooks in order to explore different uses of food-related nostalgia. Carey argues that corporate and industry advertising of meat, dairy, and eggs makes intensive use of childhood nostalgia in a way that positions "innocent pleasure" as a bulwark against recent public controversies concerning animal foods. Yet, as vegan cookbooks show, nostalgia can be framed otherwise. To illustrate, the chapter focuses on references to childhood eating in two popular vegan cookbooks by Isa Chandra Moskowitz. As Carey demonstrates, Moskowitz consistently integrates memories of food into her cookbooks, including non-vegan food; however, by emphasizing the pleasurable dimensions of eating as an imaginative and creative continuum between remembered meals and present vegan recipes, Carey argues Moskowitz discursively breaks down the presumed political barrier between pleasure and veganism.

In her contribution, Margaret Robinson turns from the mainstream discourse of veganism and food politics of the two previous chapters to focus on a localized indigenous setting and to analyze the potentially colonizing force of consumer patterns. In her chapter, "Is the Moose Still My Brother if We Don't Eat Him?" Robinson takes a speculative indigenous philosophy approach to explore the tension between new food technologies and traditional indigenous food practices based on, in the case of the Mi'kmaq, a relationship to the moose. Her chapter focuses on how the meaning of animal consumption within Mi'kmaq philosophy, ceremony, or practice might change with the advent of mass-produced cultured meat. Robinson shows that animals perform a significant cultural role for the Mi'kmaq, the indigenous people of North America's northeast coast. Traditionally, the Mi'kmaq diet was high in meat, and animals were framed as relatives whose conscious personal sacrifice enabled the survival of the people. Traditional practices emphasized the need to treat animal bodies with respect, lest the animals become offended and refuse to be caught by hunters. Mi'kmaq hunting, gathering, and eating practices changed considerably in the wake of colonization and the widespread destruction of natural habitats. Yet the recognition of all life as interconnected, expressed in the Mi'kmaq phrase "M'sit No'kmaq," which means

"All My relations," has remained constant. In this context, Robinson's chapter takes a unique turn in that it examines how the contemporary Mi'kmaq relationship to the moose may be explored through a growing knowledge of *epigenetics*, by which eating moose in the future may not require the *death of the moose*. Furthermore, as Robinson asks, how might changes in food production impact how others see Mi'kmaq and how they see themselves as a people? And what relationships are possible for Mi'kmaq communities if their bond with other animals is no longer one of dependence?

Many of the essays collected in this volume trace changes in dietary trends with a view to new technologies. Leading the fourth and final section, *New Media Is the Message*, Ophélie Véron's chapter explores a new configuration of mainstream media, as she looks at French food blogs and the digital dissemination of veganism. She shows how the blogosphere has led veganism to develop far beyond its traditional public of activists and be increasingly accepted in French society. Véron argues that vegan food blogs have played a major role in changing the way vegan food was commonly perceived in France. By revisiting traditional dishes, highlighting the culinary delights offered by vegan cuisine, and presenting it as a healthy and delicious alternative to meat-based food, these blogs have increased awareness of veganism among people outside of their usual readership and have thus helped expand acceptance of veganism in French society. The author goes on to argue that this mounting prevalence of veganism as a new, somewhat fashionable, trend has not weakened the radical impetus of veganism as a politics, but reinforced its impact, as the popularity of vegan food has raised new awareness on animal welfare and animal rights.

In their coauthored piece, Brittni MacKenzie-Dale and Juawana Grant take us into the world of award-winning television programs *The Simpsons* and *Roseanne* and explore how they introduced us to Lisa Simpson and Darlene Conner, two pop icons who brought vegetarian ecofeminism into America's living rooms. This chapter looks closely at the intermingling of their feminist and vegetarian identities, particularly during moments of conflict in their fictional communities. The characters, Lisa and Darlene, barrelled into primetime television, unapologetic about their views, ask their families and two large, loyal fan bases tough questions about carnism.

MacKenzie-Dale and Grant read the characters through the lens of Sara Ahmed's theory of feminist killjoys. As sleeper ecofeminists and prototypical vegans, Lisa and Darlene serve to disrupt both patriarchy and "carnism"—the invisible ideology of meat eating—clueing in viewers to the link between the two. Darlene and Lisa tacitly ask the question: Can you be a feminist and still eat meat?

In the next chapter, "The Carnivorous Mission of the Celebrity Chef," Francesco Buscemi focuses on the support for meat consumption offered by celebrity chefs. The study analyses four chefs ostensibly supporting other health campaigns, but continuing to represent meat eating in a positive light. Buscemi's work draws on meat studies and bio-semiotics to interrogate the separation between nature and culture. The chapter demonstrates that celebrity chefs represent eating meat in order to revive nationalistic and paternalistic elements and values that are perceived to be in decline. In our global and postindustrial times, meat links us to reconstructed versions of the nation, the authentic past, and masculinity; meat consumption, Buscemi argues, is touted as the medicine for the decline of traditional values. His chapter concludes by examining how the focus on meat consumption by celebrity chefs promotes carnism as the means of upholding the nature and culture divide upon which patriarchal masculinity depends.

Taking a semiotic approach as well, Sarah Lingo, Alexis Priestley, and Peter Royal argue in their chapter "The Worst Offense Here is the Misrepresentation: *Thug Kitchen* and Contemporary Vegan Discourse" that as conversations about food, and veganism in particular, become more popular in mainstream American culture, the tropes imbricated in food objects hold increasing political significance. Following Eivind Jacobsen, this chapter explores the "performative consequences" of tropes within popular vegan food discourses. Accordingly, the authors examine the racialized rhetoric of the blog *Thug Kitchen* to unmask the political implications and tangible consequences that extend beyond the act of consuming food for sustenance. While *Thug Kitchen* potentially encourages veganism through its use of humour, this chapter importantly concludes that the blog's rhetoric perpetuates harmful assumptions about class, race, and food access.

Although veganism can be a powerful optic for recognizing and analyzing exclusionary processes of ontological, ethical, and aesthetic import, veganism itself might in some instances reproduce inequalities. In her chapter, "Decolonizing Veganism: On Resisting Vegan Whiteness and Racism," Jennifer Polish also provides a discussion of *Thug Kitchen* to demonstrate that whiteness and mainstream vegan praxis in the USA is undeniably linked: from the gentrification driven by vegan advertising and food practices to PETA's uncritical equation of black people with factory-farmed animals and the literary blackface of website and recipe book *Thug Kitchen,* veganism is often understandably maligned as being too influenced by whiteness to matter to people of color. By centralizing the work of scholars of color like Sylvia Wynter and A. Breeze Harper, the final chapter of the collection explores the intimate relationship between racism and speciesism, ultimately arguing that only attentiveness to these interconnections can decenter whiteness and dismantle both anthropocentrist and racist violence.

As the editors of food for thought and long-time vegans, the work that has gone into putting this collection together has been particularly rewarding. Going vegan today has become a lot easier than it was ten years ago in terms of food product availability. In Denmark, the native land of one of the editors, it was hard to find soy milk in grocery stores a decade ago, but, now, the country, one of the world's largest producers of bacon, is experiencing a fast-growing popularity of veganism; recently, a major Danish news outlet even covered this development in exceedingly positive terms. But with mainstream approval comes a new set of concerns. The commercialization of veganism is expanding, and, invariably, the main focus is on food goods. Even though we recognize dietary vegans have contributed to the increase of food options, we understand veganism as being about more than what we put in our mouths. For this reason, we end this introduction and begin this volume by calling attention to what Joy and Tuider refer to as "an increasing worldwide sensitivity," which heralds nothing less than a paradigm shift in relation to the more-than human world. In this spirit, we invite the reader to reflect upon a call for change that takes us beyond carnism.

Part I

Ethics, Politics & Philosophy

Veganisms

Robert C. Jones

Introduction

Those of us living in affluent consumer culture under late capitalism, where plant-based alternatives to meat and dairy are readily available, are morally obligated to adopt vegan practice. The source of this obligation is grounded in a widely held belief, namely, that—all else being equal—unnecessary suffering and premature death are bad things and that acting with relatively minimal cost to oneself to contribute to

A substantial portion of the content of this chapter comes from Lori Gruen and Robert C. Jones, "Veganism as an Aspiration" in *The Moral Complexities of Eating Meat*, eds. Ben Bramble and Bob Fischer (UK: Oxford University Press, 2015), 153–171. I would like to thank Lori Gruen for her encouragement to expand on that essay in the writing of this current chapter. I also benefited greatly from conversations with Gunnar Eggertsson and Mark Balaguer, as well as members of the Chico Animal Rights Education Series (CARES). Thanks also to Annie Chen for helpful editing suggestions.

R.C. Jones (✉)
Philosophy, CSU Chico, Chico, CA, USA
e-mail: rcjones@mail.csuchico.edu

J. Castricano, R.R. Simonsen (eds.), *Critical Perspectives on Veganism*,
The Palgrave Macmillan Animal Ethics Series,
DOI 10.1007/978-3-319-33419-6_2

a decrease in violence, objectification, domination, exploitation, and oppression is something we should all aspire to.[1] When I say that we[2] are obligated to adopt vegan practice, not just any type of "vegan practice" will do, so I want to argue for a specific type of veganism I call *political veganism*. I will do that toward the end of this chapter since I first want to establish that it is morally wrong for the vast majority of us living in high-income, highly industrialized, consumer cultures—such as the majority of us living in the Global North—to consume animal[3] products.

To clarify, the argument is not an argument for some kind of universal veganism; that is, I will not argue that every human being on the planet is morally obligated to become vegan. Not because I don't believe it—I do—but because (a) the question of whether an indigenous Inuit subsistence hunter must stop consuming all animal products is complicated and not my focus in this chapter, and (b) I prefer to focus my argument on those of us living in Western societies.

Instead, I offer what I call a "localized version" of the argument for veganism. That is, it's an argument that applies only locally, not universally, the scope of which is directed, as I said, toward those of us living in consumer culture under late capitalism where plant-based alternatives to meat and dairy are readily available. Central to the argument is the claim that both factory farming and so-called "humane" farming are morally problematic. However, describing the treatment of nonhuman animals used in food production and why these practices are unnecessary and immoral is not the focus of this chapter. Knockdown arguments for why both methods of animal farming are morally wrong are successfully made elsewhere.[4] Instead, I want to focus on a challenge generated by the central premise of the argument for veganism, a challenge that, *prima facie*, threatens to undermine the obligation to embrace veganism for those who believe that going vegan decreases the suffering and death of sentient beings (which, I imagine, is the reason why a majority of ethical vegans go vegan in the first place). In answering this objection, I discuss a number of solutions that I believe, jointly, meet the challenge. Finally, I explore various kinds of veganisms and advocate for political veganism. But for now, let's first have a look at the localized argument for veganism.

An Argument for Ethical Veganism[5]

1. It is wrong to cause suffering and/or premature death unless there is good enough reason.[6]
2. The production of animal products causes animals suffering and/or premature death.
3. Consumption of animal products increases the production of animal products.
4. With minimal hardship (if any), a vast majority of those of us living in high-income, highly industrialized, consumer cultures (such as those of us living in the Global North) can flourish without consuming animal products.
5. A vast majority of those described in (4) consume animal products not because such products are physiologically or nutritionally necessary but for convenience or taste preference.
6. Convenience or the satisfaction of taste preference are not good enough reasons to justify the harm that the consumption of animal products causes to animals.
7. Therefore, it is morally wrong to consume animal products (1–6).
8. Therefore, a vast majority of those living in high-income, highly industrialized, consumer cultures ought to stop consuming animal products.

The argument is pretty straightforward and compelling (to me, at least), but the argument for veganism faces what some see as a serious factual and conceptual challenge to the central premise of the argument, namely (3), consumption of animal products increases the production of animal products.[7]

A Puzzle About Ethical Veganism: The Causal Impotence Objection

To argue that the raising and commodification of other-than-human animals for consumption is morally bad is one thing; to argue that individual consumers ought not to purchase animal products is quite another. The reason being that there's a bit of a puzzle—located in (3) of the argument—that few vegans address.

Recall premise (3) of the argument: Consumption of animal products increases the production of animal products. The relation implicit in (3) is a *causal* relation. That is, the idea behind (3) is that my consuming ("consuming" in the sense of my acting as a market *consumer*) animal products creates demand for animal products, and thus, (indirectly) *causes* an increase in the production of animal products, and thus, an increase in animal suffering. The assumption behind ethical veganism—and most likely the central reason why a vast majority of vegans go vegan in the first place—is that going vegan decreases animal suffering. By going vegan, according to the argument, you somehow contribute *directly* to decreasing suffering on both small ranches and factory farms.

I recently ate at the Southern California vegan fast-food chain Native Foods, where, after ordering at the counter, I was handed a placard with my order number on it. The placard read, "Crispy Battered Native Chicken Wings: One order saves three chickens!" What exactly does this mean? It can't mean that there are three chickens somewhere who are waiting to be slaughtered on a factory farm whose lives are spared when I order the Crispy Battered Native Chicken Wings.

Maybe what the placard means then is something like this: Three chickens won't be born, won't come into existence, and won't suffer the horrible lives and deaths of factory-farmed chickens because I order the Crispy Battered Native Chicken Wings. But how exactly does that work? It can't mean that I thwart the plan of some egg producer, who is waiting for the thumbs-up to hatch another three chickens, by ordering this particular vegan dish. Besides, you cannot save a nonexistent being, so it can't mean that. I think the charitable read is something like this: When consumers—as a *group*—order Crispy Battered Native Chicken Wings instead of actual chicken wings, the demand for chicken decreases, causing the chicken market to produce fewer chickens. Translating this market decrease into number of chickens actually "saved," and dividing by the number of consumers who order the Crispy Battered Native Chicken Wings, you get the average number of chickens that each individual consumer saves when ordering the Crispy Battered Native Chicken Wings, which in this case is three. But is that really what is intended by the claim on the placard? And even if it is, is it all really that simple? The answer seems to be no.

Critics argue that this kind of linear causal story connecting individual consumer choice to changes in market supply gets the real-world facts all wrong. Markets like the chicken market are too massive to be sensitive to the purchasing behaviors of any single consumer. And since the overwhelming majority of individual consumers have nothing at all to do directly with agribusiness, or the raising or killing of "livestock," an individual consumer's choice to refrain from the purchase or consumption of animal products makes *no difference at all* in decreasing the number of animals suffering and dying on factory farms. This is known as the *causal impotence objection* to ethical veganism.

One might object on the grounds that this kind of challenge is too abstract and that it's obvious that purchasing meat causes animal suffering and death; hence, annoying "hypothetical" puzzles like this should be dismissed out of hand as so much philosophical sophistry. However, it would be too fast a dismissal.

First, it's easy to imagine someone in the real world reasoning in the following way: Whether or not I order the chicken won't change anything. Regardless of what I do, the ag industry will do what it's going to do and the animal rights movement will do whatever it's going to do, so what I do makes no difference. So I guess I'll just order the chicken.

Second, it's certainly true that, (a) *collectively*, consumers of animal products (e.g., meat eaters) cause harm to animals.[8] However, from the truth of (a), it does not follow that (b) a particular consumer of animal products causes harm to animals. An inference from (a) to (b) would be fallacious since it's possible for (a) to be true while (b) is false.[9] But let's look more closely at the claim made by the Native Foods placard. It would seem that such claims make a kind of simplistic assumption, namely, that supply is sensitive to demand. But imagine the following case. I decide to prepare chicken for dinner, so I head to my local supermarket and purchase a frozen chicken. As Robert Bass points out, this purchase has "no effect on the killing, packaging, freezing and shipping of that chicken a week or two earlier…the decision weeks earlier to raise a certain number of broilers from eggs, or the decision months or years earlier to operate the chicken house where the chicken spent her life. Nothing I do brings it about that one chicken more or less is raised for food."[10]

At this point, you might think that my purchasing that one chicken reflects an increase in demand for chicken, and that an increase in demand will lead to a *future* increase in supply, and thus, one more chicken will be slaughtered as a result of my purchase. But you would be wrong for a few reasons. First, supermarkets order more chickens than they expect to sell since waste and spoilage are built into the ordering process. Second, supermarkets in particular, and agribusiness more generally, are so huge that the chicken market is insensitive to individual consumer decisions. But if this is so, why is it wrong for individuals to purchase or consume animal products such as frozen chicken? Just how responsible are we in causing suffering and harm to other animals when we consume their bodies produced in the industrialized system, and what difference might we make as individuals? It seems that individual consumers are powerless *qua* individuals to cause change in such an enormous market. If so, then (3) is false, and individual vegans make virtually no difference whatsoever in decreasing animal suffering; therefore, ethical vegans who believe that their individual purchases have direct causal efficacy on the lives of nonhuman animals (as the Native Foods placard suggests) are, at best, confused. This causal impotence objection stands as a challenge to the obligation for individuals to go vegan.

Solutions

Vegans have a number of responses to the causal impotence objection. I want to discuss just a few solutions and argue that these solutions can, in concert, answer the challenge.

The first response is simply to deny the claim of causal impotency and ask: How can an individual make no difference if *together* we make a difference? If collective action has causal impact (which it does), then at least *some* individual instances must have causal impact. Collective action is not some kind of spooky "metaphysical" occurrence, but a combination of individual actions that can each have a variety of impacts. Though seemingly imperceptible, there is nonetheless some impact (albeit, very small) that, when combined with the very small impacts of other consumers, results in causal effect.

For example, it may be that my action serves as a trigger or threshold. Suppose that the butcher only makes a call to order more chickens when the 100th chicken breast is purchased, or the poultry industry only reduces production when a threshold of 10,000 people stop purchasing chicken. It may seem that if you are not the one who purchases the 100th chicken breast or are not the 10,000th person who gave up chicken products, your refraining from such purchases makes no difference. However, your refraining affects the timing of slaughter or the cessation of slaughter. This is an impact, even if it is not a direct impact on any particular individual. So buying or not buying animal bodies does make a difference. Further, no matter what the causal impact of your refraining from consuming animal products, what is certain is that your not going vegan is practically certain to delay any threshold event happening, and therefore, practically certain to result in excess animal suffering.[11]

A second response revolves around the notion of role modeling. Many involved in vegan practice influence others who, in turn, influence others, and so on. This kind of role modeling may be understood as a species of the broader phenomenon of *social contagion* in which an action of a particular type makes another action of that type more likely. Thus, veganism can increase the probability that others become vegan, which increases the probability that the collective action of the aggregate more quickly brings about a reduction in the number of animals produced for food and other consumer goods, decreasing animal suffering and bringing about a decrease in violence, exploitation, and domination.[12]

With regard to private actions like eating leftover chicken when no one else is around (or will ever witness or even find out about), doing so may actually increase the chance that one may, in the future, eat more chicken. Veganism urges us to conceptualize chicken or pig bodies, for example, as "not food," much the way many in Western cultures think of dog bodies as "not food."[13] As people begin to view the corpses of others as inedible, the probability that they will want to consume "leftover" bodies is lowered. Someone aspiring to be the kind of person who acts to minimize suffering and oppression will thus adopt strategies that will stabilize their ability to act on their values and refrain from consuming animal products even in the case of private consumption.[14]

A third response expands on a notion mentioned above, namely, conceptualizing animal bodies as food. Lori Gruen[15] argues persuasively that the very act of ontologizing animals as food, of putting animals in the category of the edible, strips them of their individual personalities and interests. Animals have interests beyond not suffering that matter—for example, being allowed to live their lives with their family members and not being killed simply to satisfy someone else's culinary desires. Being categorized as edible, in industrial societies, renders these beings as disposable and consumable. When we place nonhuman animals in the category of "things," commodities to be bought and sold, we change both the relationships we have with them and how we think of those relationships. As humans, we understand ourselves as not in the category of the edible, and this understanding, in part, shapes how we construct our relations with each other and the ways of life we share. If we were instead to think of our bodies and other people's bodies as food, the value of our bodies and ourselves changes.

In response, it might be argued that since both human and nonhuman animals are, in fact, consumable, the problem is not that we ontologize animals as food, but that we ontologize animals as meat. Val Plumwood[16] argues that while "meat" represents reductionism, domination, alienation, and commodification, "food" suggests an acknowledgment of our ecological selves. But Plumwood conflates the fact that we are all consumable with the fact that we categorize some bodies as "edible" and others as "nonedible." That humans could be consumed as prey in certain contexts is distinct from the social categorization of certain others as edible. To be vegan is not to deny ecological entanglement, but to suggest a reconceptualization of animals in their living bodies as fellow creatures with whom we can be in empathetic relationship and for whom we must have deeper respect.[17]

A fourth response involves not the notion of individual causal efficacy, but relies instead on the notion of complicity, group causation, and group function. Elizabeth Harman,[18] for example, argues that actions can sometimes be morally wrong, not because they make any difference to the amount of suffering in the world, but because they involve a kind of joint causation, which is neither necessary nor sufficient for an effect. As per Harman's view, one need not make a causal difference to have good reason to refrain from participating in collective wrongs. For example, it is wrong to participate in group bullying even if your joining the bullying makes no difference to how badly the victim is hurt. For example, imagine a case where the bullying victim is so upset that he is not paying attention to who is actually verbally bullying him, and so it would make no difference whether you join the bullying. In other words, the causal story for the harm perpetrated upon the bullied is overdetermined by the number of the bullying group. For Harman, though refraining from individual acts of meat consumption may have little or no effect on decreasing animal suffering, it may still be wrong to (a) participate as a joint cause in such a collective wrong, or (b) fail to participate as a joint cause in a collective good. Thus, Harman identifies two moral reasons for individuals to adopt vegan practice independent of whether doing so has any direct or indirect causal effect on decreasing animal suffering. By consuming animal products, one is (a) participating as a *joint cause* in practices that cause animal suffering and/or premature death and (b) failing to participate in a movement that can do a lot of moral good.[19]

Expanding on the notion of complicity and group function, Adrienne Martin[20] argues that even if adopting vegan practice makes no causal difference to decreasing animal suffering, not doing so makes the consumer complicit in animal suffering in that the consumer shares responsibility for the direct harms perpetrated by meat, egg, and dairy producers. Martin's notion of complicity hinges on the notion of role-taking and group function. For Martin, individual consumers of animal products are complicit in the harm and suffering experienced by animals not because they contribute directly or indirectly to such harm, but because they willingly participate as members of a consumer group that has the function of signaling demand. According to Martin, such a collectivized liability account of responsibility is eminently plausible:

> Everyone who voluntarily joins [in the bullying] thereby participates in a cooperative project aimed at making the victim suffer, and it is surely right that each individual participant is thereby liable to be blamed for the victim's suffering, even if the suffering would be just as bad if the ringleader (say) were the only tormentor. What matters here is not whether there is some chance that an individual will make a difference to the suffering, or even that each

> individual is a joint cause. What matters is that the individual willingly adopts the role of participant in a group, knowing or at least suspecting that the group has the function of making its victim suffer. The individual is thereby liable to be blamed for what other group members do qua participants, including succeeding in the group's purpose; this liability stands even if the individual does not actually contribute to the victim's suffering.[21]

Likewise, the non-vegan willingly adopts a role as a participant in a *consumer group* that one knows (or ought to know) serves a function of signaling increased demand to meat, egg, and dairy producers. And in this way, consuming animal products makes one complicit in the animals' suffering and/or premature death. Conversely, in order not to be complicit, one must (at the very least) refrain from the consumption of animal products, regardless of whether such refraining is causally efficacious in reducing animal suffering and/or death.

Though none of these responses individually provides a knockdown rejoinder to the causal impotence objection to veganism, it is clear that taken as a group they do in fact adequately provide a rational basis to adopt vegan practice. Just what the vegan practice should look like is the focus of the following section.

Veganisms

Clearly, there are strong grounds for accepting the localized argument for veganism. As a rational and effective response to hierarchical, systemic, speciesist human violence perpetrated against nonhuman animals, veganism plays an indispensable role in dissolving such violence. However, there are a number of ways of conceiving of veganism.

Identity Veganism

In adopting vegan practice, a number of ethical vegans see veganism primarily as an *individual lifestyle choice*, an expression of their commitment to decreasing (and ultimately ending) the suffering and death that accompanies the commodification of sentient nonhuman beings.

Since many ethical vegans may believe (wrongly) that no animals are harmed in the production of their vegan consumer goods and foodstuffs, this ethical vegan "lifestyle" may sometimes be accompanied by a sense of ethical purity, a belief that once one adopts a vegan lifestyle, one then has "clean hands" and may carry on one's consumerism with a clear conscience. Seen as a kind of litmus test of one's commitment to social justice for animals, veganism may sometimes be thought as the "moral baseline" for those seeking to end the suffering and domination of other-than-human animals. Though there are debates among vegans about questions of purity and commitment, there appears to be a growing public perception of vegans, a kind of *vegaphobia*[22]—which may be based in fact, prejudice, or more likely a combination of both—that vegans see themselves as better than and morally superior to non-vegans; that they may sometimes appear to be "preachy"; and that they may exhibit a kind of self-righteous zealotry, acting as the "vegan police" who promulgate veganism as the universal, one-and-only way to fight systemic violence against animals.[23] It was perhaps proponents of identity veganism that prompted philosopher Val Plumwood to describe vegans as

> crusading [and]…aggressively ethnocentric, dismissing alternative and indigenous food practices and wisdom and demanding universal adherence to a western urban model of vegan practice in which human predation figures basically as a new version of original sin, going on to supplement this by a culturally familiar methodology of dispensing excuses and exemptions for those too frail to reach their exacting moral norms of carnivorous self.[24]

Such vegans are sometimes perceived—rightly or wrongly—as judging non-vegans (including ovo-lacto vegetarians) as shirking their responsibility or being self-indulgent or simply cruel.[25] This view, that the only ethical way to live is to adopt a vegan lifestyle, is called *identity veganism* by Gruen and Jones.[26] What distinguishes identity veganism from other kinds of veganism is that identity veganism is more about the practitioner's self-image, state of mind, and attitude (particularly regarding themselves vis-à-vis non-vegans) than about consumer behavior. Though, *qua* consumers, the behavior of identity vegans may be indistinguishable from that of other types of vegans, it is a kind of deluded self-righteousness of some identity vegans that distinguishes them from other kinds of vegans. If followed strictly and universally, identity veganism is thought to confer clean hands and a clean conscience. As the name implies, this sort of veganism is often thought of as an identity, or individual lifestyle choice, and is sometimes characterized—again, rightly or wrongly—as exuding an air of moral certitude and superiority.[27]

However, there are at least two reasons why identity veganism is not a kind of veganism to be endorsed. First, identity veganism is, at best, naïve and Pollyannaish and, at worst, a way to insulate oneself from a terribly inconvenient truth. For in late-capitalist consumer culture, even vegans cannot escape the cycle of state-supported, systemic, industrialized violence and destruction of animals and their habitats. Vegan or not, we all have blood on our hands. Try as they might to believe otherwise, identity vegans must face the fact that regarding our contributions to the objectification of animals and the consumption of animal products, there is no "moral sainthood."[28]

Second, since the central focus of identity veganism practice is the rejection of and abstention from the consumption of *nonhuman animal* products, identity vegans may fail to attend to the lives of other sentient beings who may suffer to produce their consumer goods—specifically, *human* sentient beings. For example, workers of the Global South exploited to produce identity vegans' nonanimal product–containing consumer goods may not be considered in the equation relating personal consumer choice with a reduction or elimination of suffering. Neither may identity vegans dedicate their practice to the environmental costs of their vegan consumerism. The circumstances driving their "clean hands" self-image may exclude damage to habitat that the production of vegan foodstuffs may (and often do) incur. This discussion leads me to the next kind of veganism I wish to address, boycott veganism.

Boycott Veganism

Like identity veganism, the guiding principle behind *boycott veganism*[29] is a rejection of the purchase and consumption of all animal products with less (or no) consideration for the human or even environmental costs. Yet, *unlike* identity vegans, boycott vegans may very well accept that a by-product of the web of production of even vegan foodstuffs may involve the harming of individual sentient nonhuman animals. However, as identity veganism is not to be endorsed, neither is boycott veganism.

First, boycott veganism (like identity veganism) sees vegan practice as a kind of individual lifestyle choice, ignoring the larger social, cultural, economic, and political contexts in which systematized, institutional violence, suffering, exploitation, domination, objectification, and commodification of both human and nonhuman animals are required to produce consumer goods of all kinds, including "vegan" consumer goods. As Jenkins and Stanescu make powerfully clear:

> [B]oycott veganism conflates conspicuous consumption with ethical action and political change....Simply replacing animal with plant-based products only transfers capital to global corporations through different mechanisms; boycott veganism serves to reinforce capitalist institutions and neoliberal social structures that promote the commodification of life and disguise market forces as neutral, amoral means of distributing social goods.[30]

Some identity and boycott vegans (e.g., "Taco Bell vegans") either tacitly or actively condone the continued existence of the very same exploitative consumer-capitalist structures that produce things such as the milk found in milk chocolate (which they refuse to consume), and the cacao produced using child slave labor (which they may willingly or perhaps unknowingly consume),[31] or palm oil, a ubiquitous ingredient found in a large number of prepared "vegan" foods produced by clear-cutting which devastates endangered (and non-endangered) species' habitats.[32]

Second, by reducing veganism to individual consumer choices, boycott vegans unwittingly reinforce the belief that by "voting" with your vegan dollars you can make real moral progress and effect political change, leaving the exploitation of human and nonhuman animals and the unprecedented catastrophic global destruction of the natural environment and animal habitats to the will of consumer-capitalist markets.[33]

Importantly, boycott vegans fail to acknowledge that a vegan lifestyle, particularly in the Global North, can be an environmentally high-impact lifestyle. For example, the packaging from vegan food doesn't take up less space in the landfill or consume fewer resources just because the food is vegan.[34] Additionally, boycott vegans overlook the fact that in terms of net suffering, harm, and destruction, being a high-consuming vegan can, in some contexts, be more damaging than being a meat eater.[35] With regard to behaviors that most impact global climate change, much attention is paid to the ways that people's home energy use, travel, food choices, and other routine activities affect their emissions of carbon dioxide (CO_2) and, ultimately, their contribution to global warming. However, the reproductive choices of an individual are rarely incorporated into calculations of their personal impact on the environment. Yet, the extra emissions of fossil CO_2 that an average individual causes when he or she chooses to have children far exceed the lifetime emissions produced by the original parent. In the USA, for example, each child adds about 9,292 tons of CO_2 to the carbon legacy of an average female, 5.7 times her own lifetime emissions. Even more startling is the fact that the potential savings from reduced reproduction are huge compared to the savings that can be achieved by changes in lifestyle. For example, a woman in the USA who adopted six basic, nonreproductive lifestyle changes would save about 486 metric tons of CO_2 emissions during her lifetime, but, if she were to have two children, this would eventually add nearly 40 times that amount of CO_2—an astonishing 18,584 metric tons—to the earth's atmosphere.[36] Again, my point here is that boycott vegans may overlook the fact that in terms of net suffering, harm, and destruction, being a high-consuming vegan can, in some contexts, be more damaging than being a meat eater. It's probable that a Michael Pollanesque omnivore who has no children, doesn't own a car, rides her bike everywhere, and doesn't travel by plane nor shop at Walmart can have a less-damaging welfare and environmental footprint (or hoofprint) than a conscientious boycott vegan who produces two children, drives a Prius, often travels by plane, and purchases vegan products at Walmart. Additionally, it is the poor and impoverished of the Global South who will take the brunt of climate change in the coming decades.[37] Clearly, a different kind of vegan practice is called for, one that engages with, rather than ignores, the global devastation to which even a vegan practice can contribute.

Revisionary Political Veganism

The kind of veganism that I advocate I call revisionary political veganism (or just political veganism, for short). Political veganism has three virtues: It is (a) revisionary, (b) aspirational, and (c) intersectional and inclusionary.[38]

Political veganism is in part a *revisionary* project in that it calls for a rejection of the conventional concept of veganism as an individual lifestyle or consumer choice. Political veganism reappropriates the term "vegan" to include a moral and political commitment to active resistance against institutional and systemic violence, exploitation, domination, objectification, and commodification directed against all sentient beings—human and nonhuman—as well as the natural environment that supports and sustains them.[39] In this sense, veganism becomes a kind of stance—an anticonsumer-capitalist stance—toward economic and political structures of violence and oppression.

Political veganism—in fact, *all* veganisms—can be only *aspirational.* The belief that abstaining from animal products allows one to avoid complicity in harming other animals ignores the complex dynamics involved in the production of consumer goods of all kinds, global entanglements we engage with each time we purchase and consume food of all sorts. Living today, even for vegans, involves participating in the deaths of sentient individuals. Vegan diets have welfare footprints in the form of widespread indirect harms to animals, harms often overlooked or obscured by advocates of identity and boycott veganism. Though vegans have attended to the tragedy that farmed animals experience, few pay much attention to the harms other animals suffer in the production of vegan foodstuffs. For example, the raising and harvesting of crops in industrialized agriculture harms and kills a large number of sentient field animals such as mice, voles, rabbits, and birds in the production of fruits, vegetables, and grains produced for human (not livestock) consumption.[40] Even if some vegans can practice "veganic" farming, few of us can afford to create food in this way.

All aspects of consumption in late capitalism involve harming others, human and nonhuman. When we live with companion animals, other animals will have to die, most obviously to feed those animals. But even if they are vegan, dogs and cats will kill and eat other animals if they get a chance.[41] While neither ignoring nor resigning oneself to these realities, as political vegans we acknowledge our complicity in these institutional vices, while doing the best we can to minimize them. Not to do so would be bad faith.[42] Political veganism commits us to striving for a moral goal, as something one works at rather than something one is. Of course, there is overlap between identity and boycott vegans on the one hand and political vegans on the other. In different contexts, someone who recognizes that veganism can be but an aspiration may also express her commitments in ways that make it seem more like a lifestyle.

However, to see that veganism is only aspirational is not to see veganism as *merely* an aspiration. To call oneself a political vegan while continuing consciously and without necessity to act in ways that condone animal exploitation (e.g., proclaiming your aspirations to vegan commitments while ordering a cheeseburger at your favorite fast-food restaurant) would be to disingenuously appropriate the language of veganism and, again, be inauthentic and act in bad faith. Despite wanting it to be otherwise, vegan or not, we cannot live and avoid killing, even if only indirectly. Given all this, veganism can be but an aspiration, and imagining otherwise is an illusion. Political veganism incorporates this fact into practice, imagining and earnestly trying to actualize—to the best of one's ability—a world in which there is no violence, exploitation, or oppression, while working at the individual, political, cultural, and structural levels to reduce harm and foster a vegan world, while fully recognizing that, even as vegans, we are complicit in this cycle of violence.

Finally, the greatest virtue of political veganism is that it is intersectional and inclusionary. Political veganism acknowledges the connections between and rejects the structures of oppression—such as human exceptionalism, speciesism, racism, sexism, ableism, and militarism—while emphasizing the relationships between the consumption of animal products and environmental destruction. Thus, political vegans reject the notion of a meat-eating environmentalist, feminist, or queer advocate. Such binaries are not aligned with the goals of dismantling speciesism and eradicating the commodification and consumption of nonhuman animals.[43]

Political veganism is wide in scope and limited not only to a rejection of the consumption of animal products but also to a rejection of the structures and institutions that link the commodification and exploitation of animals, vulnerable human populations, and the environment. Thus, political veganism is not a personal dietary lifestyle choice, but rather an active and engaged worldview dedicated to an inclusion of nonhuman animals in social justice theory and practice.[44]

Political veganism acknowledges the link between structural violence and exploitation, and the consumer-capitalist structures that drive demand for vegan foodstuffs and other "vegan" consumer goods. These include the experiences and sufferings of nonhuman animals and human workers in slaughterhouses, the trafficking and slavery of farmworkers who grow and pack vegan foodstuffs, and the impoverishment of Bangladeshi children who are beaten and forced to work 20-hour shifts, 7 days a week, for pennies to produce clothing containing no animal products for retailers such as Walmart—to point out just a few.[45] Political vegans also recognize the role that state-sponsored subsidies of agribusiness play in the dietary racism that results when such subsidies make available high-fat, cheap, animal-based foods in impoverished neighborhoods truly in need of healthful, whole, plant-based foods. Given that political veganism can be but aspirational, sincere political vegans do their best to decrease their contribution to global suffering by actively opposing these industries and the fetishizing of commodity consumer culture.[46]

If taken seriously, political veganism has some interesting—if not counterintuitive—consequences. For example, on the one hand, someone in the Global North with disposable income who eats an exclusively plant-based diet solely for reasons of personal health or who abstains from eating animal products out of concern for "animal rights" but who purchases "vegan" (e.g., non-leather) consumer goods from Walmart while cognizant of the conditions under which those kinds of items are produced would not be vegan in the sense that I am characterizing political veganism. Conversely, I can imagine a "fellow traveler" who earnestly and sincerely aspires to political veganism, but who lacks the resources, income, or employment (e.g., a freegan and perhaps a poor, vulnerable single parent) and "dumpster dives," or in some other way chooses to take in animal bodies or their by-products for sustenance, who could constitute a political vegan in the sense that I am articulating. Rather than seeing these seemingly odd consequences as a deficiency, they instead act to highlight the virtues of political veganism, illustrating why political veganism is both revisionary and aspirational.

Conclusion

I have argued that those of us living in affluent consumer culture under late capitalism, where plant-based alternatives to meat and dairy are readily available, are ethically obligated to adopt vegan practice. I have provided an argument for localized veganism and answered a number of objections to it. Further, I have identified a number of veganisms and advocated for one, namely, political veganism. I ultimately argue that vegans are obliged to actively engage with and resist those power structures built on speciesism, violence, oppression, exploitation, domination, objectification, and commodification of all sentient beings—human and nonhuman—and their habitats. I see political veganism not merely as a theoretical construct, but as a call to action and engagement by those of us in the Global North to retreat from our destructive consumer-capitalist ontologies and use our privilege to reduce and ultimately eliminate suffering, while forging moral and just relations of care, compassion, and respect.

Notes

1. I present here an ethical argument for veganism. The argument for veganism is even more compelling when we consider the environmental argument for veganism (i.e., the horrendous environmental destruction caused by the industrial–grain–oilseed–livestock complex. See Tony Weiss, *The Ecological Hoofprint: The Global Burden of Industrial Livestock* (London: Zed Books, 2013).

2. I am assuming that the target audience of this collection of essays is those of us living in the affluent Global North in consumer culture under late capitalism. That's who I mean here by "we."
3. The term "animal" is fraught and troublesome. The use of the term acts only to reinforce *human exceptionalism*, a paradigm of division and oppression that perpetuates the dangerous and misguided notion that those sentient beings, commonly referred to as "human beings"—who are normatively and operationally interpreted as metaphysically distinct from and morally superior to so-called "animals"—are outside and "above" membership in the "animal kingdom," a distinction that has served the interests of the dominant species at the expense of those oppressed species. However, rather than entirely repudiating this linguistic convention in this chapter, for ease of the reader I will instead use the terms "animals," "nonhuman animals," and "other-than-human animals" to refer to so-called nonhuman "animals."
4. See, for example, James Rachels, "The Basic Argument for Vegetarianism" in *Food for Thought: The Debate Over Eating Meat*, ed. Steve F. Sapontzis (Amherst, NY: Prometheus Books, 2004), 70–80, Dan Imhoff, *The CAFO Reader: The Tragedy of Industrial Animal Factories*, (Berkeley: Watershed Media, an imprint of U California Press, 2010), Mark Hawthorne, *Bleating Hearts: The Hidden World of Animal Suffering*, (UK: Changemakers Books, 2013), and Hope Bohanec, *The Ultimate Betrayal: Is There Happy Meat?*, (iUniverse, 2013).
5. The basic infrastructure of this kind of argument is found in Mylan Engel, "Why YOU Are Committed to the Immorality of Eating Meat" in *Social and Personal Ethics*, 4th edition, ed. William Shaw (Belmont, CA: Wadsworth Publishing Company, 2002), 212–21, and James Rachels, "The Basic Argument for Vegetarianism" in *Food for Thought: The Debate Over Eating Meat*, ed. Steve F. Sapontzis (Amherst, NY: Prometheus Books, 2004), 70–80.
6. Within the context of the relation of speciesism to the treatment of nonhuman animals, reasons that are "good enough" are varied and, some believe, not uncontroversial. I'm not interested in those debates as I believe the notion as I am using it is pretty straightforward. All I need to run this argument is for you to agree that—*ceteris paribus*—satisfying my taste or desire for bacon is not good enough reason to slaughter a pig. There are many arguments against this conclusion, but none that I know of that are not speciesist.

7. See Robert Bass, "What Can One Person Do? Causal Impotence and Dietary Choice," unpublished manuscript (2014), for a clear and thorough treatment of this challenge to vegan practice, namely, the *casual impotence objection.*
8. By "products," in the term *consumers of animal products*, I intend the products of animals—such as meat, eggs, and dairy—produced by the livestock industry. I say this because it may be possible—at least, in theory—to produce animal products that do not cause harm to animals (e.g., eggs from rescued hens, who are loved, protected, and well cared for), but these examples are so extremely rare and are such a minuscule fraction of all animal products produced and consumed as to be negligible.
9. To conclude that a *particular individual* causes x because consumers *as a whole* cause x is an error of logical scope. Bass in Robert Bass, "What Can One Person Do? Causal Impotence and Dietary Choice," unpublished manuscript (2014), sees this as an instance of the fallacy of division. Either way, it's a bad inference.
10. Robert Bass, "What Can One Person Do? Causal Impotence and Dietary Choice," unpublished manuscript (2014), 3.
11. Alastair Norcross, "Puppies, Pigs, and People: Eating Meat and Marginal Cases," *Philosophical Perspectives* 18, no. 1 (2004), 232.
12. Ben Almassi, "The Consequences of Individual Consumption: A Defence of Threshold Arguments For Vegetarianism and Consumer Ethics," *Journal of Applied Philosophy*, 28, no. 4 (2011), 404–7.
13. See Melanie Joy, *Why We Love Dogs, Eat Pigs, and Wear Cows: An Introduction to Carnism* (San Francisco: Conari Press, 2011) for a nice book-length discussion of these kinds of cases and the speciesist and carnist worldviews underlying them.
14. Interestingly, when we consider that role-modeling behavior can have both positive and negative aspects and recognize that some "negatively contagious" actions (so-called "backfire" role modeling) can affect others' behavior such that it increases the probability that an observer will engage in behaviors *opposite* to the role modeler, we have evidence that perceptions of vegans as self-righteous zealots may very well push non-vegans *away* from veganism and toward meat consumption.
15. Lori Gruen, *Ethics and Animals: An Introduction*, (UK: Cambridge University Press, 2011), 101–4.

16. Val Plumwood, "Integrating Ethical Frameworks for Animals, Humans, and Nature: A Critical Feminist Eco-Socialist Analysis," *Ethics and the Environment* (2000), 296.
17. Lori Gruen, *Entangled Empathy: An Alternative Ethic for Our Relationships with Animals*, (New York: Lantern Press, 2015).
18. Elizabeth Harman, "Eating Meat as a Morally Permissible Moral Mistake" in *Philosophy Comes to Dinner: Arguments on the Ethics of Eating*, eds. Andrew Chignell, Terence Cuneo, and Matthew Halteman (NY: Routledge, 2015) 215–31.
19. Ibid.
20. Adrienne M. Martin, "Consumer Complicity in Factory Farming" in *Philosophy Comes to Dinner: Arguments on the Ethics of Eating*, eds. Andrew Chignell, Terence Cuneo, and Matthew Halteman (NY: Routledge, 2015) 203–14.
21. Ibid., 210.
22. Matthew Cole and Karen Morgan, "Vegaphobia: Derogatory Discourses of Veganism and the Reproduction of Speciesism in UK National Newspapers," *The British Journal of Sociology* 62, no. 1 (2011), 134–53.
23. See Lori Gruen, "Facing Death and Practicing Grief" in *Ecofeminism: Feminist Intersections with Other Animals and the Earth*, eds. Carol J. Adams and Lori Gruen (New York: Bloomsbury, 2014) 127–41, and Matthew Cole and Karen Morgan, "Vegaphobia: Derogatory Discourses of Veganism and the Reproduction of Speciesism in UK National Newspapers," *The British Journal of Sociology* 62, no. 1 (2011), 134–53.
24. Val Plumwood, "Integrating Ethical Frameworks for Animals, Humans, and Nature: A Critical Feminist Eco-Socialist Analysis," *Ethics and the Environment* (2000), 286.
25. A parody of this kind of vegan has even found its way into popular culture via *The Simpsons*. In the episode "Lisa the Tree Hugger" (S4E12), Lisa's earnest proclamation to environmentalist and animal rights activist Jesse that she is a vegetarian is met with a chuckle as Jesse condescends to Lisa, "I'm a level 5 vegan. I won't eat anything that casts a shadow."
26. Lori Gruen and Robert C. Jones, "Veganism as an Aspiration" in *The Moral Complexities of Eating Meat*, eds. Ben Bramble and Bob Fischer (UK: Oxford University Press, 2015), 153–171.
27. It was pointed out to me that to avoid any potential vagueness (given that this section is rife with "many" and "some,") I should point the reader to one of these identity vegans. And so, if I must, I will. I have in mind someone like vegan superstar Gary Yourofsky.

28. See Susan Wolf, "Moral Saints" *The Journal of Philosophy* 79, no. 8 (1982), 419–39, for a brilliant and pointed rejection of moral sainthood as a virtue.
29. "Boycott veganism" as I am characterizing it, is—to my knowledge—introduced, and sharply dissected in Stephanie Jenkins and Vasile Stănescu's, "One Struggle," in *Defining Critical Animal Studies: An Intersectional Social Justice Approach for Liberation*, eds. Anthony J. Nocella II, John Sorenson, Kim Socha, and Atsuko Matsuoka (NY: Peter Lang Publishing, 2014), 74–85. At least that's the first place that I came upon the term.
30. Stephanie Jenkins and Vasile Stănescu, "One Struggle," in *Defining Critical Animal Studies: An Intersectional Social Justice Approach for Liberation*, eds. Anthony J. Nocella II, John Sorenson, Kim Socha, and Atsuko Matsuoka (NY: Peter Lang Publishing, 2014), 78.
31. Lauren Ornelas's Food Empowerment Project (F.E.P.) is a model organization encouraging consumers to recognize the connection between food choices and animal abuse, the depletion of natural resources, unfair working conditions for produce workers, and the unavailability of healthy foods in low-income areas. At their website http://www.foodispower.org/, you will find the F.E.P. list of slave chocolate producers, many of whom produce "vegan" chocolate products.
32. Mark Hawthorne, "The Problem with Palm Oil," *VegNews*, (March 22, 2013), retrieved January 15, 2016 at http://vegnews.com/articles/page.do?pageId=5795&catId=1.
33. Stephanie Jenkins and Vasile Stănescu, "One Struggle," in *Defining Critical Animal Studies: An Intersectional Social Justice Approach for Liberation*, eds. Anthony J. Nocella II, John Sorenson, Kim Socha, and Atsuko Matsuoka (NY: Peter Lang Publishing, 2014), 74–85.
34. Warren Oakes, "Why freegan?" Freegan.Info, retrieved January 15, 2016, from http://freegan.info/what-is-a-freegan/freegan-philosophy/why-freegan-an-attack-on-consumption-in-defense-of-donuts/.
35. Thanks to Patrick Newman for pointing out this criticism of boycott veganism.
36. Paul A. Murtaugh and Michael G. Schlax, "Reproduction and the Carbon Legacies of Individuals," Global Environmental Change 19, no. 1 (2009), 14–20.
37. Vicente R. Barros, Christopher B. Field, et al., eds., "Climate Change 2014: Impacts, Adaptation, and Vulnerability, Part B: Regional Aspects," *IPCC*, retrieved January 15, 2016, from https://ipcc-wg2.gov/AR5/images/uploads/WGIIAR5-PartB_FINAL.pdf.

38. Political veganism integrates the central ideas found in *engaged veganism* (Stephanie Jenkins and Vasile Stănescu, "One Struggle," in *Defining Critical Animal Studies: An Intersectional Social Justice Approach for Liberation*, eds. Anthony J. Nocella II, John Sorenson, Kim Socha, and Atsuko Matsuoka (NY: Peter Lang Publishing, 2014), 74–85.), *aspirational veganism* (Lori Gruen and Robert C. Jones, "Veganism as an Aspiration" in *The Moral Complexities of Eating Meat*, eds. Ben Bramble and Bob Fischer (Oxford University Press, 2015), 153–171.), and *freeganism* (Warren Oakes, "Why freegan?" *Freegan.Info*, retrieved January 15, 2016, from http://freegan.info/what-is-a-freegan/freegan-philosophy/why-freegan-an-attack-on-consumption-in-defense-of-donuts/).
39. Marti Kheel, "Vegetarianism and Ecofeminism: Toppling Patriarchy with a Fork" in *Food for Thought: The Debate Over Eating Meat*, ed. Steve F. Sapontzis (Amherst, NY: Prometheus Books, 2004), 327–41.
40. Joel MacClellan, "Animal Agriculture and Welfare Footprints," *Encyclopedia of Food and Agricultural Ethics*, eds. Paul B. Thompson and David M. Kaplan (NY: Springer, 2014), 140–143.
41. Lori Gruen, "Facing Death and Practicing Grief" in *Ecofeminism: Feminist Intersections with Other Animals and the Earth*, eds. Carol J. Adams and Lori Gruen (New York: Bloomsbury, 2014) 127–41.
42. See John Sanbonmatsu, "The Animal of Bad Faith: Speciesism as an Existential Project" in *Critical Theory and Animal Liberation*, ed. John Sorenson (NY: Rowman & Littlefield Publishers, 2011), 29–45, for a brilliant analysis of the role of bad faith in speciesist commitments.
43. Stephanie Jenkins and Vasile Stănescu, "One Struggle," in *Defining Critical Animal Studies: An Intersectional Social Justice Approach for Liberation*, eds. Anthony J. Nocella II, John Sorenson, Kim Socha, and Atsuko Matsuoka (NY: Peter Lang Publishing, 2014), 74–85.
44. See Robert C. Jones, "Animal Rights is a Social Justice Issue," Contemporary Justice Review, 18, no. 4 (2015), 467–482, for greater discussion of the relationship between animal liberation and social justice.
45. See http://www.globallabourrights.org and http://slaveryfootprint.org for details of the horrors in the production of many "vegan" consumer goods.
46. Stephanie Jenkins and Vasile Stănescu, "One Struggle," in *Defining Critical Animal Studies: An Intersectional Social Justice Approach for Liberation*, eds. Anthony J. Nocella II, John Sorenson, Kim Socha, and Atsuko Matsuoka (NY: Peter Lang Publishing, 2014), 77–81.

References

Almassi, Ben. 2011. The consequences of individual consumption: A defence of threshold arguments for vegetarianism and consumer ethics. *Journal of Applied Philosophy* 28(4): 396–411.

Barros, Vicente R., Christopher B. Field, et al., eds. 2014. Climate change 2014: Impacts, adaptation, and vulnerability, Part B: Regional aspects. *IPCC*, retrieved January 15, 2016, from https://ipcc-wg2.gov/AR5/images/uploads/WGIIAR5-PartB_FINAL.pdf

Bass, Robert. 2014. What can one person do? Causal impotence and dietary choice. Unpublished manuscript.

Bohanec, Hope. 2013. *The ultimate betrayal: Is there happy meat?* Bloomington: iUniverse.

Gruen, Lori. 2011. *Ethics and animals: An introduction*. Cambridge: Cambridge University Press.

Gruen, Lori. 2014. Facing death and practicing grief. In *Ecofeminism: Feminist intersections with other animals and the earth*, ed. Carol J. Adams and Lori Gruen, 127–141. New York: Bloomsbury.

Gruen, Lori. 2015. *Entangled empathy: An alternative ethic for our relationships with animals*. New York: Lantern Press.

Gruen, Lori, and Robert C. Jones. 2015. Veganism as an aspiration. In *The moral complexities of eating meat*, ed. Ben Bramble and Bob Fischer, 153–171. New York: Oxford University Press.

Harman, Elizabeth. 2015. Eating meat as a morally permissible moral mistake. In *Philosophy comes to dinner: Arguments on the ethics of eating*, ed. Chignell Andrew, Cuneo Terence, and Halteman Matthew, 215–31. New York: Routledge.

Hawthorne, Mark. 2013a The problem with palm oil. *VegNews*, March 22. http://vegnews.com/articles/page.do?pageId=5795&catId=1

Hawthorne, Mark. 2013b. *Bleating hearts: The hidden world of animal suffering*. Winchester: Changemakers Books.

Imhoff, Dan. 2010. *The CAFO reader: The tragedy of industrial animal factories*. Healdsburg: Watershed Media/University of California Press [distributor].

Jenkins, Stephanie, and Vasile Stănescu. 2014. One struggle. In *Defining critical animal studies: An intersectional social justice approach for libecbrsration*, ed. Anthony J. Nocella II, John Sorenson, Kim Socha, and Atsuko Matsuoka, 74–85. New York: Peter Lang Publishing.

Jones, Robert C. 2015. Animal rights is a social justice issue. *Contemporary Justice Review* 18(4): 467–482.

Kheel, M. 2004. Vegetarianism and ecofeminism: Toppling patriarchy with a fork. In *Food for thought: The debate over eating meat*, ed. Steve F. Sapontzis, 327–341. Amherst: Prometheus Books.

MacClellan, Joel. 2014. Animal agriculture and welfare footprints. In *Encyclopedia of food and agricultural ethics*, ed. Paul B. Thompson and David M. Kaplan, 140–143. New York: Springer.

Martin, Adrienne M. 2015. Consumer complicity in factory farming. In *Philosophy comes to dinner: Arguments on the ethics of eating*, 203–214. New York: Routledge.

Murtaugh, Paul A., and Michael G. Schlax. 2009. Reproduction and the carbon legacies of individuals. *Global Environmental Change* 19(1): 14–20.

Norcross, Alastair. 2004. Puppies, pigs, and people: Eating meat and marginal cases. *Philosophical Perspectives* 18(1): 229–245.

Oakes, Warren. Why freegan? *Freegan.Info*. Retrieved November 19, 2015, from http://freegan.info/what-is-a-freegan/freegan-philosophy/why-freegan-an-attack-on-consumption-in -defense-of-donuts/

Plumwood, Val. 2000. Integrating ethical frameworks for animals, humans, and nature: A critical feminist eco-socialist analysis. *Ethics and the Environment* 5: 285–322.

Rachels, James. 2004. The basic argument for vegetarianism. In *Food for thought: The debate over eating meat*, ed. Steve F. Sapontzis, 70–80. Amherst: Prometheus Books.

Sanbonmatsu, John. 2011. The animal of bad faith: Speciesism as an existential project. In *Critical theory and animal liberation*, ed. John Sorenson, 29–45. New York: Rowman & Littlefield Publishers.

Weis, Tony. 2013. *The ecological hoofprint: The global burden of industrial livestock*. London: Zed Books.

Wolf, Susan. 1982. Moral saints. *The Journal of Philosophy* 79(8): 419–439.

Speciesistic Veganism: An Anthropocentric Argument

A.G. Holdier

In the cold, hard lands of the Emyn Muil, it is easy to lose one's way; for a hobbit like Frodo Baggins—pursued by enemies and burdened with a heavy purpose—his chance confrontation with the creature Gollum in those dread hills offered him a merciful opportunity to choose cooperation over conflict. By sparing the life of the pathetic creature, even against the advice of his friend Samwise Gamgee, Frodo managed to make common cause with his adversary, convincing Gollum to help guide them on their quest even as the creature continued to disagree with them about the One Ring. By the end of Tolkien's *The Lord of the Rings*, circumstances made those philosophical differences re-erupt into conflict, but for a time, Frodo's quest to defeat evil was aided by his enemy.

Vegans would do well to learn from Frodo.

Given that a key concern for vegan movements is the protection of animals who would otherwise be mistreated and eaten, philosophical and

A. Holdier (✉)
Colorado Technical University, Colorado Springs, CO, USA
e-mail: agholdier@gmail.com

J. Castricano, R.R. Simonsen (eds.), *Critical Perspectives on Veganism*, The Palgrave Macmillan Animal Ethics Series,
DOI 10.1007/978-3-319-33419-6_3

political questions of animal rights or animal welfare can be practically (and temporarily) sublimated to the more immediate danger faced by factory-farmed creatures. Much like Frodo relying on the temporary assistance of a philosophical opponent to achieve a pragmatic end, vegans should consider shifting their immediate attention away from any philosophical disagreements with carnists to find a common cause that can more directly benefit the well-being of all conscious creatures.[1] Indeed, a more expedient route to the preservation of would-be slaughter victims lies in an argument based on a premise that most carnists already affirm:

1. Human flourishing should generally be promoted.

Regardless of any ethical problems that may or may not exist with this sort of anthropocentrism, if such a line of thinking could be pragmatically co-opted into the service of vegan goals, then tangible goods could still be accomplished when lives are nonetheless saved. Therefore, what follows seeks to show how a de facto form of veganism grounded on a rejection of large-scale food production industries can accomplish the anthropocentric flourishing of (1) in a way that simultaneously, if coincidentally, defends the lives of nonhuman creatures.

Altogether, each of the four different lines of unsettling evidence provides good reason to criticize standard Western animal-processing industries (APIs), given that

2. If an industry does not generally promote human flourishing, then that industry should not be supported.

"Generally promote" assumes that the costs to human flourishing are outweighed by any simultaneous benefits, whereas "support" includes such actions as the purchasing and consuming of the industry's products. Given that human factory workers are regularly and severely physically compromised in animal-processing plants, human entrepreneurs are frequently victimized and disenfranchised by monopolistic business practices in the API, human community members are physically endangered by the presence of meatpacking factories in their neighborhoods, and human communities worldwide are threatened by the overall effect of concentrated animal feeding operations (CAFOs) on climate change, there is considerable evidence that

3. The Western APIs (of meat, eggs, and dairy products) contribute in several ways to behaviors that undermine human flourishing.

Any one of these claims is sufficiently significant to give grounds for abstention from the consumer chain that funds such consequences; therefore, the sum total of the evidence for (3) indicates that the simplest course of action to simultaneously undermine each is to adopt a vegan diet—regardless of one's views on the metaphysical or moral status of nonhuman animals. The remainder of this chapter aims to develop four lines of evidence for (3) before analyzing (3) against possible benefits of APIs in light of (2); in short, this chapter seeks to determine whether speciesistic veganism might turn out to be a useful stopgap measure to reduce suffering more effectively and pragmatically while philosophical debates continue.

Employee Safety

Worker endangerment in factory farms and slaughterhouses appears in two primary forms, given that employees of the meat, egg, and dairy industries suffer both physically and psychologically from their involvement in the modern system of animal processing. Together, there is more than sufficient evidence to conclude that

3a. APIs create dangerous and deadly working conditions for human employees.

With more than 60 billion pounds of meat processed in a normal month,[2] the economic focus on manufacturing speed and production streamlining in APIs leads to increased rates of accidents to the workforce; overall, of the half-million workers in US slaughterhouses,[3] more than one-quarter are injured each year to an extent that requires more than simple first aid.[4] A recent report from the Southern Poverty Law

Center discovered that nearly 75% of workers in Alabama poultry factories suffered some significant form of workplace injury:

> In spite of many factors that lead to undercounting of injuries in poultry plants, the U.S. Occupational Safety and Health Administration (OSHA) reported an injury rate of 5.9 percent for poultry processing workers in 2010, a rate that is more than 50 percent higher than the 3.8 percent injury rate for all U.S. workers. Poultry workers often endure debilitating pain in their hands, gnarled fingers, chemical burns, and respiratory problems—tell-tale signs of repetitive motion injuries, such as carpal tunnel syndrome, and other ailments that flourish in these plants.[5]

Unfortunately, these numbers are by no means out of the ordinary for other forms of the API.

While some of these injuries heal with few complications and at least some might be compensated for via employee insurance (though this is no guarantee), chronic maladies characteristic of processing factory jobs, such as repetitive motion disorder, have seen incidence rates 30 times higher than comparable industries.[6] A recent study by the Centers for Disease Control found that 57% of interviewed participants from poultry-processing plants reported at least one sustained adverse musculoskeletal symptom,[7] and the rate at which cumulative trauma injuries are sustained in meatpacking plants is roughly 33 times higher than the national average.[8] Regardless, United States Department of Agriculture (USDA) regulations were updated in 2014 via the HACCP Inspection Models Project (HIMP) to further increase allowable line speeds in poultry-processing factories by 25%, from 140 to 175 birds per minute, while simultaneously decreasing funding for federal inspectors by up to 75%—despite the fact that the Inter-American Commission on Human Rights heard testimony concerning the potential dangers of the changes[9] and a petition pleading for the White House to reconsider the new policy garnered nearly 220,000 signatures.[10] Because injury rates were already abnormally high under previous conditions, it seems axiomatic that they should only be expected to further increase under the new, higher-stress, more risk-adverse conditions.

It is also worth noting that a 2005 Human Rights Watch Report on the US meat and poultry industry charged that companies "administer their workers' compensation programs by systematically failing to recognize and report claims, delaying claims, denying claims, and threatening and taking reprisals against workers who file claims for compensation for workplace injuries,"[11] meaning that workers' compensation for injuries is by no means guaranteed. Given that a sizable portion of factory employees are undocumented foreign laborers who are less likely to complain about working conditions lest they be deported,[12] the so-called "Climate of Fear" concerning speaking out against these sorts of working conditions is unsurprising.[13] In terms of both injuries (ranging from tendonitis to amputations) and fatalities, various APIs routinely rank as providing some of the most dangerous jobs in the USA.[14]

While the reasons for these dangerous conditions are complex, the extremely high industry-wide employee turnover rate each year only enhances this problem, as plants are staffed with largely inexperienced workers.[15] A 2005 Government Accountability Office study reported, "Labor turnover in meat and poultry plants is quite high, and in some worksites can exceed 100 percent in a year as workers move to other employers or return to their native countries."[16] Kandel and Parrado reported the same year that "estimates of annual employee turnover in the meat processing industry range from 60 to 140 percent or in some cases significantly higher."[17] In the last decade, employee replacement has become steadily more frequent, with rates as high as 200% being common in slaughterhouses, given certain parameters.[18] Altogether, when hazardous conditions are compounded by extremely fast-speed expectations and untrained employees, high rates of injury are bound to result; this is precisely what the available data indicates.

However, physical effects are not the only harms to workers that must be considered; exposure to, and participation in, the violence of this workplace also leads to profound psychological damage, to which anyone with anthropocentric concerns must attend. The stress to maintain production speeds already discussed is often unbearable, and illegal drug use is not unheard of as a supplement to try and meet an employer's demands.[19] Even more significantly, though, is that the work itself has disturbing psychological effects, including anxiety, depression, paranoia, personality disintegration, and dissociation as a result of a variety of unhealthy coping mechanisms.[20] According to the testimony of one poultry factory worker, there is indeed much to cope with:

> You are murdering helpless birds by the thousands (75,000 to 90,000 a night). You are a killer…Out of desperation you send your mind elsewhere so that you don't end up like those guys that lose it. Like the guy that fell on his knees praying to God for forgiveness. Or the guy they hauled off to the mental hospital that kept having nightmares that chickens were after him. I've had those, too.[21]

Or consider this story from a "sticker" on a kill line in a slaughterhouse in Iowa whose job it was to kill pigs and drain them of their blood:

> The worst thing, worse than the physical danger, is the emotional toll. If you work in that stick pit for any period of time, you develop an attitude that lets you kill things but doesn't let you care. You may look a hog in the eye that's walking around down in the blood pit with you and think, God, that really isn't a bad-looking animal. You may want to pet it. Pigs down on the kill floor have come up and nuzzled me like a puppy. Two minutes later I had to kill them—beat them to death with a pipe. I can't care.[22]

And though anecdotal evidence can be a shaky foundation for an argument, the prevalence and commonality of stories like these is suggestive of a widespread problem. Stephen Thierman has pointed out that working conditions in slaughterhouses are heavily predicated on additional dehumanizing psychological pressures brought about through the partitioning of the workforce based on features like race and socioeconomic status,[23] and Jennifer Dillard has cataloged many pertinent examples of the physical consequences of such an environment in her work to seek financial compensation for workers subjected to these sorts of conditions.[24] Perhaps the most damning piece of evidence in this regard, however, may well be a 2012 study of Turkish workers which concluded that "butchers, especially those who work in slaughterhouses, have [statistically demonstrable] higher levels of psychological disorders than the office workers" to whom they were compared.[25] Similar pressures were evidenced at slaughterhouses in Denmark by a 1991 study that not only observed a differentially higher proportion of stress-induced symptoms in workers holding positions on the kill line (vs. those in the stables, for example) but also concluded with the suggestion that the abnormally high strike rates in Danish slaughterhouses (compared to other industries) might be related to the relative lack of coping mechanisms for such a stressful environment.[26] Still, at this point, the Occupational Safety and Health Administration's regulations for the meatpacking industry include no considerations for the psychological well-being of the employees.[27]

Altogether, considering the high rate of physical injuries to human workers in animal-processing facilities or the suspiciously strong connection of this form of employment to psychological disorders, the supposition of (3a) is well founded.

Employee Victimization

To consider dangers of a different sort, the reality of the misanthropic dangers of APIs expressed in (3) is likewise undergirded by standard business practices within the largely monopolistic API that contribute heavily to the frequent abuse and marginalization of laborers and businesspersons located at the lower end of the socioeconomic spectrum. Indeed, it is not hard to conclude that

3b. APIs contribute to the economic disenfranchisement of human workers and entrepreneurs.

Not only is this the case for migrant workers, as discussed in the previous section, but it is also for farmers who are forced into manipulative contractual relationships functionally similar to indentured servitude.

As mentioned above, contemporary industrialized husbandry practices rely heavily on migrant and illegal immigrant workers to maintain staffing in the subpar working conditions of many slaughterhouses. In addition to physical dangers, this work can also lead disproportionally to common abusive economic arrangements for minority workers, including those of legal working status. Consider, for instance, how a 2012 study of Latino meatpackers in Nebraska determined that roughly 50% of workers had not heard of the Nebraska Meatpackers Bill of Rights but had received negative information regarding unions, roughly a third had failed to receive information about workers' compensation during their orientation, and at least 12% were unaware of their hours or their pay rate until after they had begun working.[28] Recent governmental approaches to immigration policy have only encouraged worker abuses of this sort, given that "The single-minded focus on immigration enforcement without regard to violations of workplace laws has enabled employers with rampant labor and employment violations to profit by employing workers who are terrified to complain about substandard wages, unsafe conditions, and lack of benefits, or to demand their right to bargain collectively."[29] Effectively, this sort of exploitation amounts to a contemporary rebirth of age-old silencing and slavery-type practices based on the disempowerment of the human labor force.

Surprisingly, this is also the case for business owners themselves, given the manipulative character of many of the contracts that farmers are expected to sign in order to do business with large conglomerates. Particularly prevalent among chicken farmers (though similar contractual arrangements have become increasingly common with pig farmers and, to a lesser extent, cow ranchers),[30] the nature of contract-farming arrangements means that production is closely coordinated with the integrating firm (such as Tyson Foods or Perdue Farms) that will eventually process the animal into a marketable consumable product; what this leads to is an arrangement where the farmer will "provide capital (housing and equipment), utilities, and labor. They receive chicks, feed, transportation, veterinary services, and technical guidance from integrators, who pay contract fees to the growers to raise the chicks to market weights."[31] Essentially, the farmer owns the equipment and does the job of raising the chickens, but must comply with the strict regulations laid out by his or her integrator because the farmer does not own the chickens themselves.

The effect of this arrangement is twofold. The primary effect of this arrangement is that the farmer must bear most of the unexpected costs of raising the chickens to market weight as well as pander to any additional requirements levied by the integrator as a condition of renewing a contract (including, as detailed in Jonathan Shepard's 2010 documentary *The Sharecroppers*, expensive equipment upgrades that drive farmers further into debt, thereby deepening their reliance on their relationship with the integrating firm). As Mary Hendrickson and Harvey S. James point out, "bucking the integrating firm's production standards is not an option for farmers stuck with ten-year loans on buildings that are a quarter of a million dollar investment"[32]—an investment, it is worth noting, for which the multimillion-dollar chicken-processing company is not financially responsible but from which the company reaps the majority of its profits.

Not only does the debt burden fall to the individual farmer, but the farmer's compensation from this complicated system is also typically drawn from a "tournament-style" payment structure where farmers from a given geographical area compete against each other in an annual ranking system designed to reward low-cost production.[33] Based on the inconsistencies of year-to-year farming and the impossibilities of predicting output rates of animal weight (much less predicting one's own annual output in tandem with one's neighbors), it is next to impossible for farmers to engage in any real long-term financial planning. As Dudley Butler, former administrator at the Department of Agriculture said in Alice Brennan and Connie Fossi Garcia's 2015 documentary *Cock Fight*, "all the tournament system is, is a cost-controlling device for the companies. Sure, they give a bonus to somebody over here, but then they give a discount to somebody over here." In short, the farmers must bear the majority of the costs while reaping a minority of the profits; it should come as little surprise, then, that in 2001, 71% of US farmers who grew only chickens lived at or below the federal poverty line with little demographic improvement since.[34]

The secondary effect of this unnaturally complex economic arrangement between animal farmers and the companies that own and process the animals themselves is the degradation of farmers' moral compasses as a result of their economic instability. As James and Hendricks discovered in 2007, "perceived economic pressures are correlated with a greater willingness of farmers to tolerate unethical conduct," including the mistreatment of "the land, animals or the food they produce."[35] Although a purely anthropocentric argument might ignore this nonetheless interesting fact, vegans concerned about animal rights or welfare would do well to pay attention to this human-centered harm and its spillover effects to other species.[36] Even apart from increased toleration of unethical conduct, economic pressures on laborers from the cradle to the grave of a food animal's life give a strong indication that claim toward economic disenfranchisement expressed in (3b) is sound.

Community Safety

Notably, workers are not the only human beings to experience adverse effects in connection with standard practices of the API. To take a wider perspective on the scope of (3) and its ramifications, community members in neighborhoods surrounding slaughterhouses and other meat-processing facilities are negatively affected by APIs to a degree that is increasingly confirmatory of the next premise:

3c. APIs contribute to the physical endangerment of neighboring community members.

This can be demonstrated in at least two ways: firstly, through common practices that can lead to contaminated products, and, secondly (and more problematically), via crime rates that statistically increase in areas adjacent to abattoirs.

Not only do increased line speeds raise risk factors for employee safety but they also simultaneously lower reasonable quality expectations for the end result of the production chain as unavoidable inspection oversights impact a greater percentage of workflow output in the system, which prompted Joy to remark that "it appears that in our nation's meatpacking plants, contaminated meat is the rule, rather than the exception."[37] When inedible contaminants and pathogens are accidentally introduced into the production line, faster speeds make it more difficult for inspectors to catch each mistake; for example, rates of food poisoning cases associated with contaminated meat products have increased at rates roughly comparable to chain speed rate increases inside meat production facilities.[38] In the words of a USDA inspector for a pork production facility testing the pilot HIMP program (previously mentioned in connection with poultry line speed increases), "contamination such as hair, toenails, cystic kidneys, and bladder stems has increased under HIMP. Line speeds don't make it any easier to detect contamination. Most of the time they are running so fast it is impossible to see anything on the carcass."[39] This same report indicated that the plant in question was, at least at times, processing in excess of 1200 animal carcasses per hour (one carcass every three seconds). Not only is this sort of contamination thoroughly unsurprising, given the working conditions inside the factory, but it also has wide-ranging effects on the eventual health of the human consumers of the animal meat products.

However, on the local level, an even more troubling side effect of the meat production industry is indicated by the results of a study published in 2009 on the spillover effects that slaughterhouses have on the communities in which they are located. Even when controlling for variables like unemployment or demography, "the findings indicate that slaughterhouse employment increases total arrest rates, arrests for violent crimes, arrests for rape, and arrests for other sex offenses in comparison with other industries."[40] The correlation between abattoir employment and sexual crimes, particularly rape, was especially strong, leading the researchers to suggest that their data may imply that "the work done within slaughterhouses might spillover to violence against other less powerful groups, such as women and children," a point that feminist care ethicists like Carol Adams have been arguing for years.[41] Dubbed the "Sinclair Effect"—after the author of the landmark novel *The Jungle*, which detailed the dismal working conditions in the American meatpacking industry of the early twentieth century—the product of this phenomenon is of acute anthropocentric concern regardless of one's views on the morality of animal abuse itself.

Notably, popular-level considerations have already begun to take slaughterhouse employment into consideration during courtroom deliberations in criminal trials. Dillard reports how in two cases from the early 2000s "the murders at issue were performed in a manner similar to the way in which an animal at the defendant's former place of employment would be slaughtered,"[42] making the habits of the defendant connected to his profession particularly relevant. Given that noninstitutionalized forms of animal abuse have long been recognized as carrying implications for similarly violent attitudes toward human beings,[43] such a conclusion is hardly a large leap. Similarly, with increased risks of disease and localized violence evidently connected with current abattoir realities, the case for (3c) is likewise a short jump.

Community (and Global) Victimization

Although also problematic on a local level, the most widespread anthropocentric consequence of the contemporary industrialized animal husbandry paradigm is the significant set of contributions made by APIs to global climate change. While relatively small-scale environmental effects have been evidenced in areas directly around large-scale animal processing facilities, global-level concerns about land degradation and deforestation, air and water pollution, and subsequent biodiversity instability make for a convincing case:

3d. APIs contribute significantly to anthropogenic climate change.

It is worth noting that, if true, the potential ramifications of this consequence of APIs could affect entire human populations, even if they abstained from consuming animal products raised in any format.

On the smallest level, in this regard, Fitzgerald has documented a plethora of studies concerning the degradation of the immediate environment following the development of large-scale CAFOs, largely due to the high amount of manure that is necessarily produced in industrialized farms with thousands of animals inside.[44] On average, a CAFO must process roughly 50 pounds of liquid and solid waste matter from each of its steers on a daily basis[45]; standard industry practice is to deposit the manure into large, frequently open-air "lagoons" where it is stored until it can be recycled as fertilizer, posing a significant environmental risk in the interim period.[46] Disease-causing microbes flourish in such systems, and a 2001 report from the Natural Resources Defense Council and the Clean Water Network detailed many of the possible ways that lagoon systems can fail and contaminate local neighborhoods' air and water supplies:

> People living near factory farms are placed at risk. Hundreds of gases are emitted by lagoons and the irrigation pivots associated with sprayfields, including ammonia (a toxic form of nitrogen), hydrogen sulfide, and methane. The accumulation of gases formed in the process of breaking down animal waste is toxic, oxygen consuming, and potentially explosive, and farm workers' exposure to lagoon gases has even caused deaths. People

> living close to hog operations have reported headaches, runny noses, sore throats, excessive coughing, respiratory problems, nausea, diarrhea, dizziness, burning eyes, depression, and fatigue.[47]

And even if health risks and environmental concerns were set aside, the aesthetic experience of CAFO exposure is more than mildly unpleasant; as Eric Schlosser eloquently describes the hometown of one of the nation's largest CAFOs, "You can smell Greeley, Colorado, long before you can see it. The smell is hard to forget but not easy to describe, a combination of live animals, manure, and dead animals being rendered into dog food. The smell is worse during the summer months, blanketing Greeley day and night like an invisible fog."[48] Altogether, it should not be surprising that a variety of movements have sprung up to challenge the encroachment of large-scale operations into rural community life.[49]

On a wider scale, the polluting side effects of CAFO-style farms spread far beyond the local communities where the factories themselves are located. Though estimates of the overall quantities of greenhouses gases (GHGs) produced by industrialized farming operations vary, two conclusions do not seem to be in dispute: firstly, that meat and dairy operations account for the majority of food-related GHG emissions (in most cases at least 50%), and, secondly, that livestock operations are one of the largest single industries that contribute to GHG emissions internationally—ranging from 18 to 20% of overall GHG emissions in both the USA and Europe.[50] A single cow can produce more volatile organic compounds that contribute to methane and ammonia emissions than do many small cars, and New Zealand's cattle and sheep industry, for one example, is responsible for 43% of the country's overall GHG emissions.[51] Given that global demand for meat and milk products is not only increasing but expected to double by 2050, atmospheric conditions unfortunately show no sign of benefiting from a potential downturn in the livestock industry that would reduce the level of pollutants in the air.[52] As has been detailed in a wide variety of other settings, potential consequences of the greenhouse effect are already affecting human livelihoods around the world.

Further environmental concerns over contemporary animal-processing methods are found in second-order impacts such as the necessary land used to facilitate standard industry practices. The thousands of animals in industrialized farms require large amounts of food, typically in the form of cereal grains; it has been estimated that more than a third of the world's cereal output is dedicated to farm animal feed,[53] despite the fact that "it would be much more efficient for humans to consume cereals directly since much of the energy value is lost during conversion from plant to animal matter."[54] Even though livestock already occupy 20% of terrestrial animal biomass[55] and 80% of anthropogenic land use overall,[56] the continuous need to expand growing operations to meet factory demands has been a significant motivation for deforestation in places like the Brazilian Amazonian region.[57]

Not only does the destruction of habitats in this way release stored carbon reserves into the atmosphere at higher rates,[58] but that devastation also poses a significant risk to global biodiversity. As humans continue to encroach on wild habitats, native species are continually put at risk—not only has the World Wildlife Fund listed livestock as a potential threat for 37% of its listed terrestrial ecoregions, but 23 of Conservation International's 35 emergency-level global hotspots for biodiversity have been reportedly affected by livestock and livestock-related projects.[59] Finally, limited resource consumption is characteristic of the rather inefficient meat-processing industry; as Matsuoka and Sorenson summarize, "Producing meat is more energy-consumptive than producing vegetables for consumption, requiring far higher amounts of water, at least 16 times as much fossil fuel, and producing 25 times as much carbon dioxide emissions."[60]

Given the myriad impacts on global climate change to which these industries contribute, not only is the soundness of (3d) easy to defend, but it is also a fourth example of a problematic consequence of the industry poignantly affecting human beings.

The Counterpoint

However, it might well be the case that, as significant as they are, these costs could be superseded by sufficient benefits resultant from the animal production industry. A comprehensive anthropocentric analysis must consider both benefits and harms to human populations, and, broadly construed, this means:

4. APIs contribute in several ways to actions that do not undermine human flourishing.

This would be defensible, firstly, insofar as this means:

4a. APIs create jobs for human workers.

Secondly (and most prominently), (4) is also strengthened by the following point:

4b. APIs create popular animal products that create pleasure for human consumers.

In the USA alone, the meat- and poultry-processing industries provide jobs for nearly a half-million human beings.[61] Considering that only roughly 7% of the US population identifies as either vegetarian or vegan, nearly 296.5 million consumers in the USA alone enjoy some form of animal-based food regularly—often because the taste of the meal is described as enjoyable.[62] Though gainful employment and pleasant aesthetic experiences might ultimately be outweighed by sufficiently significant concerns, they are factors that cannot, in principle, be ignored.

Speciesistic de facto Veganism

Recall the arguments to this point:

1. Human flourishing should generally be promoted.
2. If an industry does not generally promote human flourishing, then that industry should not be supported.
3. The Western APIs (of meat, eggs, and dairy products) contribute in several ways to actions that undermine human flourishing.

   3a. APIs create dangerous and deadly working conditions for human employees.

3b. APIs contribute to the economic disenfranchisement of human workers and entrepreneurs.
3c. APIs contribute to the physical endangerment of neighboring community members.
3d. APIs contribute significantly to anthropogenic climate change.

4. APIs contribute in several ways to actions that do not undermine human flourishing.

4a. APIs create jobs for human workers.
4b. APIs create popular animal products that create pleasure for human consumers.

Firstly, we are now at the point where the costs of (3) can be considered in light of the benefits of (4) to determine what the sum effect of the API is on human welfare and the potential for human flourishing. Given that an instance of employment is not an inherent good (because of any number of possible workplace injustices that could, in fact, damage a life to a greater degree than a paycheck would assist it), the benefits of (4a) are directly countermanded by the nature of the jobs provided as detailed in (3a): in more than a few cases, workers' physical and mental ailments are sufficiently debilitating such that it likely would have been better for the worker in question to have continued looking for a different job rather than settling for a job at the trauma-inducing slaughterhouse.

Secondly, it seems unlikely that the simple aesthetic pleasure of taste on which (4b) is grounded will ever overrule the harmful total weight of (3a–d). Granting for the sake of argument that meat is, in fact, aesthetically pleasurable, the noncompulsory nature of at least most aesthetic pleasures makes such a benefit irrelevant in light of the significant ethical problems that cause the experience in question.[63] If no physical pleasures are taken to rationally predominate over concerns as drastic as those listed above, then it cannot be the case that an optional, fleeting pleasure outweighs a collection of substantial, long-lasting harms. That is to say that, regardless of how tasty animal meat is for some humans, the painful experiences, financial corruption, physical endangerment, and climate-based catastrophes to which that meat contributed in its production chain are, in fact, more significant. So much so that

5. The sum benefits of APIs as listed in (4) are ethically outweighed by the sum costs of APIs as listed in (3).

Which, rephrased in light of (2), means

6. Therefore, APIs do not generally promote human flourishing.

And, if human flourishing is indeed something to value as proposition (1) indicates, then we must conclude that

7. Therefore, APIs should not be supported.

Which is precisely to say that the products of APIs—in this case, the collection of Western industries that raise, process, harvest, and slaughter animals via concentrated, industrialized methods—should neither be purchased nor consumed.

Admittedly, this argument does not require a principled vegetarian or vegan conclusion, but rather a systematic rejection of animal outputs produced commercially in the most common Western method. Raising and butchering one's own meat (or, similarly, eggs or dairy products) in one's own backyard for one's own consumption would not be open to criticism on these grounds—additional arguments not restricted to purely anthropocentric concerns would be required for the condemnation of such activities—but the sheer rarity of homegrown (and home-killed) options make this objection essentially irrelevant for most consumers.[64]

Importantly, one can ignore nonhuman animal rights and well-being entirely and still recognize, based on the argument presented here, that the current standard system of industrialized animal husbandry leads to human suffering. Consequently, even the most devout speciesist could still conclude, on the sole basis of his or her concern for *homo sapiens*, that a de facto vegan diet is morally obligatory in most Western contexts (wherever conditions [3a–d] sufficiently obtain). Therefore, much like Frodo temporarily making use of the pitiful Gollum to reach his goal of destroying the ring in the fires of Mount Doom, a speciesistic attitude can still be beneficially appropriated in the service of vegan goals. Consequently, if it contributes to the expedited prevention of creaturely slaughter, then animal welfarists can rest somewhat more comfortably on the pragmatic beachhead of this anthropocentric argument.

Notes

1. Following Melanie Joy, *Why We Love Dogs, Eat Pigs, and Wear Cows: An Introduction to Carnism* (San Francisco: Conari Press, 2010), 28–30, a "carnist" is an individual who, on the basis of some ideology, chooses to eat meat.
2. Economic Research Service, "Livestock and Meat Domestic Data," United States Department of Agriculture. Accessed 1/12/2016, http://www.ers.usda.gov/data-products/livestock-meat-domestic-data.aspx#26056.
3. Carey Biron, "Meatpacking Workers Fight "Unacceptable And Inhumane" Conditions," *Mintpress News*, March 27, 2014, http://www.mintpressnews.com/meatpacking-workers-fight-unacceptable-and-inhumane-conditions/187409/.
4. Eric Schlosser, *Fast Food Nation: The Dark Side of the American Meal* (New York: Houghton Mifflin, 2002), 173 and 185.
5. Tom Fritzsche, "Unsafe at These Speeds," *The Southern Poverty Law Center*, February 28, 2013, https://www.splcenter.org/20130301/unsafe-these-speeds#summary.
6. Roger Horowitz, "Government, Industry Play the Numbers Game on Worker Safety in Meatpacking Plants," *LaborNotes*, June 13, 2008, http://labornotes.org/2008/06/government-industry-play-numbers-game-worker-safety-meatpacking-plants.
7. US Dept. of Health and Human Services, Centers for Disease Control and Prevention, and the National Institute for Occupational Safety and Health, "Evaluation of Musculoskeletal Disorders and Traumatic Injuries Among Employees at a Poultry Processing Plant," by Kristin Musolin, et al. http://www.cdc.gov/niosh/hhe/reports/pdfs/2012-0125-3204.pdf, i.
8. Schlosser, *Nation,* 173.
9. Lydia Zuraw, "Inter-American Commission on Human Rights Hears Testimony on Poultry Worker Safety," *Food Safety News*, March 27, 2014, http://www.foodsafetynews.com/2014/03/iachr-hearing/#.Vh9E7PlVhBd.

10. Biron, "Meatpacking".
11. Lance Compa, *Blood, Sweat, and Fear: Workers' Rights in U.S. Meat and Poultry Plants* (Humans Rights Watch, 2004), 57.
12. Atsuko Matsuoka and John Sorenson, "Human Consequences of Animal Exploitation: Needs for Redefining Social Welfare," *Journal of Sociology and Social Welfare* XL, no. 4 (2013): 15.
13. Fritzsche, "Unsafe".
14. For example, in 2013 "Farmers, Ranchers, and Other Agricultural Managers" ranked as the 9th most dangerous industry with 21.5 fatalities per 100,000 full-time workers and "Fishers and Related Fishing Workers" ranked 2nd with 75 fatalities per 100,000 full-time workers, with preliminary data from 2014 indicating that fishing industry fatality rates are further increasing. See U.S. Dept. of Labor, *National Census of Fatal Occupational Injuries in 2014 (Preliminary Results)* 2015, http://www.bls.gov/news.release/pdf/cfoi.pdf; Joy, *Introduction to Carnism*, 79–81.
15. Gail Eisnitz, *Slaughterhouse: The Shocking Story of Greed, Neglect, and Inhumane Treatment Inside the U.S. Meat Industry* (Amherst: Prometheus, 2007), 62.
16. U.S. Government Accountability Office 2005, 7.
17. William Kandel and Emilio A. Parrado, "Restructuring of the US Meat Processing Industry and New Hispanic Migrant Destinations," *Population and Development Review* 31, no. 3 (2005): 458.
18. Amy J. Fitzgerald, "A Social History of the Slaughterhouse: From Inception to Contemporary Implications," *Human Ecology Review* 17, no. 1 (2010): 64.
19. Schlosser, *Nation*, 174.
20. Matsuoka and Sorenson, "Human Consequences," 16–17; Joy, *Introduction to Carnism*, 82–85.
21. Sandor Ellix Katz, *The Revolution Will Not Be Microwaved: Inside American's Underground Food Movements* (White River Junction: Chelsea Green, 2006), 258.
22. Eisnitz, *Slaughterhouse*, 87.
23. Stephen Thierman, "Apparatuses of Animality: Foucault Goes to a Slaughterhouse," *Foucault Studies* 9 (2010): 103–104.
24. Jennifer Dillard, "Slaughterhouse Nightmare: Psychological Harm Suffered by Slaughterhouse Employees and the Possibility of Redress through Legal Reform," *Georgetown Journal on Poverty Law and Policy* 15, no. 2 (2008): 395–396.

25. Abdurrahim Emhan et al., "Psychological Symptom Profile of Butchers Working in Slaughterhouse and Retail Meat Packing Business: A Comparative Study," *The Journal of the Faculty of Veterinary Medicine* 18, no. 2 (2012): 319.
26. Tage Kristensen, "Sickness Absence and Work Strain Among Danish Slaughterhouse Workers: An Analysis of Absence from Work Regarded as Coping Behavior," *Social Science and Medicine* 32, no. 1 (1991): 24.
27. U.S. Dept. of Labor, Occupational Safety and Health Administration. Safety and Health Topics, "Meat Packing Industry—OSHA Standards," Accessed January 12, 2016. https://www.osha.gov/SLTC/meatpacking/standards.html
28. María Teresa Gastón, "Meatpacking Workers' Perceptions of Working Conditions, Psychological Contracts and Organizational Justice," (MA Thesis, University of Nebraska-Omaha, 2011): 30 and 47–49, http://digitalcommons.unomaha.edu/studentwork/9/.
29. R. Smith, A. Avendaño, and Martínez Ortega, *Iced Out: How Immigration Enforcement has Interfered with Workers' Rights* (Washington, DC: AFL-CIO, 2009), 5 http://digitalcommons.ilr.cornell.edu/laborunions/29/. The whole of the report is worth reading as it details several examples of corrupt practices commonly used to arrest abused workers and leave abusive conditions unresolved rather than to address the abuse itself.
30. Mary Hendrickson and Harvey S. James, "The Ethics of Constrained Choice: How the Industrialization of Agriculture Impacts Farming and Farmer Behavior," *University of Missouri Agricultural Economics Working Paper* (2004), 9. Available at SSRN: http://ssrn.com/abstract=567423 or http://dx.doi.org/10.2139/ssrn.567423.
31. James MacDonald, *Financial Risks and Incomes in Contract Boiler Production* (The United States Department of Agriculture Economic Research Service, August 4, 2014), http://ers.usda.gov/amber-waves/2014-august/financial-risks-and-incomes-in-contract-broiler-production.aspx#.Vh9PXPlVhBf.
32. Hendrickson and James, "Constrained Choice," 13.
33. MacDonald, "Financial Risks."
34. The Pew Campaign to Reform Industrial Animal Agriculture 2013.
35. Harvey S. James and Mary Hendrickson, "Perceived Economic Pressures and Farmer Ethics," *University of Missouri Agricultural Economics Working Paper* (2007), 16–18. Available at SSRN: http://ssrn.com/abstract=1007080 or http://dx.doi.org/10.2139/ssrn.1007080.

36. For example, standard operating procedures inside large-scale egg farms require non–egg-laying male chicks to be killed, typically by being suffocated in a plastic bag (See Mylan Engel, "The Immorality of Eating Meat," in *The Moral Life: An Introductory Reader in Ethics and Literature*, ed. Louis P. Pojman (New York: Oxford University Press, 2001), 884) or by being thrown into a meat grinder while still alive (Elisabeth Braw, "The Short, Brutal Life of Male Chickens," *Al Jazeera America*, February 20, 2015, http://america.aljazeera.com/articles/2015/2/20/the-short-brutal-life-of-male-chickens.html); neither of these are actions that would be easily promotable among the general public, but the economic necessity of standard factory farm processes lead to their acceptance among farmers and workers.
37. Joy, *Introduction to Carnism*, 76.
38. Fitzgerald, "Social History,"64.
39. Food Integrity Campaign, "WTF Hormel?!—Affidavit #2" *The Government Accountability Project*, accessed January 12, 2016, http://www.foodwhistleblower.org/campaign/wtf-hormel/#affidavits.
40. A.J. Fitzgerald, L. Kalof, and T. Dietz, "Slaughterhouses and Increased Crime Rates: An Empirical Analysis of the Spillover From "The Jungle" Into the Surrounding Community," *Organization & Environment* 22, no. 2 (2009): 158.
41. Fitzgerald, Kalof, and Dietz, "Spillover," 175.
42. Jennifer Dillard, "Slaughterhouse Nightmare: Psychological Harms Suffered by Slaughterhouse Employees and the Possibility of Redress through Legal Reform," *Georgetown Journal on Poverty Law and Policy* 15, no. 2 (2008): 400.
43. Arnola Arluke et al., "The Relationship of Animal Abuse to Violence and Other Forms of Antisocial Behavior," *Journal of Interpersonal Violence* 14, no. 9 (1999): 963–975; Catherine Miller, "Childhood Animal Cruelty and Interpersonal Violence," *Clinical Psychology Review* 21, no. 5 (2001): 735–749.
44. Fitzgerald, "Social History," 66.
45. Schlosser, *Nation,* 150.
46. Discussing one of the largest CAFOs in the country (located outside of Greeley, Colorado), Schlosser indicates that just two feedlots "produce more excrement than the cities of Denver, Boston, Atlanta, and St. Louis—combined." See Schlosser, *Nation*, 150.

47. Robbin Marks, *Cesspools of Shame: How Factory Farm Lagoons and Sprayfields Threaten Environmental and Public Health* (National Resources Defense Council and the Clean Waters Network, 2001), 1.
48. Schlosser, *Nation*, 149.
49. Fitzgerald, "Social History," 63.
50. Hope W. Phetteplace, Donald E. Johnson, and Andrew F. Seidl, "Greenhouse Gas Emissions from Simulated Beef and Dairy Livestock Systems in the United States," *Nutrient Cycling in Agroecosystems* 60 (2001): 99; Tara Garnett, "Livestock-Related Greenhouse Gas Emissions: Impacts and Options for Policy Makers," *Environmental Science and Policy* 12 (2009): 491; Jessica Bellarby, et al., "Livestock Greenhouse Gas Emissions and Mitigation Potential in Europe," *Global Change Biology* 19, no. 1 (2013): 1.
51. Matsuoka and Sorenson, "Human Consequences," 14.
52. Garnett, "Emissions," 491.
53. Bellarby, et al., "Mitigation Potential," 1.
54. Garnett, "Emissions," 494.
55. Henning Steinfeld, et al., *Livestock's Long Shadow: Environmental Issues and Options* (Rome: The Food and Agriculture Organization of the United Nations, 2006), xxiii. Available: http://www.fao.org/docrep/010/a0701e/a0701e00.HTM.
56. Bellarby, et al., "Mitigation Potential," 1.
57. Garnett, "Emissions," 494.
58. Garnett, "Emissions," 494.
59. Steinfeld, et al., *Livestock's Long Shadow*, xxiii.
60. Matsuoka and Sorenson, "Human Consequences," 14.
61. "The United States Meat Industry at a Glance," *The North American Meat Institute*, accessed January 12, 2016, https://www.meatinstitute.org/index.php?ht=d/sp/i/47465/pid/47465.
62. Frank Newport, "In U.S., 5% Consider Themselves Vegetarians," *Gallup.com* July 26, 2012, accessed January 12, 2016, http://www.gallup.com/poll/156215/consider-themselves-vegetarians.aspx.
63. Admittedly, some aesthetic pleasures might well be genuinely necessary for an individual's flourishing existence, but it seems remarkably unlikely that someone would defend the animal-based products of APIs discussed here on such grounds. Because space constraints prevent a more comprehensive consideration of this potential objection, I will simply assert that this is not the case and trust that my boldness is uncontroversial.

64. Though difficult to determine concretely, estimates based on USDA census data indicate that CAFO-style farms account for more than 99% of farmed and slaughtered animals in the United States. See "Ending Factory Farming," *Farm Forward*, accessed January 12, 2016, https://farmforward.com/ending-factory-farming/.

References

Arluke, A., et al. 1999. The relationship of animal abuse to violence and other forms of antisocial behavior. *Journal of Interpersonal Violence* 14(9): 963–975.

Bellarby, Jessica, et al. 2013. Livestock greenhouse gas emissions and mitigation potential in Europe. *Global Change Biology* 19(1): 3–18.

Biron, Carey L. 2014. Meatpacking workers fight "Unacceptable and inhumane" conditions. *Mintpress News*, March 27. http://www.mintpressnews.com/meatpacking-workers-fight-unacceptable-and-inhumane-conditions/187409/. Accessed 14 Jan 2016.

Braw, Elisabeth. 2015. The short, brutal life of male chickens. *Al Jazeera America*. February 20. http://america.aljazeera.com/articles/2015/2/20/the-short-brutal-life-of-male-chickens.html. Accessed 14 Jan 2016.

Brennan, Alice, and Connie Fossi Garcia. 2015. *Cock fight*. Film Documentary. Youtube.com and Fusion Interactive, also http://interactive.fusion.net/cock-fight/. Accessed 14 Jan 2016.

Compa, Lance. 2004. *Blood, sweat, and fear: Workers' rights in U.S. meat and poultry plants*. New York: Human Rights Watch.

Dillard, Jennifer. 2008. Slaughterhouse nightmare: Psychological harm suffered by slaughterhouse employees and the possibility of redress through legal reform. *Georgetown Journal on Poverty Law and Policy* 15(2): 391–408.

Economic Research Service. 2015. *Livestock and meat domestic data*. United States Department of Agriculture. http://www.ers.usda.gov/data-products/livestock-meat-domestic-data.aspx#26056. Accessed 14 Jan 2016.

Eisnitz, Gail A. 2007. *Slaughterhouse: The shocking story of greed, neglect, and inhumane treatment inside the U.S. meat industry*. Amherst: Prometheus.

Emhan, Abdurrahim, et al. 2012. Psychological symptom profile of butchers working in slaughterhouse and retail meat packing business: A comparative study. *The Journal of the Faculty of Veterinary Medicine, University of Kafkas* 18(2): 319–322.

Ending Factory Farming. 2016. *Farm forward*. https://farmforward.com/ending-factory-farming/. Accessed 14 Jan 2016.

Engel Jr., Mylan. 2000. The immorality of eating meat. In *The moral life: An introductory reader in ethics and literature*, vol. 1, ed. Louis P. Pojman, 856–890. New York: Oxford University Press.

Fitzgerald, Amy J. 2010. A social history of the slaughterhouse: From inception to contemporary implications. *Human Ecology Review* 17(1): 58–69.

Fitzgerald, A.J., L. Kalof, and T. Dietz. 2009. Slaughterhouses and increased crime rates: An empirical analysis of the spillover from "The Jungle" into the surrounding community. *Organization & Environment* 22(2): 158–184.

Food Integrity Campaign. 2016. WTF Hormel?! – Affidavits. *The Government Accountability Project*, 2014. http://www.foodwhistleblower.org/campaign/wtf-hormel/#affidavits. Accessed 14 Jan 2016.

Fritzsche, Tom. 2013. Unsafe at these speeds. *The Southern Poverty Law Center*, February 28. https://www.splcenter.org/20130301/unsafe-these-speeds#summary

Garnett, Tara. 2009. Livestock-related greenhouse gas emissions: Impacts and options for policy makers. *Environmental Science and Policy* 12: 491–503.

Gastón, María Teresa. 2011. Meatpacking workers' perceptions of working conditions, psychological contracts and organizational justice. Master's thesis, University of Nebraska. Omaha. http://digitalcommons.unomaha.edu/studentwork/9/

Hendrickson, Mary, and Harvey S. James. 2004. The ethics of constrained choice: How the industrialization of agriculture impacts farming and farmer behavior. *University of Missouri Agricultural Economics Working Paper*. No. AEWP 2004–3. Available at SSRN: http://ssrn.com/abstract=567423 or http://dx.doi.org/10.2139/ssrn.567423

Horowitz, Roger. 2008. Government, industry play the numbers game on worker safety in meatpacking plants. *LaborNotes*, June 13. http://labornotes.org/2008/06/government-industry-play-numbers-game-worker-safety-meatpacking-plants. Accessed 14 Jan 2016.

James, Harvey S., and Mary Hendrickson. 2007. Perceived economic pressures and Farmer Ethics. *University of Missouri Agricultural Economics Working Paper*. Available at SSRN: http://ssrn.com/abstract=1007080 or http://dx.doi.org/10.2139/ssrn.1007080

Joy, Melanie. 2010. *Why we love dogs, eat pigs, and wear cows: An introduction to carnism*. San Francisco: Conari Press.

Kandel, William, and Emilio A. Parrado. 2005. Restructuring of the US meat processing industry and new hispanic migrant destinations. *Population and Development Review* 31(3): 447–471.

Katz, Sandor Ellix. 2006. *The revolution will not be microwaved: Inside America's underground food movements*. White River Junction: Chelsea Green.

Kristensen, Tage S. 1991. Sickness absence and work strain among Danish slaughterhouse workers: An analysis of absence from work regarded as coping behavior. *Social Science and Medicine* 32(1): 15–27.

MacDonald, James. 2014. *Financial risks and incomes in contract Broiler production*. The United States Department of Agriculture Economic Research Service. August 4. http://ers.usda.gov/amber-waves/2014-august/financial-risks-and-incomes-in-contract-broiler-production.aspx#.Vh9PXPlVhBf. Accessed 14 Jan 2016.

Marks, Robbin. 2001. *Cesspools of shame: How factory farm lagoons and sprayfields threaten environmental and public health*. Washington, DC: National Resources Defense Council and the Clean Waters Network.

Matsuoka, Atsuko, and John Sorenson. 2013. Human consequences of animal exploitation: Needs for redefining social welfare. *Journal of Sociology and Social Welfare XL* 4: 7–32.

Miller, Catherine. 2001. Childhood animal cruelty and interpersonal violence. *Clinical Psychology Review* 21(5): 735–749.

Newport, Frank. 2012. In U.S., 5% consider themselves vegetarians. *Gallup.com*. July 26. http://www.gallup.com/poll/156215/consider-themselves-vegetarians.aspx. Accessed 14 Jan 2016.

Phetteplace, Hope W., Donald E. Johnson, and Andrew F. Seidl. 2001. Greenhouse gas emissions from simulated beef and dairy livestock systems in the United States. *Nutrient Cycling in Agroecosystems* 60: 99–102.

Schlosser, Eric. 2002. *Fast food nation: The dark side of the American Meal*. New York: Houghton Mifflin.

Shepard, Jonathan. 2010. *The Sharecroppers*. Film Documentary. Youtube.com and http://thesharecroppers.blogspot.com/. Accessed 14 Jan 2016.

Smith, R., A. Avendaño, and J. Martínez Ortega. 2009. *Iced out: How immigration enforcement has interfered with workers' rights*. Washington, DC: AFL-CIO. http://digitalcommons.ilr.cornell.edu/laborunions/29/.

Steinfeld, Henning, et al. 2006. *Livestock's long shadow: Environmental issues and options*. Rome: The Food and Agriculture Organization of the United Nations. http://www.fao.org/docrep/010/a0701e/a0701e00.HTM.

The Pew Campaign to Reform Industrial Animal Agriculture. 2013. *The business of Broilers: hidden costs of putting a chicken on every grill*. Washington, DC: Pew Environment Group. http://www.pewtrusts.org/en/research-and-analysis/reports/2013/12/20/the-business-of-broilers-hidden-costs-of-putting-a-chicken-on-every-grill. Accessed 14 Jan 2016.

The United States Meat Industry at a Glance. 2016. *The North American Meat Institute*. https://www.meatinstitute.org/index.php?ht=d/sp/i/47465/pid/47465. Accessed 14 Jan 2016.

Thierman, Stephen. 2010. Apparatuses of animality: Foucault goes to a slaughterhouse. *Foucault Studies* 9: 89–110.

U.S. Department of Health and Human Services, Centers for Disease Control and Prevention, and the National Institute for Occupational Safety and Health. 2014. *Evaluation of Musculoskeletal disorders and traumatic injuries among employees at a poultry processing plant*, by Kristin Musolin, et al. http://www.cdc.gov/niosh/hhe/reports/pdfs/2012-0125-3204.pdf

U.S. Department of Labor. 2015. *National Census of fatal occupational injuries in 2014 (Preliminary Results)*. http://www.bls.gov/news.release/pdf/cfoi.pdf

U.S. Department of Labor. Occupational Safety and Health Administration. *Safety and health topics: Meat packing industry – OSHA standards*. https://www.osha.gov/SLTC/meatpacking/standards.html. Accessed 14 Jan 2016.

U.S. Government Accountability Office. 2005. *Workplace safety and health: Safety in the meat and poultry industry, while improving, could be further strengthened*. GAO-05-96. Washington, DC. http://www.gao.gov/products/GAO-15-477

Zuraw, Lydia. 2014. Inter-American Commission on Human Rights Hears Testimony on Poultry Worker Safety. *Food Safety News*, March 27. http://www.foodsafetynews.com/2014/03/iachr-hearing/#.Vh9E7PlVhBd. Accessed 14 Jan 2016.

Human Rights are Animal Rights: The Implications of Ethical Veganism for Human Rights

Jeanette Rowley

Introduction

Veganism is currently surging in popularity[1] in response to concern and compassion for nonhuman[2] others, involving a growing social interest in environmental and humanitarian issues and the health benefits of a plant-based diet. The popularity of veganism has generated much positive press coverage replacing historically negative perceptions[3] and brought about a broad range of vegan-friendly consumables, for which there is significant demand. The original concept of veganism as a lifestyle adopted to express moral integrity concerning the appropriation and suffering of nonhumans[4] has, in recent times, come to be regarded as strategically instrumental in countering wider oppressive structural forces

Jeanette Rowley is a Doctoral student at Lancaster University, United Kingdom.

J. Rowley (✉)
Lancaster University, Bailrigg, UK
e-mail: j.rowley2@lancaster.ac.uk

J. Castricano, R.R. Simonsen (eds.), *Critical Perspectives on Veganism*,
The Palgrave Macmillan Animal Ethics Series,
DOI 10.1007/978-3-319-33419-6_4

that result in the abuse of nonhumans. On this basis, veganism is also a lifestyle choice adopted by some feminists, anticapitalists, humanitarians, and "green" campaigners who promote the intersecting nature of oppression.[5] The original rationale of veganism, however, as a means of elevating the moral standing of nonhuman others in their own right, remains significant in the normalization of the vegan lifestyle. This specific meaning of veganism, commonly referred to as ethical veganism,[6] is recognized in Europe and the UK as qualifying for legal protection in human rights and equality frameworks.[7] Under human rights and equality law, the practical manifestation of the beliefs of ethical vegans—those whose lifestyles revolve around respect for the moral standing of nonhuman others—is accommodated and protected. As such, ethical veganism is significant in human rights discourse and in the postmodern call for a reorientation of the moral grounds for protective rights.

The institution of human rights is already under stress and may be on the verge of granting legal rights to some nonhumans[8] on the basis that they possess characteristics that are sufficiently close to those alleged innate to autonomous human identity.[9] While the sentience[10] of nonhuman beings, although recognized in law,[11] is deemed insufficient to confer moral standing, their sapience—the recognition of complex cognition or intelligence—commands more attention. In a postmodern era that has seen a rapid expansion of human rights, some nonhuman animals may, therefore, be deemed to satisfy the criteria for moral standing and may soon be included in the category of beings that are allegedly in possession of autonomy and, subsequently, dignity and rights. This view of rights is congruent with what is known as the autonomy orthodoxy.[12] It is the traditional concept of a human rights enterprise that, as Kelly Oliver[13] has observed, is significantly problematic in the effort to dismantle the anthropocentric human–animal boundary.

There is, however, an alternative view of the moral basis of rights that challenges the notion of human identity central to the autonomy orthodoxy and, in so doing, has positive implications for the moral standing of nonhumans. This view is developed by human rights scholars in light of Emmanuel Levinas's ethics of alterity and prioritizes responsibility intrinsic to the innate human capacity to welcome the vulnerable, sentient other.[14] This view holds that the ethical grounds for rights ought to promote responsibility to others because human society is created through relationships with others, not through the actions of autonomous individuals. On this view, the significance of intertwined compassionate relationships supersedes autonomy as fundamental to human Being, and it is on this basis that critical human rights scholars call for a reorientation of human rights. As a way of life that primarily emphasizes responsibility and extends an ethic of care and compassion, I argue that the otherness of ethical veganism represents this notion of human identity.

In a human rights system that protects ethical veganism, two questions are raised. The first questions the notion of human identity operational in human rights, and the second questions the moral principle of justice, on which human rights operate. In response to these questions, I argue that ethical veganism brings to human rights an identity of otherness and that otherness is, therefore, operational as a principle of justice. The otherness of ethical veganism brings law to a transformational threshold, at which the moral standing of nonhuman others is recognizable in human rights. As such, ethical veganism mobilizes both the form of human identity and otherness as the transcendental principle of justice sought by critical human rights scholars. On this view, the ethical foundation of rights also concerns nonhuman others. As such, ethical veganism in human rights strikes at principles central to the speciesist and exceptionalist notion of human identity, honored in the long-standing autonomy orthodoxy of human rights.

The argument that human rights are also the rights of nonhuman animals is developed in the context of and value for natural moral rights and three additional foundational ideas. These foundational ideas concern the implicit recognition of the moral standing of nonhuman others in a body of critical human rights scholarship that calls for the rejection of autonomy and reason as the moral baseline for protective rights, Emmanuel Levinas's ethics of alterity, and the increasing popularity of veganism and its protection under human rights law. These foundational ideas are used to illustrate that nonhumans have *a priori* rights that are not dependent on subject-specific characteristics enabling a sovereign of autonomous agency. Rather, they highlight the moral standing of nonhuman others and the way it is transported to human rights by the telos of ethical veganism. From this perspective, it is argued that an alternative, compassionate human identity and the moral standing of nonhumans are both recognized and accommodated by human rights law. As such, ethical veganism is significant in the existing call for a reorientation of the moral grounds for protective rights and urges the logical explicit inclusion of all nonhumans.

To explain this theory, I will firstly outline the moral basis of human rights and how current claims for the rights of some nonhuman animals are brought into the existing philosophical framework. I will then highlight a branch of human rights scholarship that rejects the existing moral basis of human rights in favor of the adoption of Levinas's ethical philosophy of otherness and explain how the general rationale of this body of literature implicitly concerns nonhuman others. Following this, I will portray ethical veganism as an identity in a simplified overview of Levinas's philosophy[15] to explain its application to critical human rights scholarship. The aim is to illustrate the moral standing of all nonhumans, how veganism operates as a transcendental mechanism in human rights, and why ethical veganism necessitates the explicit provision of protective rights for all species.

The Nature of Human Rights

Article 1 of The Universal Declaration of Human Rights[16] states: "All human beings are born free and equal in dignity and rights. They are endowed with reason and conscience…" Reason and conscience are, thus, the innate characteristics that define and determine human identity and a moral entitlement to protective rights. Consequently, rights arguments focus on the subject. They seek to identify the intrinsic dignity-inducing characteristics that determine entitlement to protection. These characteristics are established and entrenched in the ontology of human rights and concern the subject as an individual with self-determining agency and reason. They are also the anthropocentric, exclusive, supremacist and exceptionalist, Kantian-inspired foundations of a protective rights enterprise that place an impossible burden of proof on nonhumans to demonstrate their moral standing.

Originally, human rights were limited by modernist notions of capability. They were restricted to men, but not granted to all men. Women, children, the colonized, barbarians, the impoverished, the insane, and other minority groups, were also excluded.[17] Such groups of human beings were deemed incapable of autonomy and reason, considered not fully human and, thus, unworthy of human rights.

The postmodern expansion of human rights, in a new light of universality, gives way to inclusion: to women, to different cultures, to the disabled, and to other marginalized groups, such as those in same-sex relationships whose oppression and exclusion have found legitimate expression. In this context, advocates for nonhuman others are gaining ground. The Great Ape Project seeks rights for "close relatives"[18]; Steven Wise, spearheading the Nonhuman Rights Project,[19] seeks rights for some nonhumans on the basis of complex cognition, and India, a nation that has recognized the intelligence of dolphins, suggests that they should be considered nonhuman persons with their own specific rights.[20] In the postmodern expansion of human rights then, the inclusion of some nonhumans appears within reach. In these circumstances, the status of nonhuman Being is firmly on the political agenda, especially for those privileged enough to come within the scope of the exceptionalist framework of worthy identity as a result of their ability to signify a level of cognition similar to that of human Being, deeply cemented in the autonomy orthodoxy.

The autonomy orthodoxy is the Kantian-inspired validation for humanity's moral claim to protective rights. The claim lies in the certainty that autonomous reason is an ontological fact of human Being: an *a priori* human attribute that has long differentiated the human from the animal kingdom. The idea, however, that human beings are governed by autonomous will, is challenged by Emmanuel Levinas, whose thesis on ethics details how human Being is constituted in and through relationships with others. In Levinas's philosophy, the other has an innate affective presence, to which the subject willingly responds with a compassionate welcome. This idea, which is that the other is *a priori* the subject, dissolves any notion of the primacy of autonomy. For Levinas, the other is pre-ontological: an *a priori* entity that transcends the subject's autonomy.

In its search for ethical foundational principles, the postmodern and posthuman[21] debate in human rights concerns these notions of human Being. It explores the extent to which human beings are in possession of an autonomous construction of morality, or whether, indeed, they are subject to an alternative phenomenon that constitutes their ethical Being. This exploration follows a particular view of human rights that has identified Kantian values as promoting an unethical humanity that espouses a focus on self at the expense of others and community. The starting point for this debate is the observation that the project of human rights has, at worst, failed and, at best, operates unethically.

The Inadequacy of the Autonomy Orthodoxy of Human Rights

Disillusioned by the failure of human rights, Costas Douzinas,[22] Upendra Baxi,[23] and William Paul Simmons[24] observe that rights abuses are prevalent across the globe and that despite many decades of the existence of a universal declaration on human rights and fundamental freedoms, suffering is widespread and immense. These scholars indicate that a significant factor in the failure of human rights is the false ontological assumptions about human identity and morality that are entrenched at the heart of human rights. This branch of scholarship emerged from what has come to be known as the Protest School[25] of human rights. Protest scholarship develops a "suffering narrative" which emphasizes that the mission of the human rights project is the amelioration of suffering and not the protection of the individual autonomous ego. Before exploring further the attraction of Levinas's theory for critical rights scholars, it will be helpful to explain why they object to current values, entrenched by human rights, and illustrate how their suffering narrative relates to notions of human identity and the question of the animal.[26]

Postmodern critical human rights scholar, Costas Douzinas, argues fervently against the current ontology of human rights. Human rights, for Douzinas, deny intertwined human Becoming and do little but serve homogeneity and maintain a Kantian egoistic same-for-self ideology. As such, current human rights are individualistic and an ineffective means to eliminate suffering. What Douzinas means by "same-for-self" is the reduction of difference to sameness through a homogenizing process that rejects the plurality of difference in vulnerability and suffering. Douzinas urges that the obligation of human rights is to recognize intertwined human Becoming and heterogeneity, give priority to unique singular others, attend to suffering, and empower oppressed victims. Douzinas claims that, currently, human rights legitimize Kantian same-for-self morality rather than facilitate emancipation from suffering. Maintaining that the essence of human Being is compassion to different vulnerable others, Douzinas argues that current human rights betray humanity.

For Douzinas, current human rights betray humanity because they function as a mechanism of personal and social delusion. They reduce heterogeneity and promote human Being as self-orienting in its Becoming rather than as coexisting in the context of a community of uniquely different others. Human rights betray individuals who are denied their true nature as ethical beings constituted in an intersubjective context. As a consequence, human beings bear the burden of isolation. In constructing human beings as essentially isolated, autonomous and atomized, human rights have encouraged same-for-self claim rights at the expense of compassionate duty for the unique, vulnerable, suffering other.

Rejecting autonomous Being, and in consideration of one's absolute duty to the other, Douzinas asserts that "before my right and before my identity as organised by rights, comes my obligation, my radical turn towards the claim to respect the dignity of the other."[27] Since rights have meaning only in this context, then "the right of the other always and already precedes mine."[28] What Douzinas means by this is that in rejecting the autonomy orthodoxy, the moral basis for human rights concerns the phenomenological experience of the primacy of the significant, ethically inducing other. What Douzinas wants in a framework of protective rights is a transcendental ethical principle to be entrenched in human rights. Such a principle will promote the primacy of the duty to the unique and vulnerable other.

Douzinas, therefore, argues that human rights ought to transcend the subject and reclaim their purpose as functional in a community of unique others, whereby their main end is to resist domination and oppression. In this regard, Douzinas calls for a rejection of autonomy and reason as the moral basis of human rights and, instead, looks to the utility of Levinasian ethics of alterity—a philosophy of how ethics arise in the face of the absolutely different other—as a robust moral principle for grounding human rights.

Similarly, Upendra Baxi attacks the Kantian values entrenched in the foundational architecture of human rights and urges the uptake of

Levinas's ethics of duty to unique others. For Baxi, the proclamation of rights for autonomous and rational subjects has proved to be a disastrous ontological construction that caused a hierarchical system and excluded many. Baxi emphasizes that human rights must entrench the right of the other to be different and make explicit the essence of human being as that which prioritizes compassion for vulnerability and suffering. For Baxi, human rights have an obligation to promote humanity's compassionate welcome to otherness in all its difference and to create the conditions in which empathy overrides self and the subject, putting first and foremost the experience of the other before acting in self-prioritization.

Baxi is a realist who believes in embracing plurality. He believes that human rights are dynamic and able to constitute new contexts as the institution evolves to eradicate all oppression. In achieving these new contexts, Baxi argues that bearers of human rights are obliged to critically evaluate their interests and that critical and positive morality requires a continual process of "de-/re-/valuation."[29] Baxi reiterates the importance of acknowledging that difference[30] is the fundamental principle of a project to furnish those who suffer with protective rights. In this context, difference means the absolutely dissimilar Being that is "non-interchangeable, incomparable and unique."[31] Baxi applies this sentiment to human rights, upholding them as representing recognition, respect and protection of difference, not as entitlements for homogenizing sameness. With regard to expanding the heterogeneity of Being, Baxi emphasizes that what human rights can be remains to be seen in the context of its dynamic nature and its ability to exist in new and unimagined paradigms.

Deconstructionist William Paul Simmons also advocates the adoption of Levinas's philosophy. Simmons believes that the ontology of current human rights assimilated all humanity as being in possession of autonomous agency but is blind to difference and unable to hear the voices of marginalized groups. For Simmons, human rights create homogeneity, which he regards to be the original violence of the creation of human rights. Simmons argues that this original violence of homogeneity "cauterizes"[32] the marginalized other. By this, he means, firstly, that human rights is an institution which declares others as rightless; secondly, that human rights dismisses the voice of the other; and, finally, that the voice of the other is powerfully portrayed as unimportant, facilitating a functional dumbing down of feelings associated with compassion for the other's suffering.

In this theoretical context of incongruent foundational concepts to true human identity and an absence of otherness in human rights, Simmons presents an antihegemonic operational principle of justice for human rights practice in the principle of the saturated marginalized other. The most saturated, marginalized other is the other who presents to the subject an ego-overwhelming hyper-presence. This principle is based on a combination of phenomenological principles and the Levinasian concrete, nonabstract ethical encounter with the other as that which affects the autonomy of the subject. Emphasizing that philosophy has an important role to play in deconstructing unethical homogeneity in the existence and operation of human rights, Simmons adopts Levinas's ethics of alterity as a foundational principle for its invigoration. As such, Simmons reiterates the failure of the autonomy orthodoxy, the need to recognize and entrench heterogeneity, and the primacy of the suffering of the unique, absolutely different other.

In the interests of ethics and the enhancement of the operational success of human rights, the ideas of Emmanuel Levinas are utilized by critical human rights theorists in their call for a reorientation of the ethical foundations of human rights. Protest scholars are influenced by Levinas's philosophy because it emphasizes intersubjectivity between unique beings and a compassionately responsive human identity as constituting the very essence of ethics itself. Ethics for Levinas emerge in the context of a compassionate human nature manifest in a concrete, nonabstract, dutiful relationship with the autonomy-limiting other. As such, critical scholars believe that the primacy of the vulnerable unique other, rather than an autonomous individual, should be paramount and explicit in the ethics and normative operation of human rights. This scholarship argues that if Levinas's philosophy was adopted in the moral foundation of human rights, then the notion of human identity, operational in human rights, would be that which willingly succumbs to the call of the other before it concerns itself with rights-for-the-self: human rights would be a human rights of the other, in recognition of duty to vulnerability and for the alleviation of the other's suffering.

Levinas's Ethics of Alterity

Levinas's ethics center on his notion of human identity. As mortal subjects at the mercy of the life process, he identifies human beings as, essentially, vulnerable and weak. For Levinas, the "extreme precariousness"[33] of Being is the basis for universal awareness of shared mortality, suffering and vulnerability. In this context, Levinas describes the identity of the subject as a being with an innate response-ability to welcome the vulnerable other. The vulnerable, mortal other, on the other hand, automatically, without effort or intention, takes the subject hostage through a gentle harnessing of their Being. In a relationship of intertwined becoming, the subject has response-ability and desires to offer an asymmetrical, compassionate welcome. For Levinas, ethics begin with this automatic recognition of the vulnerable other. In oversimplified terms, Levinas's ideas about human Being can be thought of as a phenomenological event of cause and effect. The call of the other "causes" the ethical becoming of the subject, but the subject willingly submits to the call.[34]

For Levinas, ethics arise because of an affective presence intrinsic to Being. This presence is revealed by the expression of what he calls, the "face."[35] Being in possession of face, as explained by Levinas, is being in possession of an innate power that impacts on the subject such that a response is willingly offered. The face, for Levinas, is not the physical face but is an expression of Being that can be presented through, for example, "the whole body—a hand or a curve of the shoulder."[36] Levinas makes it clear that the face is the expression of a presence that represents a "wordless vocalisation of suffering."[37] Ethics arise in the encounter with the face of the other because it reveals an infinite, unknowable presence and generates the moral rule "thou shalt not kill."[38]

In this context, the other and the subject are bound together in an intertwined existence. This asymmetrical welcome, which is freely offered to the other who expresses their Being, is, for Levinas, ethics itself. The presence of the other induces an ethical response-ability in inescapable, intertwined relationships, in which the subject is bound by the effect of the other's *a priori* metaphorical "speaking." In this way, the other expresses an innate, affective presence that enables the subject to extend a compassionate welcome. The subject is, thus, responsible to the other.

In very simple terms, Levinas regards suffering, vulnerability, caring and compassion as essential human attributes. His philosophy constructs a human identity that is vulnerable to external forces of mortality and, as such, is always and already universally predisposed to welcome vulnerable others in all their difference. In the presence of the other emerges the ethical event and the uncertainty of total autonomous agency. Levinas thus argues that the autonomy of human Being is overridden by external forces.[39] Human Being is not autonomous but formed and dependent on an intertwined existence with others. It is in this context of intertwined relationships that ethics arise. For Levinas, ethics and ethical humanity emerge in and, importantly, because *of* relationships with others, not as a result of a freethinking autonomous will.

With regard to nonhuman others, Levinas gave little thought to whether or not they were capable of generating an ethical response.[40] His thesis, however, has attracted significant and sustained attention,[41] culminating in the (perhaps common sense) affirmation that the vulnerability and suffering of nonhumans are both widely recognized and an acknowledged reality to which human beings respond. As such, the ethics of alterity directly concern nonhuman others and, in this regard, ethical veganism can be seen as the ultimate expression of a response to their moral standing.

Though none of the protest scholars mentioned above address the moral standing and suffering of nonhuman others, the suffering narrative of human rights emphasizes the rejection of the autonomy orthodoxy and that its mission is the recognition of the vulnerable other and the alleviation of suffering. As such, turning to the philosophy of Emmanuel Levinas for foundational ethical principles for the existence and operation of human rights, the contested animal–human distinction is brought to the debate of moral standing in the ontology of human rights.

Critical Human Rights and the Implicit Recognition of the Moral Standing of Nonhumans

Critical human rights scholars theorize the orthodoxy of autonomy as unethical for humanity. They expose it as betraying true compassionate human identity and as cauterizing and silencing the Being of vulnerable others. As such, the autonomy orthodoxy of human rights is prominent at the contested boundary of human and nonhuman Being. In terms of the question of the animal, the observations of these scholars are significant in the context of "accepted wisdom, that the distinctions between humans and animals are conceptually and materially indecisive."[42] In this regard, the suffering narrative of human rights implicitly recognizes that a human rights law that espouses the autonomy orthodoxy also upholds and protects a notion of human integrity at the expense of nonhuman others. Whereas human beings bear the burden of isolation and personal and social delusion, nonhuman others bear the violent consequences of such atomizing.[43] On this view, human rights function as an oppressive mechanism manifest as a moral obstacle to the dissolution of the species boundary and an ontological constraint on human ethical Becoming.

The moral standing of nonhuman others is, thus, implicitly recognized in human rights theory on the basis of its rejection of the autonomy orthodoxy, its target being the amelioration of suffering, and in its utilization of an alternative philosophy of human Being which emphasizes duty to the unique, suffering other as paramount and significant in human Becoming. In addition to the implicit inclusion of the moral standing of nonhuman others in critical human rights theory, the telos of ethical veganism is significant in its search for a transcendental identity, in which saturated otherness is operational both in human Becoming and as a principle of justice in human rights.

Critical Human Rights and the Need for a Transcendental Identity

In their rejection of the autonomy orthodoxy, critical human rights scholars speak of a human identity that is formed, not on the basis of sovereign qualities but of intertwined Being, influenced by and responsive to external stimuli related to the presence of others. They emphasize a human nature willing to accommodate a duty to suffering and argue that the ethics of human rights require explicit reference to intersubjectivity rather than to the human rights construction of isolated, atomized individuals.

In this regard, Levinasian otherness is the transcendental principle of justice sought and the preferred concept for a reinvigorated notion of human identity in human rights. It is a principle that is recognized as transcendental because, in operation, it overthrows the autonomy orthodoxy and entrenches compassionate relations as the basis of ethics. For Douzinas, Baxi and Simmons, the utility of Levinas's philosophy is that it accommodates duty to vulnerability and suffering and recognizes unique singularity and heterogeneity. In this context, ethical veganism is significant in itself and in facilitating otherness by transporting the moral standing of nonhuman others to practical human rights.

Ethical Veganism as Levinasian Otherness

Living a life of ethical commitment[44] and indistinction[45] regarding the moral status of nonhuman others, from which there is apparently no escape,[46] ethical vegans represent a notion of human identity that is significant in the call for a transcendental identity of otherness to operate in human rights. In the spirit of otherness, ethical vegans respond to the unique, singular, unknowable other through "an all-encompassing ethic that infuses every aspect of a person's being"[47] and one that "cannot be removed."[48] This is evidenced in literature[49] commenting on the essence of veganism as representing a culture grounded in morality and ethics. Espousing the development of a human identity that extends asymmetrical, compassionate consideration and welcome for other-than-human life, vegan identity illustrates the capacity of human Being to espouse a transcendental ethics of compassion that is extended to nonhuman others at a level beyond normative behaviors, such as those concerning loving "pets." On this view, the telos of vegan ethics operates on a posthuman level of Being.

Rather than representing an autonomous, atomized ego, veganism as Levinasian identity represents the other. In a profound, asymmetrical, self-for-other relationship with nonhumans, ethical vegans respond to the signification of Being and moderate their lifestyle choices to accommodate and promote the moral standing of the ethically inducing other. The ethical demand to welcome and care for the absolutely different other and to be partly constituted by this asymmetrical relationship is met willingly and freely by ethical vegans. This asymmetrical relationship allows compassion, the essence of human Being—as represented by the innateness of the ethical obligation in Levinasian philosophy—to flow freely in the face of infinite difference and response-ability. In this regard, ethical veganism, as responsive to a Levinasian infinite demand, is supported by the fact that the life choices of ethical vegans are made in recognition of the moral standing of millions of beings they will never meet or know.

Instead of a homogeneous same-for-self human identity, in which personal subjectivity and ego supersede ethical integrity, ethical vegan human identity is open to absolute difference and is saturated[50] with the prerequisite *a priori* ethical obligation. Going beyond the humanist and posthuman call to respect and welcome human difference, ethical veganism is the lived expression of a human identity, imbued with Levinasian infinite compassionate duty to and responsibility for the unique, dissimilar, noninterchangeable other: the ultimate in unknowable difference, the nonhuman other.

Ethical Veganism as Transcendental in Human Rights

On the basis of rights for the self, the postmodern expansion of human rights has included various groups of marginalized others—for example, women, same-sex couples and disabled people. This expansion of human rights assimilates a homogenous group of beings-for-the-self who are allegedly in sovereign command of autonomy. As such, current human rights have reduced the infinity of heterogeneity of Being and limited the potential of protective rights.

The recognition of ethical veganism[51] in human rights, however, promotes a different notion of human identity. Rather than an identity of being-for-self, ethical veganism presents the idea of a human rights of otherness and with it the idea that protective rights can operate according to a human desire for an intertwined community of Being that is concerned with human rights as primarily being the expression of compassion for the vulnerable, mortal other. Rather than limit the transformational potential of rights,[52] ethical vegan human rights represent, in the practical application of human rights law, a transcendental principle of justice that is based on a positive, infinite response to the call of the nonhuman other. This is because ethical vegans do not claim the same-for-self personal rights for the individual ego in atomistic Kantian terms, but, instead, put forth claims made in the spirit of otherness: an otherness that welcomes the *a priori*, asymmetrical, ethical responsibility, and compassionate duty to the vulnerable, suffering other. When an ethical vegan comes to human rights, she does so, not to claim self-for-same ego-driven rights, but with an *a priori* ethical demand to welcome, care for, and protect the absolutely different other. Rather than same-for-self claims, ethical vegan claims manifest themselves in the form of responsibility and entrench otherness as the sought-after postmodern principle of justice. In this justice, nonhuman others find the moral grounds for protective rights.

This identity of otherness brought to human rights by ethical veganism initiates a broader positive response to the moral standing of nonhumans. Ethical vegan human rights create practical duties[53] on wider society that require participation in upholding the moral status of nonhumans. The implication is that human rights validate otherness and concern the moral standing of nonhumans. As such, ethical veganism animates otherness as a transcendental principle of justice.

The validation and operation of ethical veganism in human rights practice, therefore, indicate that human rights are on the verge of a transformative threshold at which its ontology incorporates the moral standing of nonhuman life. By recognizing ethical veganism,[54] law accommodates, protects, and advances much more than ethical vegan values. In this intersection where human rights meet the moral status of nonhumans, law facilitates the operational Levinasian transcendental principle of justice that is based on infinite responsibility to profound otherness. The legal rights of ethical vegans represent the ethics of alterity and the moral, social, and political responsibility to the absolutely different nonhuman other.

On this basis, unlike other postmodern identities that represent Kantian autonomy in law, ethical vegan identity and rights are transformational in their orientation toward a paradigm of inclusivity that dissolves the species boundary.[55] This inclusivity honors universal suffering as that which denotes moral standing, and the subsequent duty to the mortal other. An ethical vegan jurisprudence of human rights highlights an existing human identity of otherness that is relevant to the call for a transcendental principle of justice to operate in human rights. This justice, in which otherness and ethical affectivity are foundational principles, always and already concerns nonhumans and is operational in current human rights through the acknowledgment of ethical veganism.

Conclusion

This chapter has argued that human rights are also the rights of nonhumans based on the rejection of the autonomy orthodoxy and the development of a suffering narrative of human rights. It has described ethical veganism as a Levinasian compassionate human identity that transports the moral standing of nonhuman others to human rights law. It has done so to theorize ethical vegan identity as a transformative mechanism currently operational in human rights practice.

The chapter has explained that the moral basis of human rights is currently human dignity subsequent to Kantian-inspired autonomous agency. In this regard, the postmodern expansion of human rights is presently challenged by advocates for the autonomy and moral standing of *some* nonhuman others on the basis that they display sophisticated cognitive abilities, typically characteristic of humans.

Autonomous agency, as a definition of human identity in human rights, however, has come under attack from critical theorists who find significant value in Levinas's philosophy of otherness. Critical theorists offer a compelling alternative to the autonomy orthodoxy, advocating for a transcendental ethic of otherness to operate in the ontology of human rights to promote heterogeneity and respect for the infinity of difference that signifies Being. As such, with its emphasis on the dissolution of autonomy in favor of a human identity predisposed to a duty to suffering, contemporary critical human rights theory provokes questions about the moral standing of nonhumans and the significance of ethical veganism. It is argued that the critical postmodern account of human rights, in which the philosophy of Emmanuel Levinas features prominently, implicitly concerns nonhuman others with regard to their moral standing. Further, ethical veganism, as an emerging human identity that serves otherness and promotes the moral standing of others, is highly relevant to a contemporary human rights discourse that promotes the infinite heterogeneity of Being.

In particular, ethical veganism in human rights is observed as an alternative notion of human identity that espouses Levinasian otherness and brings the moral standing of nonhuman others to human rights. Human rights law responds by imposing duties on wider society to uphold the ethical convictions of vegans. The moral standing of nonhumans in human rights is, therefore, both theoretically represented and practically acknowledged. In this regard, ethical veganism supports the postmodern and posthuman call to reorient human rights from dignity in autonomous agency to dignity in responsibility to otherness. In this reorientation, rights are manifest in the form of compassionate responsibility and the grounds for human rights concern all animals. On this basis, ethical veganism in human rights is transformational.

Notes

1. For examples of contemporary press reporting, see Veronika Nagy, "Into the Mainstream: Why Society is Embracing Vegetarianism," Palatinate, accessed December 3, 2015, http://www.palatinate.org.uk/into-the-mainstream-why-society-is-embracing-vegetarianism/and Elizabeth Crawford, "Vegan is Going Mainstream, Trend Data Suggests," Food Navigator-USA.com, accessed December 3, 2015, http://www.foodnavigator-usa.com/Markets/Vegan-is-going-mainstream-trend-data-suggests.
2. This chapter may use the terms "animals," "nonhuman others," "nonhuman animals," and "nonhumans" interchangeably.
3. For an analysis of historical press coverage of veganism, see Matthew Cole and Karen Morgan, "Vegaphobia: Derogatory Discourse of Veganism and

the Reproduction of Speciesism in UK National Newspapers," *The British Journal of Sociology*, 62, no. 1, (2011): 134–153.

4. The original Vegan Society was formed in response to the suffering of nonhumans.
5. The original idea of intersecting oppression can be found in the work of Theodor Adorno and Max Horkheimer. It emerged in the discipline of Critical Theory and has been utilized by, for example, feminists and eco-feminists to illustrate and explain the mutual oppression of women and nonhumans under patriarchy. An example of prominent literature in this area is a book by Carol Adams, *The Sexual Politics of Meat* (Cambridge: Polity Press, 1990).
6. Ethical veganism denotes a way of life espousing an ethical commitment to the moral standing of nonhumans and a conscious and proactive avoidance of consuming products derived from their exploitation. This essay regards ethical veganism as the intention to express moral integrity toward (but not limited to) the nonhuman other.
7. Ethical veganism is protected as a "philosophical belief" under Article 9 of the European Convention on Human Rights and Fundamental Freedoms (ECHR) and is promoted as such by the Equality and Human Rights Commission of the UK. This protection stems from principles in the International Bill of Rights, concerning a paramount universal value in human rights that recognizes the need to live according to one's deep convictions, and from statements made in a case concerning ethical veganism which was presented to the European Court of Human Rights in 1991. See *C.W. v. United Kingdom* App no 18187/91 (ECtHR, 10 February 1993).
8. It is now believed that reports about the Argentinian case concerning the Orangutan known as Sandra being granted personhood were incorrectly reported.
9. See "Nonhuman Rights Project," accessed September 18, 2015, http://www.nonhumanrightsproject.org/.
10. To "feel," "perceive," or "experience". The term is used extensively to refer to the capacity of nonhumans to feel pain and suffer.
11. For example, the Treaty of Amsterdam explicitly recognizes the sentience of nonhuman animals.
12. This is an expression used by Simon Critchley to represent the dominant tradition in western philosophy regarding human identity. See Simon Critchley, *Infinitely Demanding: Ethics of Commitment, Politics of Resistance* (London: Verso, 2012).
13. Kelly Oliver, "What Is Wrong with (Animal) Rights", *Journal of Speculative Philosophy*, 22, no. 3, (2008): 214–217. See also Kelly Oliver, "Animal

Ethics: Toward an Ethics of Responsiveness", *Research in Phenomenology* 40, (2010): 267–280.

14. For the purposes of this chapter, "otherness" is used in the positive Levinasian sense in which the other is paramount. This is contrasted with literature utilizing the concept of "the other" as negative "othering" to explain marginalization and exclusion. The Levinasian other is a being with an inner world which cannot be known and is a singular, unique, absolutely different but paramount other. The other presents the external force that co-constructs ethical humanity.
15. Levinas's philosophy is prolific and complex. This chapter presents the essence of his ideas concerning human Being, Becoming, and the importance of the other in generating ethical humanity. For the essence of his thesis, see: Emmanuel Levinas, *Totality and Infinity: An Essay on Exteriority*, trans. Alphonso Lingis (Netherlands: The Hague, 1961/2000). For a very readable interpretation of its importance to the question of the animal, see Matthew Calarco, *Zoographies: The Question of the Animal From Heidegger to Derrida* (West Sussex: Columbia University Press, 2008).
16. "The Universal Declaration of Human Rights," United Nations, accessed September 18th, 2015, http://www.un.org/en/documents/udhr/.
17. For a very readable introduction to human rights, see Michael Hass, *International Human Rights: A comprehensive Introduction* (Oxon: Routledge, 2014).
18. "The Great Ape Project," accessed September 18, 2015, http://www.projetogap.org.br/en/mission-and-vision/.
19. "Nonhuman Rights Project," accessed December 3, 2015, http://www.nonhumanrightsproject.org/.
20. The official circular, released by the Indian Ministry of Environment and Forrest, states "…the unusually high intelligence; as compared to other animals means that dolphins should be seen as "non-human persons" and as such should have their own specific rights…". However, India has not granted specific rights to dolphins but merely suggested their claim to moral standing and banned their use in entertainment. See "Circular: Policy on Establishment of Dolphinarium – Regarding," Government of India, Ministry of Environment and Forests, Central Zoo Authority, accessed September 18, 2015, http://cza.nic.in/ban%20on%20dolphanariums.pdf.
21. Joseph Indaimo considers the notion of human identity espoused by human rights and the utility of Levinas's philosophy of otherness. Joseph Indaimo, *The Self, Ethics and Human Rights* (Oxon: Routledge, 2015).
22. Costas Douzinas, *The End of Human Rights* (Oxford: Hart Publishing, 2000).

23. Upendra Baxi, *The Future of Human Rights,* 3rd ed. (Oxford: Oxford University Press, 2008).
24. William Paul Simmons, *Human Rights Law and the Marginalized Other* (New York: Cambridge University Press, 2011).
25. For a discussion of a useful (but loosely grouped) categorization of human rights scholarship, see Marie-Bénédicte Dembour, *Who Believes in Human Rights? Reflections on the European Convention* (Cambridge: Cambridge University Press, 2006).
26. The question of the animal, posed by Jacques Derrida, concerns the contested human–animal boundary. See Jacques Derrida, *The Animal That Therefore I Am*, trans. D. Wills, ed. M. L. Mallet (New York: Fordham University Press, 2008).
27. Douzinas, *The End of Human Rights*, 465.
28. Ibid.
29. Baxi, *The Future of Human Rights*, 15.
30. Ibid.
31. Emmanuel Levinas, *Outside the Subject*, trans. Michael B. Smith (London: Continuum International Publishing Group, 1987/2008), 98.
32. Simmons, *Human Rights Law and the Marginalized Other*, 10–11.
33. Emmanuel Levinas, "Peace and Proximity," in *Emmanuel Levinas: Basic Philosophical Writings,* trans. P. Atterton, ed. Adrianne Peperzak, Simon Critchley and Robert Bernasconi (Bloomington: Indiana University Press, 1996), 167.
34. Levinas would not describe his philosophy as phenomenology due to the innate qualities in Being that facilitate ethical coexistence.
35. The explanation of the face is developed by Levinas in *Totality and Infinity: An Essay on Exteriority* from p. 39. Levinas refers to the "face" not merely as a physical object but as the most significant expression of the other's presence and expression. For a discussion of the Levinasian face and nonhumans, see Peter Atterton, "Levinas and Our Moral Responsibility to Other Animals," *Inquiry* 54, no. 6, (2011): 633–649, accessed September 18, 2015, doi: 10.1080/0020174X.2011.628186.
36. Levinas, *Totality and Infinity: An Essay on Exteriority,* 262; and Levinas, "Peace and Proximity," 166–167 refers again to the expression of face.
37. Judith Butler, "Precarious Life," in *Radicalizing Levinas,* ed. Peter Atterton and Matthew Calarco (New York: State University of New York Press, 2010), 7.
38. Emmanuel Levinas, *Ethics and infinity: Conversations with Philip Nemo,* trans. R.A. Cohen (Pittsburgh: Duquesne University Press, 1985), 89.

39. For Levinas, the power of the other is not essentially over the subject but cooperates with an innate, compassionate, welcoming mechanism within the subject. The debate in philosophical and phenomenological discourse centers on whether or not human Being is internally or externally constituted.
40. Levinas was specifically interested in theorizing, in theological terms, the destiny of humankind. With regard to nonhumans, Levinas felt that they remained basic beings, while humans were exceptional on the basis of their ethical becoming in the face-to-face event. As ethical beings, Levinas theorized that human beings transfer the idea of suffering onto nonhuman Being. This is highly contested, as is his general theological approach. For a discussion about the limitations of Levinas's original ethical theory for a workable principle of justice, see Simmons, *Human Rights Law and the Marginalized Other*. Generally, Levinas's thesis has been developed by contemporary scholars, including Simon Critchley and Judith Butler, who prefer to remove limiting factors. In this regard, the idea of transference is also disregarded in the present discussion in the light of more persuasive evidence in science concerning nonhuman Being.
41. Discussions cover several controversial aspects of Levinas's philosophy. See for example, and for further references: Jacques Derrida, *The Animal That Therefore I Am*; Dianne Perpich, *The Ethics of Emmanuel Levinas* (California: Stanford University Press, 2008); Peter Atterton, "Levinas and Our Moral Responsibility to Other Animals," *Inquiry* 54, no. 6, (2011): 633–649, accessed September 18, 2015, doi: 10.1080/0020174X.2011.628186; Christian Diehm, "Facing Nature: Levinas Beyond the Human," *Philosophy Today* 44, no. 1, (2000): 165–198; William Edelglass, Jim Hatley and Christian Diehm ed., *Facing Nature: Levinas and Environmental Thought* (Pennsylvania : Duquesne University Press, 2012); David Clark, "On Being the last Kantian in Nazi Germany: Dwelling With Animals After Levinas," in *Animal Acts: Configuring the Human in Western History,* Jennifer Ham and Matthew Senior ed., (Routledge, New York, 1997),165–198; John Llewelyn, "Am I obsessed by Bobby? (Humanism of the other animal)," in *Re-reading Levinas,* Robert Bernasconi and Simon Critchley ed., (Indiana University Press, 1991), 234–245; Tamra Wright, Peter Hughes and Alison Ainley, "The Paradox of Morality: An Interview With Emmanuel Levinas," trans. A. Benjamin and T. Wright, in *The provocation of Levinas: Rethinking the Other,* Robert Bernasconi and David Wood ed, 168–180. (Oxfordshire: Routledge, 1988/2014), 168–180.
42. Anat Pick, *Creaturely Poetics: Animality and Vulnerability in Literature and Film* (Columbia University Press, 2011), 1.

43. I thank Jodey Castricano for making this more explicit point in response to an earlier draft of this chapter.
44. Bob Torres, *Making a Killing: The Political Economy of Animal Rights* (Edinburgh: AK Press, 2007), 134.
45. Matthew Calarco, "We Are Made Of Meat: Interview With Matthew Calarco," Interviewer Leonardo Caffo, *Relations: Beyond Anthropocentrism* 1, no. 2, (2013): accessed September 18, 2015, http://www.ledonline.it/index.php/Relations/issue/view/3.
46. Joanne Stepaniak, *Being Vegan: Living with Conscience, Conviction, and Compassion* (Lowell House, Illinois: NTC/Contemporary Publishing Group, Inc. 2000), 112.
47. Ibid.
48. Ibid.
49. See for example: Matthew Cole, "'The Greatest Cause on Earth': The Historical Formation of Veganism as Ethical Practice," in *The Rise of Critical Animal Studies: From the Margins to the Centre,* Nik Taylor and Richard Twine ed., (Routledge, Oxon, 2014), 203–224; Barbara McDonald, "Once You Know Something, You Can't Not Know It: An Empirical Look At Becoming Vegan," *Society & Animals*, 8, no. 1, (2000), 1–23; Rachel, MacNair, "McDonalds "empirical look at becoming vegan"," *Society & Animals* 9, no. 1, (2001), 63–69; Victoria Moran, *Compassion The Ultimate Ethic: An Exploration Of Veganism,* 3rd ed., (New Jersey, NJ. The American Vegan Society, 1991); Stepaniak, *Being Vegan: Living with Conscience, Conviction, and Compassion.* See also on the subject of vegetarianism as a way of Being: Kenneth Joel Shapiro, "I am a vegetarian": Reflections on a Way of Being', *Society & Animals*, 2, no 2, (2014), 128–147.
50. Simmons, *Human Rights Law and the Marginalized Other*, uses the idea of phenomenological saturation in his development of a political principle for justice which is developed from a combination of the ethics of alterity and phenomenological theory.
51. Comments made by the Equality and Human Rights Commission of the United Kingdom and in *C.W. v. United Kingdom* App no 18187/91 (ECtHR, 10 February 1993).
52. Douzinas, *The End of Human Rights*, and Indaimo, *The Self, Ethics and Human Rights*, overlooking the potential of ethical veganism in human rights, believe that human rights have reduced the heterogeneity of the infinity of alterity and the ethics of otherness to the *genus* of sameness in the signification of being.
53. In accommodating ethical vegans, providers of goods and services are required to participate in upholding the ethical convictions of vegans by

sourcing and supplying consumables that are not derived from nonhuman others. The moral standing of nonhumans is, therefore, upheld by nonvegans and wider society as a legal requirement in the protection of ethical veganism.

54. The UK Equality and Human Rights Commission literature frequently refers to veganism as a protected lifestyle under the ECHR, the Human Rights Act 1998 and The Equality Act 2010. Public Authority provisions for vegans include food, work wear (uniform items, footwear, accessories), and educational items in school (such as items not made from the skin of other species or containing substances derived from other species). The Equality Act 2010 imposes a duty to accommodate veganism. This is the interpretation of provisions in the International Bill of Rights concerning the human right to live according to one's deep convictions.
55. For a discussion about balancing the rights of equal beings, see Doug Halls, "Agency, vulnerability and Societas: Toward a Levinasian Politics of the Animal," in *Facing Nature: Levinas and Environmental Thought,* ed. William Edelglass, Jim Hatley and Christian Diehm (Pennsylvania: Duquesne University Press, 2012): 41–65.

Bibliography

Adams, Carol. 1990. *The sexual politics of meat.* Cambridge: Polity Press.

Atterton, Peter. 2011. "Levinas and our moral responsibility to other animals." *Inquiry* 54(6): 633–649. doi:10.1080/0020174X.2011.628186. Accessed 18 Sept 2015.

Baxi, Upendra. 2008. *The future of human rights,* 3rd ed. Oxford: Oxford University Press.

Butler, Judith. 2010. "Precarious life." In *Radicalizing Levinas,* edited by. Peter Atterton and Matthew Calarco, 3–19. New York: State University of New York Press.

C.W. v. United Kingdom App no. 18187/91 European Court of Human Rights. http://hudoc.echr.coe.int/eng#{"fulltext":["vegan"],"article":["9"],"documentcollectionid2":["GRANDCHAMBER","CHAMBER","DECISIONS","COMMUNICATEDCASES"],"itemid":["001-1503"]}

Calarco, Matthew. 2008. *Zoographies: The question of the animal from Heidegger to Derrida.* West Sussex: Columbia University Press.

Calarco, Matthew. 2013. We are made of meat: Interview with Matthew Calarco. Interviewer Leonardo Caffo. *Relations: Beyond Anthropocentrism* 1(2). http://www.ledonline.it/index.php/Relations/issue/view/3. Accessed 18 Sept 2015.

Circular: Policy on Establishment of Dolphinarium. 2013. Government of India, Ministry of Environment and Forests, Central Zoo Authority. http://cza.nic.in/ban%20on%20dolphanariums.pdf

Clark, David. 1997. On being the last Kantian in Nazi Germany: Dwelling with animals after Levinas. In *Animal acts: Configuring the human in western history*, edited by Jennifer Ham and Matthew Senior, 165–198. New York: Routledge.

Cole, Matthew. 2014. The greatest cause on Earth: The historical formation of veganism as ethical practice. In *The rise of critical animal studies: From the margins to the centre*, edited by Nik Taylor and Richard Twine, 203–224. Oxon: Routledge.

Cole, Matthew, and Karen Morgan. 2011. "Vegaphobia: Derogatory discourse of veganism and the reproduction of speciesism in UK national newspapers." *The British Journal of Sociology* 62(1): 134–153.

Crawford, Elizabeth. 2015. "Vegan is going mainstream, trend data suggests." *Food Navigator-USA.com,* March 17. http://www.foodnavigator-usa.com/Markets/Vegan-is-going-mainstream-trend-data-suggests. Accessed 3 Dec 2015.

Critchley, Simon. 2012. *Infinitely demanding: Ethics of commitment, politics of resistance*. London: Verso.

Dembour, Marie-Bénédicte. 2006. *Who believes in human rights? Reflections on the European convention*. Cambridge: Cambridge University Press.

Derrida, Jacques. 2008. *The animal that therefore I am*, edited by Marie-Louise Mallet. Translated by D. Wills. New York: Fordham University Press.

Diehm, Christian. 2000. "Facing nature: Levinas beyond the human." *Philosophy Today* 44(1): 51–58.

Douzinas, Costas. 2000. *The end of human rights*. Oxford: Hart Publishing.

Edelglass, William, Jim Hatley, and Christian Diehm (eds.). 2012. *Facing nature: Levinas and environmental thought*. Pennsylvania: Duquesne University Press.

Equality and Human Rights Commission. http://www.equalityhumanrights.com/your-rights/human-rights/what-are-human-rights%3F/the-human-rights-act/freedom-of-thought-belief-and-religion. Accessed 9 Dec 2015.

European Convention on Human Rights. Council of Europe. https://ec.europa.eu/digital-agenda/sites/digital-agenda/files/Convention_ENG.pdf. Accessed 9 Dec 2015.

Great Ape Project (GAP). http://www.projetogap.org.br/en/. Accessed 11 Oct 2015.

Halls, Doug. 2012. "Agency, vulnerability, and societas." In *Facing nature: Levinas and environmental thought*, edited by William Edelgalss, James Hatley, and Christian Diehm, 41–65. Pittsburgh: Duquesne University Press.

Hass, Michael. 2014. *International human rights: A comprehensive introduction*, 2nd ed. Oxon: Routledge.

Indaimo, Joseph. 2015. *The self, ethics and human rights*. Oxon: Routledge.

Levinas, Emmanuel. 1961/2000. *Totality and infinity: An essay on exteriority*. Translated by A. Lingis. Netherlands: The Hague.

Levinas, Emmanuel. 1966. "Peace and proximity." In *Emmanuel Levinas: Basic philosophical writings*, edited by Adriaan T. Peperzak, Simon Critchley and Robert Bernasconi'. Translated by P. Atterton. 161–170. Bloomington: Indiana University Press.

Levinas, Emmanuel. 1985. *Ethics and infinity: Conversations with Philip Nemo*. Translated by R.A. Cohen. Pittsburgh: Duquesne University Press.

Levinas, Emmanuel. 2008. *Outside the subject*. Translated by Michael B. Smith. London: Continuum International Publishing Group.

Llewelyn, John. 1991. "Am I obsessed by bobby? (Humanism of the other animal)." In *Re-reading Levinas*, edited by. Robert Bernasconi and Simon Critchley, 234–245. Bloomington: Indiana University Press.

MacNair, Rachel. 2001. McDonalds 'empirical look at becoming Vegan'. *Society & Animals* 9(1): 63–69.

McDonald, Barbara. 2000. Once you know something, You can't not know it: An empirical look at becoming vegan. *Society & Animals* 8(1): 1–23.

Moran, Victoria. 1991. *Compassion the ultimate ethic: An exploration of veganism*, 3rd ed. New Jersey: The American Vegan Society.

Nagy, Veronika. 2015. "Into the mainstream: Why society is embracing vegetarianism." *Palatinate*, December 1. http://www.palatinate.org.uk/into-the-mainstream-why-society-is-embracing-vegetarianism/. Accessed 3 Dec 2015.

Nonhuman Rights Project. http://www.nonhumanrightsproject.org/. Accessed 11 Oct 2015.

Oliver, Kelly. 2008. "What is wrong with (animal) rights." *Journal of Speculative Philosophy* 22(3): 214–217.

Oliver, Kelly. 2010. "Animal ethics: Toward an ethics of responsiveness." *Research in Phenomenology* 40: 267–280.

Perpich, Dianne. 2008. *The ethics of Emmanuel Levinas*. California: Stanford University Press.

Pick, Anat. 2011. *Creaturely poetics: Animality and vulnerability in literature and film*. West Sussex: Columbia University Press.

Shapiro, Kenneth J. 2014. "I am a vegetarian: Reflections on a way of being." *Society & Animals* 2(2): 128–147.

Simmons, William Paul. 2011. *Human rights Law and the marginalized other*. New York: Cambridge University Press.

Stepaniak, Joanne. 2000. *Being vegan: Living with conscience, conviction, and compassion*. Lowell House: NTC/Contemporary Publishing Group Inc.

Torres, Bob. 2007. *Making a killing: The political economy of animal rights*. Edinburgh: AK Press.

Treaty of Amsterdam, 2 October 1997. (Consolidated Version). European Union. http://www.refworld.org/docid/3dec906d4.html

United Kingdom *Equality Act 2010*. 8 Apr 2010. http://www.legislation.gov.uk/ukpga/2010/15/contents

United Kingdom *Human Rights Act 1998*. 9 Nov 1998. http://www.refworld.org/docid/3ae6b5a7a.html

Universal Declaration of Human Rights, 10 December 1948, 217 A (III). UN General Assembly. http://www.refworld.org/docid/3ae6b3712c.html

Wright, Tamra, Peter Hughes, and Alison Ainley. 2014. The paradox of morality: An interview with Emmanuel Levinas, edited by Robert Bernasconi and David Wood. Translated by A. Benjamin, and T. Wright. In *The provocation of Levinas: Rethinking the other*, 168–180. Oxfordshire: Routledge, 2014.

Hegel, Eating: Schelling and the Carnivorous Virility of Philosophy

David L. Clark

Emmanuel Levinas criticizes "the Hegelian enterprise" for ignoring the ethical significance of the "residue" that is not "reducible" to the work of "the Concept" [*der Begriff*], and for configuring that labor as an expression of the subject's sovereign potency.[1] Of the *Encyclopaedia Logic*'s pretensions to totality, Levinas writes: "[I]f philosophizing consists in assuring oneself of an absolute origin, the philosopher will have to efface the trace of his own footsteps and unendingly efface the traces of the effacing of the traces, in an interminable methodological movement of staying where it is" (*OB* 20). The heroic virility of this Sisyphean task blinds the

This essay was first published in *Cultures of Taste/Theories of Appetite: Eating Romanticism*, ed. Timothy Morton (New York: Palgrave, 2004), 115–139. It is being reproduced here by permission of the publisher.

D.L. Clark (✉)
Department of English & Cultural Studies, McMaster University, Hamilton, ON, Canada
e-mail: dclark@mcmaster.ca

J. Castricano, R.R. Simonsen (eds.), *Critical Perspectives on Veganism*, The Palgrave Macmillan Animal Ethics Series,
DOI 10.1007/978-3-319-33419-6_5

philosopher to the ungraspable remnant haunting speculative idealism's claim to absolute self-sufficiency as its interior limit. Philosophy faces there its radical opening toward an alterity that Levinas identifies with the subject's defenselessness before the hunger of the other. Yet Hegel denies any such liability with a brawny rhetoric of self-possession: "In fact one cannot think for someone else, any more than one can eat or drink for him."[2] Levinas insists that far from validating the project's presence of mind, Hegel's claim to eat and think for himself renders him infinitely vulnerable to the other's vulnerability. "Only a subject that eats can be for-the-other," Levinas argues, comparing this dietary self-unraveling to "snatching the bread from another's mouth" (*OB* 74). Levinas is offended that Hegel proceeds as if he could eat alone and leave no traces, not even of their erasure; he fails to recognize that in his "mouth there remains the word or the morsel of bread that it is impossible for me not to give to…my neighbour."[3]

Levinas's characterization of Hegel as a gluttonous, selfish eater, a supreme instance of what Jacques Derrida calls "carno-phallogocentrism,"[4] derives from Hegel's most complicated critic, F.W.J. von Schelling. In his 1809 masterwork on human freedom, Schelling had introduced into speculative idealism the troublesome notion of "the indivisible remainder" [*der nie aufgehende Rest*], the indigestible morsel resisting even the greatest application of reason.[5] Schelling had tried to no avail to share this remnant with his most important philosophical "neighbor" whose company he had enjoyed in Tübingen and Jena. But were there not analogous leavings in Hegel already? For Schelling, as for Levinas, the question is more ethical than merely epistemological or ontological: "we still do not believe…that someone can be virtuous, or a hero, or a great man at all, by means of pure reason" (*PI* 282). In refusing the remainder, Hegel represents not only the apogee of speculative thought but also the greatest threat to the project first delineated in *Of Human Freedom*: to reduce virtue to the calculations and history of reason is to nullify virtue because it annihilates the abyss of freedom "to do Good or Evil" (*PI* 256). As Slavoj Žižek proposes, "Hegel reduces Evil to the subordinated moment in the self-mediation of Idea qua supreme Good"; but for Schelling, evil "remains a permanent possibility which can never be fully sublated in and by the Good."[6] Schelling responds to this crisis of consumption by parsing Hegel for the absent presence of the unsublated and unsublatable, to show that the *Logic*'s principles are not only in error but also unjust. In their ferociously reiterated claim to self-grounding autonomy (or what Schelling calls the "hunger of selfishness" and the "desirous, hungry, and poisonous" (*PI* 263) need for remainderless independence), they are paradigmatic of "evil." Since Schelling, however, must also think and eat ("in this life," Schelling notes, "it is…necessary to interiorize everything"),[7] the question is how to think and to "eat well": how to think eating *otherwise*?[8] How to teach Hegel some table manners? This is the difficult lesson that Schelling undertakes, mimicking Hegel's own gustatory fascinations so as to turn the carnivorous language of the master against himself.

First delivered in the five years following Hegel's death in 1831, *On the History of Modern Philosophy* represents the most passionate battle Schelling fought against the academic celebrity whom he had once called friend. Significant portions were later repeated almost "verbatim" in talks at the University of Berlin,[9] where, in 1841, Schelling was installed by the King of Prussia as Hegel's successor and charged with slaying the great man's intellectual progeny.[10] A more torturously overdetermined philosophical *habitus* would be hard to imagine. Hegel had once asserted that art was a thing of the past,[11] but to make a similar claim about Hegel could only have seemed a monumentally imposing problem. What could taking Hegel's place mean when, as Schelling himself earlier conceded, to reject his work was tantamount to abandoning philosophy itself?[12] Moreover, what did it mean to labor in the void of speculative philosophy's putative conclusion of itself in Hegel? Given these convoluted psychic, professional, and intellectual conditions, it is no wonder Schelling lasted in Berlin but a few years. Still, Schelling strenuously renounced the presence of Hegel's ideas both in himself and in others, perhaps no more vividly than in his ardent claim that Hegel's *Science of Logic* "completely eats up being" (*L* 153). The unearthly meal at which Schelling places Hegel is as grim as it is vigorous, a scene of incomparable appetite and predation (who or what could devour this peerless devourer?) culminating in nothing less than the destruction of everything that is. What are we to make of this *holocaust* of being? What is "eating," and what must "being" be if Hegel's philosophy absorbs it without remainder?

What's Eating Schelling?

Schelling's lectures concentrate many of his later themes, but it is Hegel's ghost with which they are most animated; like all histories, they are at some level a work of mourning. Hardly a sentence could be said not to tremble in the awful presence of the *Logic*: all major contributions to modern philosophy—including Schelling's—swirled around the dense gravitational well of its relentlessly capacious dialectic. The Hegelian Absolute Spirit, Schelling remarks, "has the function of taking up all the preceding moments into itself as that which brings *everything* to an end" (*L* 156). But he responds to this hyperbolically strong thought with his own virile pronouncements, unabashedly figuring it not as philosophy's culmination but as the "monstrous" triumph of its "negative" mode. By violently translating actuality into form and idea, Hegel "negates all having-happened, everything historical" (*L* 159). But, for Schelling, the world is more than its rational intelligibility; it is an open-ended "question of existence" (*L* 159) and the "ecstatic" confrontation with the enigma of the particular, the contingent, the other.

Almost thirty years earlier, Schelling had made a similar case:

> [N]owhere does it appear as though order and form were original, but rather as if something initially ruleless had been brought to order. This is the incomprehensible basis of reality in things, the indivisible remainder, that which with the greatest of exertion cannot be resolved in the understanding but remains eternally in the ground. (*PI* 239)

From Hegel, however, we learn that the "Absolute Idea" is "the unity of thinking and being," the pinnacle of a self-generated but "subjectless" (*L* 155) dialectical process whose conceptual determinations it is the task of the *Logic* to enact. Ostensibly, no "indivisible remainder" haunts Hegel's system; nothing is left behind (*L* 141). Schelling himself sought a version of the "unity of thinking and being"—all German idealists did. But his perennial fascination with the question of why there is thinking and being at all enabled him to refuse the destruction of the latter by the former. The extraordinary fact that there *are* thoughts and that there *is* being (including the *that* of thought) prevents their straightforward unity, and in particular, the unproblematical—remainderless—absorption of being into thought. In Hegel, this unity is the crowning achievement of speculative idealism. For Schelling it remains "the unruly" problem at its origin, the founding difficulty toward which he spent a lifetime making more or less convincing forays.

How does Hegel's system claim to incorporate "reality" [*Wirklichkeit*] (*L* 145; *SW* 10.141), and to "present itself as the absolute philosophy, as the philosophy which leaves nothing outside of itself" (*L* 133)? At the source of Hegel's self-declared "success" is his radical reconception of "the Concept." Hegel claims to have negated all traces of the negativity attending conventional notions of the Concept. Rather than the abstracting idea of a thing (and thus *not* the thing of which it is the idea), Hegel's de-negated Concept is what a thing *is* in its essence—what it incrementally reveals itself to be through the clarifying "concretion" of mediation and reflection. Nothing lies outside mediation, only entities that have "forgotten" themselves to be mediations, and are therefore imagined as "real." The Concept is all. But as Schelling remarks in *Foundations for Positive Philosophy*, "Nothing is more conceivable than the concept…Nothing is easier to transport into pure thought" (cited in White, *Absolute Knowledge*, 97). Schelling flinches at the *Logic*'s "denigration of nature" (*L* 154), its illegitimate incursion into the ontological. The *Logic*'s conceptual map, whose explanatory power Schelling is quick to concede, had become so ambitious as to overtake the world. "Concepts as such do in fact exist nowhere but in consciousness," Schelling insists, attempting to undo Hegel's perverse reversal of thought and being: "they are, therefore, taken objectively, after nature, not before it" (*L* 145).

By "withdrawing completely into pure thinking," Hegel "has splendidly expressed the essence of the truly negative or purely rational philosophy" (*L* 145). A "truly" negative system of "complete" withdrawal—note Schelling's absolutizing adjectives—must negate its negation, so that no trace of its mediation with that from which it withdraws remains. Hegel's system eats up being, and in that self-involved act of total incorporation it consumes and refuses the radical loss that is its founding contingency. What is left is "a science in which there is no question of existence, of that which really exists" (*L* 133). Of that unthought loss, Schelling laconically notes: "philosophy should have grasped this," but instead "it put itself beyond all contradiction" (*L* 133). With nothing but logical thoughts to think, the system's universe is infinitely pliable, a docile territory suffering the strange indignity of having always already been colonized. For once one limits thinking to the thought of the Concept, nothing "real" remains to withstand the maw of the mind. In this self-sealed empire, there is only the self-out-stretching but finally empty "movement" of the Concept "in pure, i.e. unresisting ether" (*L* 146). What offends is not only the Logic's ambitiousness but also its feigned humility, its concealment of the violence that enables it. For Schelling, the Owl of Minerva is not the benign afterthought to actuality but a voracious bird of prey intent on devouring everything in its path.

Hegel will always be more complicated than Schelling makes him out to be, in part because the thinkers were bound together in ways neither fully wished to acknowledge. Xavier Tilliette characterized the relationship as "star-crossed," naming the bonds of rivalry and love with which they were betrothed.[13] In this homosocial mise-en-scène, Hegel is not merely Schelling's adversary, but a figure for a certain Hegelianism in Schelling's work and memory: a "Hegel" whose work was itself haunted by recollections of an early significant encounter with Schelling (Pinkard 110). This messily interiorized "Hegel" is caught in Schelling's work of mourning, with all the heightened acts of identification and (dis)avowal that such work invariably implies—a process made more convoluted and competitive because of Hegel's claim to have consumed "Schelling" as part of speculative idealism's incorporation of all previous systems.[14]

This anthropophagic clash suspiciously resembles Immanuel Kant's image of desirous coupling as cannibalism,[15] and would partly explain the showily affective contradictions characterizing Schelling's negotiations—his impatience with Hegel for saying too much and too little, for derailing the course of philosophy and making its succession as "positive" philosophy inevitable, for being weak-minded and "monstrously" powerful. It would explain too why Schelling wishes both to anticipate and to succeed Hegel, accusing him of plagiarizing methods he had originally invented while characterizing the master's work as entirely antithetical to his own. Even this last claim is contradictory, for in identifying his position as the "positive" contrary of Hegel's "negative" philosophy, Schelling's coarsely polarizing language paradoxically fastens him to his opponent even more closely. Playing the role of the determinate negation of the master's system, Schelling proves Werner Hamacher's observation that the reach of Hegel's *Logic* is so capacious that it hungrily anticipates every dissenting reading.[16]

All You Can Eat

Hegel once remarked that *Aufhebung* was at root a grasping gesture, and in a commentary on Kant he ironically suggested that concepts have "teeth."[17] Schelling picks up on these desublimations, as if needing us to feel the clasping and tearing that underwrites the *Logic*, the desirous will of the philosopher behind the philosophy:

> [he] wants the Absolute as the result of a science, and this science is precisely the *Logic*. Therefore the Idea continually develops throughout this whole science. By Idea Hegel also means what is to be realized, what develops, and what is wanted in the whole process; it is the Idea which at the beginning is excluded from pure being, which, as it were, *eats up being*, which happens via the determinations of the concepts which are put into being; after it has completely eaten up being and transformed it into itself, it is itself, of course, the *realised* Idea. (*L* 153)

Schelling reconfigures as ingestion the rigorous, vigorous path of the Concept. From the radical emptiness, indeterminacy, and abstraction of "Being, pure Being" with which the text famously begins, *to* the "fulfilled being" whose content is identical with its thought, and therefore, the paradigm of "the Concept that grasps itself,"[18] the *Logic* tracks the unstoppable progress of a fantastic meal. The end "result" of this gluttony is a state of plenitude, "the concrete and also absolutely intensive totality." For the philosopher of "the indivisible remainder," the fact that nothing remains, not even "nothing," is as troublesome as it is astonishing. Hence Schelling's amplifying incredulity: Hegel's system not only "eats being," it "*completely* eats being." And again: the self-mediation of the "Idea" can only mean that it "has completely eaten up being" [*das Seyn ganz aufgezehrt hat*] (*L* 155; *SW* 10.154).

What truly last supper are we here being asked to condemn? That Schelling will proceed to chastise the lifelessness of this infinitely voracious system only confirms its most ghastly resonances. Death would be among the closest analogues to Hegel's system, to whom Milton's Satan grimly promises: "ye shall be fed and fill'd/Immeasurably, all things shall be your prey."[19] Like Death, the *Logic* consumes but is never itself the object of consumption. Not so much at the top of an imaginary food chain as over the top, Hegel's system encapsulates what John D. Caputo calls "a site of a metaphysical metaphorics that transports eating to the sphere of absolute eating, of absolutely carnivorous virility."[20] Kant had once playfully warned against eating and thinking simultaneously, the great philosopher of "taste" comprehending how the two acts are not so much antithetical as competing metonymies of introjection. In Hegel Schelling sees that a thinking that thinks only "concepts" is indistinguishable from a kind of eating, a highly idealized consumption that predigests what it ingests, allowing only "ether" to pass its lips: in theory, the *Logic* takes up only concepts and leaves behind the same in the form of the text's argument, ingestion and expression functioning as virtually identical expressions of the same ruminating impulse. Nothing is therefore said here about Hegel's Concept *digesting* being, about the coils and recoils of its assimilation into the phantasmatic body of the *Logic*, much less about the dregs or remainders that might naturally be assumed to result. The eating body that Schelling's trope evokes possesses an impossible morphology: it is as if the system of the *Logic* were a mouth and nothing else. Hegel's text functions in a suspiciously "angelic" manner, as if eating were magically a matter only of tasting the world.[21] In addition to eating being that text has therefore eaten being itself, an extraordinary instance of what Derrida calls "exemplorality," the phantasmatic process that "assimilates everything to itself by idealizing it with interiority, masters everything by mourning its passing, refusing to touch it, to digest it naturally, but digests it ideally, consumes what it does not consume and vice versa."[22]

Every Breath You Take

Although Žižek contends that there is an "abundance of 'anal,' excremental innuendo in Schelling" (*Indivisible Remainder* 36), the predominant figure in his lectures is oral. The abjected and excreted are significant primarily for their absence, having seceded to modes and metonymies of emission and expression, breathing and eating—all part of an elaborate topology of the mouth. Alongside the immoderate spectacle of the *Logic*'s appetite, other oral fixations animate Schelling's lecture. For example, "the meagre diet of pure being" (*L* 138) to which Hegel's system subjects itself proves more than a mouthful, a rich source of critique that the philosopher chose only to disseminate in the form of the spoken word, his voice filling the lecture halls with this talk about Hegel before and after Hegel's own voice had fallen silent. Hegel's speculative idealism dreamt of ending philosophy by assimilating all previous systems, but the mere existence of Schelling's *Lectures on the History of Modern Philosophy* puts to us that reports of that death had been greatly exaggerated. Schelling survives Hegel, complexly embodying the indivisible remainder. Indeed, in specifically creating a history, Schelling seeks to locate Hegel's work in a larger framework as one moment in an epoch whose future is uncertain rather than foreclosed. The ironic "laughter" (*L* 154) that Schelling hears in response to the carnivorous claims of Hegel's system forms the unabashed opposite to the system's "conspicuous narrow-chestedness," the ways in which its fussy restraint about certain modes of representation (i.e., tropes borrowed from nature and natural processes) "means that it cannot speak openly and express itself and it is though breath and voice have been taken from it, so that it can murmur incomprehensible words" (*L* 162). To compare the *Logic*'s self-unfolding "Idea" to "eating" is therefore to desublimate it, to choke off its desires for "subjectless" disembodiment.

Of course, Hegel can only simulate this philosophical reticence and modesty. Schelling observes that even and especially at the beginning of the *Logic*, where Hegel claims to be saying the very least one can say so as not to contaminate the dialectic's inauguration with any unthought presupposition, massively consequential premises are already at work—enumeration and succession, for example. Contemplating the initial move from "Being" to "Nothing" to "Becoming," Schelling asks: "How do I end up, here at the farthest edge of philosophy, where it hardly dare open its mouth yet, where it finds word and expression only with great effort, using the concept of number?" (*L* 148). The *Logic* is from the start overtaken by another logic, that of the supplement: the body of philosophy cannot *not* resort to "word and expression," including numbers and numbering. Hardly opening its mouth, it is already talking expansively and is thus caught in the network of differentiated signs for which numbers are kind of pure instance. The system says little, but this quietness belies how much it has already consumed at the instant that the Concept gets under way. The *Logic* cannot eat and talk at the same time; but, for Schelling, the less it says the, more it actually consumes. To suggest that Hegel's *Logic* eats is thus not only to satirize its pretensions to totalization but also to activate what remains alive but unvoiced within it. Part of the permanent complexity of coming *after* Hegel is evident in Schelling's compulsion to speak both against and for Hegel, putting words in the mouth of his adversary who, believing that the Concept articulates itself, would rather not speak at all.

Perhaps, the most striking instance of this forced ventriloquism comes late in the lecture. Schelling puzzles out what he considers the point of maximum incoherence: how one gets from the world of the Concept to that of living creatures, the passage "into the unlogical world, indeed, into the world which is opposed to what is logical" (*L* 153). Hegel is hypocritically fastidious and evasive in characterizing the transition from Absolute Idea to Nature as a "releasing" ["*entlassen*"] (*L* 155; *SW* 10.153). "Releasing" tolls Schelling back to his sole self, condensing into one term the "common mistake of every philosophy that has existed up until now"—their coy inability to confront the enigma of creation (Bowie, "Introduction" 30). Schelling balks especially at the diction's palliative defensiveness: "The expression 'release'—the Idea releases nature—," he says with mock incredulity, turning the word over as if hearing it for the first time, "is one of the strangest, most ambiguous and thus also timid expressions behind which this philosophy retreats at difficult points" (*L* 155). Schelling responds with deliberate coarseness and schematism, like a child driven to vulgarity in the face of too much refinement. Out of the mouths of babes, then: "Jacob Boehme says: divine freedom vomits [*erbricht*] itself into nature. Hegel says: divine freedom releases nature. What is one to think in this notion of releasing?" (*L* 155; *SW* 10.153). Schelling finds Hegel's tasteful avoidance of the indivisible matter at hand unpalatable, making his insistence on "vomit"—the very figure of distastefulness—as inevitable as it is shocking. "Releasing" hides the foundational crisis of thought and being that the *Logic* should acknowledge and explore: How was it that "something initially ruleless had been brought to order?" (*PI* 238) Why is there anything? As Schelling well knew, these questions unsettlingly lack answers. They model for speculative idealism what Derrida describes as the "object" of radical disgust, the disturbing remainder that "does not allow itself to be digested, or represented, or stated—does not allow itself to be transformed into auto-affection by exemplorality." "It is an irreducible heterogeneity which cannot be eaten either sensibly or ideally," Derrida argues, "and which…by never letting itself be swallowed must therefore cause itself to be vomited" (Derrida, "Economimesis" 21). Schelling is disgusted by Hegel's refusal to think and to speak about disgust, to confront the aporia about the origin to which philosophy is most deeply summoned.

Schelling had once spoken of God's agonistic self-creation in precisely these theosophical terms: as an originary purgation, a contracting expulsion whose "motivation" may be self-purification but whose uncanny effect is ambivalently and interminably to attach the Absolute both to the abject and to the process of abjection (*Stuttgart Seminars* 208). Without necessarily recommitting himself to the terms of that genesis story, Schelling contrasts Hegel's "releasing" with Boehme's vomit because the latter re-embodies the conditions of creation whose difficulties and resistances to thought have been spirited away into the neutral and neutralizing language of relaxation by Hegel's negative philosophy. Whatever accounts positively for creation, it is not primly about "releasement" but raucously about unthought urges, forced losses, unconsciousness abysses, and the sheer messiness of being mortal. "Vomit" succinctly captures this perdurable knot of problems, while also feeding into the chain of oral tropes with which the lecture is quickened in its tensing against Hegel's maw. Schelling doubtless assumes that Hegel would have found its anachronistic theosophical connotations and anthropomorphizing naïveté to be precisely what demanded its expulsion from the body of philosophy.

Let Them Eat Flesh

If philosophy were as ascetic as Hegel seems to wish, if it were truly to disavow all presuppositions, conceptual borrowings, and rhetorical supplements, vomiting from itself even and especially "vomit", it would be dumb, the silent night in which all cows are mute. To that enforced stiltedness, to that comical image of Hegel swallowing his own words to accommodate the quietness of thought thinking itself, we might contrast the myriad ways in which Hegel's project draws on alimentary and oral tropes. Schelling's figure of eating embarrasses Hegel to confirm and vivify what he can also baldly command: "Hegel must come to reality" (*L* 154). But Schelling's gustatory tropes also recall the degree to which Hegel elsewhere dwells thoughtfully upon that elusive question: "What is eating?" More than perhaps any other modern European philosopher, Hegel grasped how "eating" is inevitably a "metonymy of introjection," a figure for a range of psychic processes of interiorization and idealization. As Hamacher argues, "the metaphorics of consuming, of sucking, of digesting structure the entire corpus of Hegel's texts just as much as the metaphorics of grasping and generating does" (*Pleroma* 234).[23] Tropes of incorporation seem most readily to hand when Hegel renounces competing philosophical positions as a way of securing his own virile preeminence. Consider the moment in the *Phenomenology* when Hegel demarcates his project from all forms of intuitionism. Schelling would have known this point only too well: it leads directly into Hegel's infamous slur about the weakened philosophy that tries "to palm off its Absolute as the night in which…all cows are black." "This is cognition naively reduced to vacuity," Hegel says, implying that plenitude (non-vacuity) is a sunlit pasture where the philosopher can plainly see what he wants to eat.[24] But who is calling the cows black? Isn't the attempt to think the empty abstraction of being so anorexically bereft of substance that it reduces cognition to vacuity? Had Hegel not told Schelling that he had completed the draft of the *Phenomenology* in the dead of night (Pinkard, 256). A significant portion of the lecture consists in Schelling's spirited defence of his interest in intuition as a form of cognition that is not merely conceptual. An "intellectual" form of intuition remained an important possibility, whereas for Hegel the term meant but one thing, the soft-headed blurring of "the differentiations of the concept." In dissolving, diffusing, and dissipating the Absolute, intuitionists seek to restore

> the feeling of essential being: in short, by providing edification rather than insight. The beautiful, the holy, the eternal, religion, and love are the bait required to arouse the desire to bite; not the concept, but ecstasy, not the

> cold march of necessity in the thing itself, but the ferment of enthusiasm, these are supposed to be what sustains and continually extends the wealth of substance. (*P* 5)

Intuitionism salivates at the ringing sound of these wonderful but empty words, "abstract form[s] ready-made" (*P* 19). Introducing them as if they were so alien to his sensibilities that they can only be exhibited as generalities devoid of real content, Hegel notes that philosophical modernity is especially susceptible to their lure. These baiting words only offer the lucidity of intoxication—the intellectual equivalent of empty calories. But philosophy must be more thoughtful than a crude stimulus–response, and more discriminating in its tastes. As with Kant's long-standing critique of the intoxicated *Schwärmerie*, the question is one of non-productivity and ill-gotten pleasure, of a certain illegality and simulation, when what is called for is a clear head and hard work. Hegel seeks to mortify the intuitionists, with all of their lofty talk and edifying objectives, by rendering them lowly and animalistic; teeth bared, cravings aroused, they reveal themselves to be too vulnerable to the charm of the pharmakon to be included in the manly company of the scientists of absolute knowledge.

The philosophers of feeling are not the only ones at Hegel's dinner table to feel his scorn. So too are the empiricists, "who assert the certainty of the reality of sense-objects" (*P* 65). Hegel asks: if I say "now" that "here is a tree," where is "here" and when is "now"? These deictics always point to some palpable thing that simply is not there. How to explain this vanishing? What has taken the place of this nothingness? Those who fail to ask these questions and continue to have faith in sense-certainty have yet "to learn the secret meaning of the eating of bread and the drinking of wine," Hegel muses:

> For he who is initiated into these Mysteries not only comes to doubt the being of sensuous things, but to despair of it; in part he brings out the nothingness of such things himself in his dealings with them, and in part he sees them reduce themselves to nothingness. Even the animals are not shut out from this wisdom, but on the contrary, show themselves to be most profoundly initiated into it; for they do not just stand idly in front of sensuous things as if these possesses intrinsic being, but, despairing of their reality, and completely assured of their nothingness, they fall to without

> ceremony and eat them up. And all Nature, like the animals, celebrates these open Mysteries which teach the truth about sensuous things. (*P* 65)

That the empiricists are superceded by the beasts in their knowledge about the being of sensuously apprehended objects relegates them to the lowest rungs of the great chain of knowing. But Hegel risks identifying his argument about sense-certainty with animals, even under the cloak of irony, reminding us of the unique privilege that he accords to eating—even animalistic *fressen* rather than *essen*—as a way of thinking about thinking.

What is the "truth about sensuous things" that animals "teach" as well as any Hegelian philosopher? Strictly speaking, it does not concern things as such: it is the "as such" of things that is under interrogation. What animals and scientists of knowledge teach comes in the form of what they do to things, or rather what they have summarily already done with them at the moment of translating them into "food." "Things" mean nothing to them. Whether as squirming and fearful prey, or as the fleeting points of the "Here" and "Now," sensuous being is only *apparently* certain and substantial, its particularity always already en route to being transmuted into what it properly is—an other that belongs wholly to the devourer's universe. So destined to be eaten are they that Hegel imagines for a moment that they willingly forfeit their lives; they partly "reduce themselves to nothingness."

Without ceremony they fall into the claws of the carnivore and are consumed, in a scene reminiscent of the idylls in early modern texts in which fish gladly jump into the nets of fishermen. Breugel the Elder's hallucinatory canvas, *The Land of Cockaigne* (1567), also comes to mind, even if the birds that he represents as willingly lying down on dinner plates serve his condemnation of the sins of gluttony and sloth. As if to remember these improbable scenes of preying and of capture, and to speak for the about-to-be-eaten, Schelling asks: "The whole world lies… in the nets of the understanding or of reason, but the question is how exactly it got into those nets, since there is obviously something other and something more than mere reason in the world, indeed there is something which strives beyond these barriers" (*L* 147). Schelling's query recalls an avaricious metaphor Hegel once used for the assimilative reach of thinking—"the diamond-nets of the understanding"—but shifts the purpose of these webs from scooping booty out of the bowels of the earth to capturing animals for slaughter and consumption.[25]

The scene in the *Phenomenology* is no peaceable kingdom but a fantastically carnivorous prospect, as glimpsed through the eyes of a predator whose monochromatic vision allows it to experience the universe in only two shadings—either as itself or as a more or less ready-to-hand extension of itself. Whatever otherness sensuous, particular things might possess, their alterity is translated and nullified at the moment they are worlded and brought within the hungry creature's ken. Completely assured of the nonbeing and non-exteriority of their prey objects, the animals "fall to without ceremony and eat them up." Scientists of knowing, like animals, do not stand idly by the world of things, either concerned about their solidity as "sceptics" or assured of their certainty and externality as empiricists. Hegelian philosophers swallow the universe whole: what is real is not sensuous being but "Reality," the matrix of universals that makes the individual parts meaningful and thus food *for* thought. "[T]o give actuality to the universal, and impart to it spiritual life" (*P* 20): this is the ambivalently generous task of the phenomenologist, who turns things over to universality by taking away their sensuous being. We could call it eating reduced to its ideal form. Consumption would then be the "open Mystery" not of nature but of spirit whose logic nature advances and rehearses, reproducing in the world of "free contingent happening" what is already happening in the world of *Geist* (*P* 492). Is the *Phenomenology* not a history of that spiritual ingestion, an account of the incrementally achieved coherence by which all things attain a substantial place within an overarching rationality?

Hegel uses animal eating to say something counterintuitive about consciousness: its content is rooted not in the false certainty of sense but in the true certainty of universals. But the metaphor only functions because he is also saying something counterintuitive about eating. For the animals' relationship with food is not a brute confrontation with an "unwilling" other, but a process of incorporation that has predigested its food at the instant it is deemed to *be* food. And from the point of view of Absolute Knowledge, *everything* is food. Animals, like philosophers, exist in a world not experienced as over and against themselves but in fact and from the beginning as an annexation of themselves, always already eaten. Actual eating, like consciousness, is therefore a kind of afterthought, made possible because of an always anterior scene of violent sacrifice. A truly resistant other would not excite the slightest desire to bite since it would not be imagined or experienced as food or even, Hegel seems to suggest, as part of the world of things. Without having been incorporated as food, this world could not be eaten; without having been conceived by and in the Concept, there would be no thought, no consciousness. The carnivorousness, the tearing of flesh and gnawing on bones, and the lip-smacking pleasures of taste and texture, all the mortal ferocity of hunting and eating is thus also elided. It is as if for eating to happen at all it must first consume itself, spirit its literal referent away into something less messy, more hygienically proper—closer to "thinking." So idealized is this scene that the animals get caught up in a curious switch. The hungry lives of creatures are mobilized to figure forth the psychic life of the philosopher. Yet the animals devour their prey with ease, soaking up the world of wiggling sensuous things rather than tearing them limb from limb, moving without resistance through an ethereal realm where things are not things at all but mere nothings. Under Hegel's ironic gaze they seem not to be eating others so much as *thinking thoughts*.

You Are What You Eat; or, Those Were Pearls That Were His Eyes

For Hegel, consumption and digestion are not merely illustrations of consciousness improperly borrowed from the naturalistic realm, but part of the underlying incorporative logic that is clarified and renewed in Hegel's discussion of "Assimilation" midway through the "Organics" section of the 1830 *Encyclopaedia*. So closely matched are these discussions that the latter reads as an extended gloss on the former, testifying to the continuity of Hegel's thinking about eating.[26] Here Hegel makes explicit and unironic what was implicit and ironic in earlier text. Animal life's relationship with externality is explained not positively as the confrontation of two discrete entities but negatively as the forcefully posited outside of the organism: "This basic division, or expulsion of the Sun and everything else, constitutes the precise standpoint of animation" (*Philosophy of Nature* 3.136).[27] There is no extra organism, no other of the organism except the *organism's* outside, that which it has posited as its exteriority or determinate negation. Externalization is in reality a negative form of *Er-Innerung* or "inwardizing," as Hegel had argued in the *Phenomenology*. And that outside realm is impressively without limit, extending to "the Sun and everything else," a phrase whose disconcerting perfunctoriness mimics animal consciousness in its indifference to the differends comprising the universe of sensuous beings.

The figure of animals eating is itself a trope for their maximally colonial mode of being-in-the-world. As if surprised by what is in truth its own possession, the world in its entirety, the organism shudders, contracts, and, in a sustained reiteration of itself in opposition to the world, seeks hungrily through its senses to augment its reach, to forge an "immediate unity of the being of the organism and that belonging to it" (*PN* 138). Therein begins a lifetime of assimilation and its discontents: "This system of living movement is the system opposed to the external organism; it is the *power* of digestion—the power of overcoming the outer organism" (*PN* 120; *W* 9.448). Such conquering is crucial to the organism's ability to replicate itself, to posit itself *as* itself and thus return to itself. To effect this self-possession, the organism's difference from externality comes to be recognized as a repetition of a difference within itself: the process by which the organism consolidates itself begins via its relationship with externality, in which digestion names the means by which the organism relates to itself as an object, as other to itself. But this is only one step in a logical sequence that concludes—or rather, since life, for Hegel, is processive and so interminably assimilative, and is *always* concluding—at the moment when that relationship is itself subject to mediation, and deemed to be an expression of a more profound kinship, the organism's relation to itself.

In positing itself as itself and for itself, retreating to the sanctity of itself as pure self-relation, "as real being-for-self" [*reales Fürsichsein*] (*PN* 163; *W* 9.491), the organism must negate its negation, consume its originating link with its outside. What is most troublesomely *other* is not the outside of the organism but the organism's *relationship* to the outside. The living creature must maintain a certain (biological) confidence and virility against the shaming threat of vulnerability, exposure, and dependence. Disgusted with itself for not exhibiting more self-confidence, Hegel suggests, the organism is filled with "loathing." But the living creature seeks a way to relieve itself of its "lack of self-reliance" in the proud accomplishment of excretion (*PN* 164). As Hegel suggests, "the significance of the excrements is merely that through them the organism acknowledges its error, and rids itself of its entanglement with external things" (*PN* 164). Excretion is therefore not the expulsion of materials that escape the masterful logic of digestion but, quite to the contrary, evidence of the organism's competence in throwing off involvement with anything but itself. Like a criminal covering his or her tracks, no trace can remain of the organism's devouring dependence on another, especially the objectifying otherness generated within itself by virtue of having to eat in the first place. In Hamacher's fine phrasing, "Thus it is not the objective thing, the food itself, which is digested, but rather the external relationship to it. What the organism digests is the process of digestion itself" (*Pleroma* 248). For Schelling, this is too weirdly coprophagous by half; what makes the remainder indivisible is precisely its radical resistance to being co-opted by the logic of digestion. Proof of the inviolability of the remnant abounds in the living world. In the *Lectures on the History of Modern Philosophy*, Schelling consequently turns to that teeming universe, too easily generalized by Hegel as "the Sun and everything else," where even in the lowliest of animate creatures he glimpses the impervious presence of what can only "unwillingly accept the concept": the "shell and casing" of mollusks demonstrate that "matter always seeks to maintain its independence" from assimilation. What for Hegel is self-evidently "excretion" is for Schelling more complicated, neither inside nor outside the organism but both simultaneously, the scene not of annihilating sacrifice but of an open-ended struggle between life and nonlife: "the inorganic, matter that lays claim to a being-itself has here already entered the service of the organism, but without being completely conquered by it" (*L* 123). What the living creature eliminates is its negation, as is the case in Hegel's *Encyclopaedia* argument about assimilation, but for Schelling this is only partly successful. In this failure lies a deeper accomplishment. The grittily inedible exterior of shellfish is to some extent its exterior, its determinate negation, yet Schelling's main point is that it is uncanny and inexplicable as well, the product of the conquering work of the negative...and something irreducibly in excess of it.

Without So Much as a Crumb

Hegel's scenes of assimilation exemplify the perfect crime, whose evidence confirms its invisibility rather than remembering its occurrence. To the thinker of "the indivisible remainder," that the logic of digestion can claim to annihilate even its own dregs, emptying them of their embarrassing excess by putting them into its service, seemed extremely improbable. For Hegel, this process constitutes life as such, enabling the living creature to ward off the inorganic. But Schelling wonders what life could be if it truly left nothing behind. For Schelling, life is saturated with personality, will, longing, and mourning and thus self-difference, incompletion, and loss. If the discussion of creaturely "assimilation" in the *Encyclopaedia* glosses the passage from the *Phenomenology*, it also encapsulates why for Schelling the *Logic* is deeply lifeless, closer to a machine—"a machine for spinning flax" (*L* 162)—than to any mortal thing that lived and breathed on the Earth. The *Logic* is dead in the way that life in Hegel is dead, closer to a perfectly efficient motor than existents whose dwelling place is the unlogical world.

We see now why, when Hegel, as Schelling says, tries "to breathe a life, an inner compulsion to the progression" into the dialectic, he does so "in vain" (*L* 144). At best, the *Logic* is a mimicry of life, secretly borrowing its self-quickening properties to account for the unfolding of the Concept, but without thereby importing anything unseemly into the dialectic—the very excesses that make life interestingly alive for Schelling. If there is evidence of life in the *Logic*, it lies in the traces of the mortal man who created it. For Heidegger, this is where Hegel fails his project's most radical ontological possibilities. In the name of sanctity of being, he complains that the *Logic* is not rigorously impersonal enough, that it remains trapped within the metaphysics of subjectivity.[28] Schelling, by contrast, insists that the problem is that its claim to "subjectless" anonymity is insincere and in particular that its effort to purge itself of the life of the philosopher is misguided and duplicitous. He sees his task as returning the remainders to Hegel's *Logic*, haunting its pristine architecture with the remnants—the "shell and casing"—of what it has incompletely disavowed.

Evidence of this excess lies in the ruses underwriting the *Logic*, which Schelling treats as the vestiges of the "'creative' author" (with a "personality and spirit," the very sources of malignancy in Schelling's universe) haunting the "reasonable" thinker as its spectral other (*L* 148). Nothing less than a "double deception" is at work in Hegel's argument: "(1) by the thought being substituted for by the *concept*, and by the *latter* being conceived of as something which moves itself, when the concept for its own part would lie completely immobile if it were not for the concept of a thinking subject… (2) by pretending that the thought is driven forward only by a necessity which lies in itself, although it obviously has a goal that it is striving towards, and this goal, however much the person philosophizing seeks to hide consciousness thereof from himself, for reason unconsciously affects the course of philosophizing all the more decisively" (*L* 138–9). In both its means and its ends, the *Logic* remains a great deal more than it appears or understands itself to be. Schelling's uncannily psychoanalytic assessment characterizes Hegel's magisterial text as riven by conceptual errors and rhetorical tricks, as well as by conflicting levels of awareness. In the face of the *Logic*'s implacably necessitarian self-representation, Schelling invites a symptomatic reading of the desires and self-differences by which Hegel's text is enlivened and troubled but to which it appears confidently blind. Hegel cannot have completely eaten up being, not when his own being-in-the-text is so fraught, layered by conscious and unconscious inclinations and deceptions. In its lucidity Hegel's system seems to possess no unconscious; but for Schelling it is precisely because it *has* an unconscious that its moments of forgetfulness are not simple lapses in conscious apprehension but active modes of remembering beyond conscious control which materially shape the argument. The "hiding" of desires is not the opposite of their frank revelation but their disclosure through negation, all the more decisive for being secreted. Schelling functions as an amanuensis not so much for what the *Logic* has forgotten but for what it remembers, in excess of itself, in the mode of forgetting.

All this talk of *Er-Innerung*, the work of interiorizing remembrances within a text that, in theory, should be a depthless surface of logical determinations and thus possess no interior life inevitably brings Schelling around to eating and other metonymies of introjection. Why doesn't being, "the most abstract and most empty thing of all" (*L* 138), remain inert? Hegel claims that out of "necessity" being logically demands "to be more full of content." But this account strikes Schelling as less about thinking logically and more about behaving like a creature who "cannot be satisfied with that meagre diet of pure being" (*L* 138). And who could that famished creature be but Hegel, or at least his phenomenological remnant in the text? Schelling's phrasing is curious: the *Logic* displays "not a necessity which lies in the concept itself, but rather a necessity which lies in the philosopher and which *is imposed upon him by his memory*" (*L* 138). The *Logic*'s underlying desirousness remembers the life of the author that its logical "restraint" would rather suppress and forget: where the *Logic* is, there Hegel shall be. But who, "Hegel"? Schelling now evokes an additional memory on the near side of the *Logic*, making its presence felt inside Hegel. The diction suggests that the philosopher is himself enlivened by an internal other, feeling the force of his own memory as if it were a kind of intrusion just as the Concept's progress in the *Logic* is irrepressibly vivified from a distance by the philosopher. Neither "Hegel" nor his text can claim magisterial authority over themselves, since they each appear to be informed from elsewhere.

Both psychic and textual memories, then, are scenes of the same unforgettable hunger. The *Logic* remembers a memory that has always already inflicted itself upon Hegel; the latter memory, "Hegel"'s memory, imposes itself upon the *Logic* via the philosopher; the former memory, the *Logic*'s symptomatic traces, imposes itself upon Hegel via the *Logic*. Together they render "Hegel" into a virtualized switch point across which the unconscious force of recollection flows at will. Hauntingly, Hegel's imposing "memory" survives the death of the author, stimulating the Concept with ferocious appetites that strictly speaking it cannot have and yet does, albeit in a figurative manner not easily described as living or dead. Finally, however, it is the circularly consumptive nature of the system that is most memorable: "the meagre diet of pure being" with which the *Logic* begins its omnivorous; if borrowed "life" is born out of a lack, it is responsible for having created. For Schelling, this is perhaps the final and most devious twist in Hegel's logic of digestion. In its desire to account fully for its own conditions, Hegel's system also invents the lack whose filling up spectrally mimics the ravenous life of the body that Schelling discerns and uses to read Hegel against the grain.

Hegel will not admit to the lived experiences upon which the quickness of the system secretly depends, and, as if shielding himself from this recognition, he ensures that the lack with which the system begins its rigorous journey is not a negative absence against which the system might be imagined to tense itself, but a positive emptiness that the system has imposed upon itself as evidence of its greatest rigor and highest reflective power. A logical lack, being purged of all determinate content, sublates a phenomenological absence, at once holding it away and relying upon it as the source of the system's repressed hunger to fill itself with "the realized Idea." Thus, the exiguity with which the system begins is not a sign hiding a lack but a sign of lack that hides Hegel's reliance on what Alan White calls "modes of experience…whose differences from conceptual thinking he refuses to acknowledge" (*L* 156). "Meagreness" is for Schelling not another name for the nothingness of pure being but the price the system has paid for withdrawing into thought. The emptiness of the system's beginning can then be said to hide in the open the disavowals that are for Schelling its truest meaning. The initial "meagerness" is the sign both of the system's inability to rid itself of its phenomenological remainder and of its unwillingness to give up trying to do so.

Indian Food

The "meagerness" of Hegel's "diet" triggers in Schelling an unexpected moment of lavishness and exorbitance. Schelling has been refuting Hegelian philosophy for "pretending at the beginning to be asking for very little, which is, as it were, not worth mentioning, as devoid of content as being itself, so that one cannot, as it were, help allowing it." Then his argument blends into the following conceit:

> The Hegelian concept is the Indian God Vishnu in his third incarnation, who opposes himself to Mahabala, the giant prince of darkness (as if to the spirit of ignorance), who has gained supreme power in all three worlds. He first appears to Mahabala in the form of a small, dwarflike Brahmin and asks him for only three feet of land (the three concepts of "being," "nothing," and "becoming"); hardly has the giant granted them than the dwarf swells up into a massive form, seizes the earth with one step, the sky with the other, and is just in the course of encompassing hell as well as the third, when the giant throws himself at his feet and humbly recognizes the power of the highest God, who for his part generously leaves to him the power in the realm of darkness (under His supreme power, of course). (*L* 148)

What to make of this staging of the icon of modern Western philosophy in the midst of a premodern, hallucinated East? The story of Vishnu and Mahabala concentrates Schelling's objections, illustrating the nexus of manliness, incorporation, mealymouthedness, knowledge, and power animating his grievance from the start. Chief among the offences Schelling charges Hegel with is philosophical modesty, the *Logic*'s insistence that its ever-widening dialectic is not an aggressively appropriative process directed by human desires but a "subjectless" sequence of determinations to which the thinker must humbly submit himself. The dialectic demands a certain "restraint," "a refusal to intrude into the immanent rhythm of the Concept" (*P* 36), to which Schelling reacts with a digression that is the opposite of "restraint" in form and content: a tale of gluttony told in the "exotic" form of a myth that feels extravagant even in the midst of passionate argumentation. The tale is a spectacular instance of what Schelling finds absent from Hegel's prose, "the bold metaphor" (*L* 143). Let us then call it an *hors d'oeuvre*, a morsel imported from outside philosophy for the Hegelians to choke on. Poaching materials from the considerable archive he was developing about the significance of Indian mythologies, Schelling notes that while appearing to practice such selflessness and ascesis, Hegel's system is all about a rationalized violence no less violent for being rationalized. The progressive derivation of the logical categories is not an immanent, self-moving, and self-contained process, but is, like Vishnu in the form of the "dwarflike Brahmin," secretly driven by the colonialist desires of the philosopher. There is thus no point at which Hegel's alien hand is not in this process, and this is never more forcefully so than when he is claiming the least, at the inauspicious start of the *Logic*. The apparently unassuming Brahmin vegetarian turns out to have the most voracious appetite of all, and in an instant he tricks the meat-eating Mahabala (otherwise known as Bali or "the strong one" in Hindu mythology) into surrendering his world.

This displaced memory of India suggests that Hegel's *Philosophy of History* looms in the background. A text more deeply invested in the carnivorous virility of philosophy, and the European worldhood for which that philosophy can function as an alibi, would be difficult to imagine. As Balachandra Rajan has argued, India is for Hegel nothing but "the home of mixed genres, hybridization, and…monstrosity," offering "to the European gaze the sterile yet edifying spectacle of a civilization profoundly unable to analyze itself."[29] Negatively exemplary in its feminized excess, lack of reflective powers, and, above all, deceitfulness, "India" forms only the raw beginning of the Occident's grand march of intellect. "India" is to be explained away, confirming the masterful superiority of Germanic thinking. If there is anything positive to be said about India's moment in world history, it is that it isn't "Africa", which for Hegel lies on the far side of humanity and history. For the philosopher, Africa is hardly worth the effort of abjection, but he savors India, an imagined realm evoking a revealing combination of repulsion and fascination. As Bahti suggests, Hegel distinguishes "India" and "Asia" from "Africa" because they are "historical, comprehensible—and digestible" (*Allegories of History* 304 n. 5). If Hegel's science of logic eats Being, it cannot be unrelated to his science of history, which devours beings, and the histories and cultures of entire peoples, in a process that Bahti describes as cannibalistic in nature (80).

Could ontophageous and anthropophageous feeding be rigorously distinguished from each other? In treating himself to one of the founding stories of the Great Puranas, Schelling cites the very Sanskrit texts that Hegel describes as the epitome of the "Orient's" fancifulness and unproductive promiscuity. In using the story of Vishnu to lay bare the willfulness and cunning of the Hegelian system, he demonstrates that "India" is *not* entirely digestible, and that there remains in the supposed wastefulness and decadence of its mythologies enough critical power to best the man who claimed imperiously to have bested it. An ancient Eastern tale of mendacious conquest speaks to an analogously assimilative violence underwriting a contemporary Western one. The fact that the ostensibly "primitive" story embarrasses the "cultured" one emanating from Berlin only reproduces the Orientalist condescension that Schelling and Hegel shared. If "India" is for Schelling not a figure for falsehood, it remains the source of a story about lying. The point is hardly to raise ancient Indian mythology to the status of a truth but to denigrate modern European philosophy, in a way roughly analogous to Hegel using the Eleusian Mysteries to condemn contemporary empiricism. This little bit of comparative anthropology is possible not only because it offers a vivid point of ironic purchase on Hegel's argument but also because of the profound failings that he saw characterizing both worlds. As Wilhelm

Halbfass has argued, "according to Schelling, neither the Indians nor Hegel were able to grasp the truly 'positive,' i.e., existence in its concreteness, the 'factuality' of the one God and of revelation."[30]

What is striking is the aplomb with which this Hindu allegory is folded into the argument. Schelling does nothing to prepare the listener or reader for the comparison and, indeed, with the exception of a brief parenthetical gloss, hardly allows it to be one at all: "The Hegelian concept" is not like "the Indian God Vishnu"; it *is* this very unlikely thing. What better way to shock the Hegelians than to spring an "exotic" analogy on them, and then heighten the disorientation by proceeding as if reaching for an ancient tale from the Puranas were the most natural thing in the world for a historian of modern philosophy? And perhaps in a way it was. Schelling's long-standing interest in India appears in an early letter to A.W. Schlegel, praising their "sacred texts" as greater in critical power even than the holy scriptures.[31] This is precisely the opposite of the position he maintains in the more than one hundred pages devoted to Indian and related Eastern traditions in his *Philosophy of Revelation*, where he asserts the supremacy of revealed Christian truth over all its shadowy types—including both "Asian" religious texts and Hegel's philosophy of reflection. Still, the fact that Schelling hybridizes his own text, freely mixing "philosophy" with "mythology," suggests that part of what he is seeking to achieve, narratively speaking, is a kind of "India" effect—this, as an affront to the supposed purism of Hegel's *Logic*.

According to one historical model, before he became an apologist for the Prussian court Schelling inaugurated post-Hegelian philosophies of finitude. We can characterize these philosophies as struggling with the question of eating well, dwelling with the knowledge that the interiorizing memory of the other is faithful to the precise extent that it is faithless, and that at the point one begins to eat being one has already been given over to that which is otherwise than being. Schelling's focus on the metonymies of introjection, his fascination with the rhetoric of carnivorous assimilation belongs to a more extensive visceral philosophical vocabulary that responds passionally and cognitively to the limits of reflection. This rhetoric, invested with certain powers of horror, evokes affective phenomena (selfish hunger, addictive longing, lacerating mutilation, unappeasable melancholy, demonic possession, incombustible waste, malignant fury, unconscious desire) that unsettle the distinctions (body and spirit, ethics and ontology, life and nonlife, ground and non-ground, freedom and necessity, negative and positive philosophy, West and East) upon which the intelligibility and ideological investments of the thetic world rest.[32] As early as 1809 Schelling had called for speculative idealism to abandon its enervating obsession purely with matters of the spirit, to cease its ferocious "war against being" (*Stuttgart Seminars* 232). Speaking as the upright opponent of philosophers who had only managed to "emasculate" themselves during the prosecution of this war, he called for more potent thinkers to join him in taking on "flesh and blood" (*PI* 236). In the lectures *On the History of Modern Philosophy*, this labor of reincarnation haunts Hegelianism with the specter of its carnivorous virility.

Notes

1. Emmanuel Levinas, *Otherwise Than Being; Or Beyond Essence*, tr. Alphonso Lingis (Pittsburgh: Duquesne UP, 1998), 18. Hereafter *OB*.
2. G.W.F. Hegel, *The Encyclopaedia Logic*, tr. T.F. Geraets et al. (Indianapolis: Hackett, 1991), 55.
3. John Llewelyn, *Emmanuel Levinas: The Genealogy of Ethics* (London and New York: Routledge, 1995), 143–4.
4. Jacques Derrida, "Force of Law: The Mystical Foundation of Authority," tr. Mary Quaintance, *Cordoza Law Review* 11.5-6 (July/August 1990), 953.
5. F.W.J. Schelling, *Philosophical Investigations into the Essence of Human Freedom and Related Matters*, tr. Priscilla Hayden-Roy, in Ernst Behler, ed., *Philosophy of German Idealism* (New York: Continuum, 1987), 239. *Philosophische Untersuchungen über das Wesen der menschlichen Freiheit* (1809), *Sämtliche Werke*, ed. K.F.A. Schelling (Stuttgart: Cotta, 1860), vol. 7, 360. Hereafter *PI*; the German is cited as *SW*.
6. Slavoj Žižek, *The Indivisible Remainder: An Essay on Schelling and Related Matters* (Verso: London and New York, 1996), 6.
7. F.W.J. Schelling *Stuttgart Seminars*, in *Idealism and the Endgame of Theory: Three Essays by F.W.J. Schelling*, tr. and ed. Thomas Pfau (Albany: State University of New York Press, 1994), 239.
8. Jacques Derrida, "'Eating Well,' or the Calculation of the Subject," tr. Avital Ronell, in *Points: Interviews, 1974–1994*, ed. Elisabeth Weber (Stanford: Stanford UP, 1995):

The question is no longer one of knowing if it is "good" to eat the other or if the other is "good" to eat, nor of knowing which other. One eats him regardless and lets oneself be eaten by him...The moral question is thus not, nor has it ever been: should one eat or not eat, eat this and not that...man or animal, but since one must eat in any case and since it is and tastes good to eat, and since there is no definition of the good [*du bien*], how for goodness' sake should one eat well [*bien manger*]? And what does this imply? What is eating? How is this metonymy of introjection to be regulated? (282)

9. Andrew Bowie, "Translator's Introduction," F.W.J. von Schelling, *On the History of Modern Philosophy* (Cambridge: Cambridge UP, 1994), 1.
10. Terry Pinkard, *Hegel: A Biography* (Cambridge: Cambridge UP, 2000), 662.
11. G.W.F Hegel, *Aesthetics: Lectures on Fine Arts*, tr. T.M. Knox, Vol. 1 (Oxford: Clarendon Press, 1975), 11.
12. Alan White, *Absolute Knowledge: Hegel and the Problem of Metaphysics* (Athens, Ohio: Ohio UP, 1983), 94.
13. Cited by David Farrell Krell in *Contagion: Sexuality, Disease, and Death in German Idealism and Romanticism* (Indianapolis: Indiana UP, 1998), 198 n.10. Tilliette's Shakespearean image of pathetic misrecognitions and bad timing masks the business of the evil meal at its heart, the scene of poisoning whose tragic consequences consume the couple in death.
14. Franz Xaver von Baader wrote to Hegel that Schelling's "early philosophy of nature was a generous, tasty steak but now he just cooks up a ragout with Christian spices." Cited in Thomas F. O'Meara, *Romantic Idealism and Roman Catholicism: Schelling and the Theologians* (Notre Dame: University of Notre Dame Press, 1982), 89.
15. Immanuel Kant, *The Metaphysics of Morals*, tr. Mary Gregor (Cambridge: Cambridge UP, 1991), 166.
16. Werner Hamacher, *Pleroma-Reading in Hegel*, tr. Nicholas Walker and Simon Jarvis (Stanford: Stanford UP, 1998), 3.
17. G.W.F. Hegel, *Aesthetics*, 511; *Hegel's Lectures on the Philosophy of Religion*, tr. and ed. E.B. Speirs and J. Burdon Sanderson, 3 vols. (London: Routledge, 1962), 3.341.
18. G.W.F. Hegel, *The Encyclopaedia Logic* (with the Zusätze): Part 1 of the *Encyclopaedia of Philosophical Sciences* with the Zusätze, tr. T.F. Geraets, W.A. Suchting, and H.S. Harris, (Indianapolis: Hackett, 1991), 307.
19. John Milton, *Paradise Lost*, II.843-5. *John Milton: Complete Poems and Major Prose*, ed. Merritt Y. Hughes (Indianapolis: Bobbs-Merrill, 1957), 252.

20. John D. Caputo, *Against Ethics: Contributions to a Poetics of Obligation with Constant Reference to Deconstruction* (Bloomington and Indianapolis: Indiana UP, 1993), 199.
21. See Denise Gigante's remarks about this "angelic" form of (non)consumption in Milton in "Milton's Aesthetics of Eating," *Diacritics* 30.2 (Summer 2000), 88–112.
22. Jacques Derrida, "Economimesis," tr. Richard Klein, *Diacritics* 11.1 (1982), 20.
23. See also Timothy Bahti, *Allegories of History* 110.
24. G.W.F. Hegel, *Phenomenology of Spirit*, tr. A.V. Miller (Oxford: Oxford UP, 1977), 9. Hereafter *P.*
25. G.W.F. Hegel, *Hegel's Philosophy of Nature*, tr. and ed. M. J. Petrey, 3 vols. (London: George Allen and Unwin, Ltd. 1970), 1.202.
26. See Mark C.E. Peterson, "Animals Eating Empiricists: Assimilation and Subjectivity in Hegel's *Philosophy of Nature*," *The Owl of Minerva* 23.1 (Fall 1991), 49–50. See also Tilottama Rajan's important work on incorporation in Hegel in "Framing the Corpus: Godwin's 'Editing' of Wollstonecraft in 1798," *Studies in Romanticism* 39 (2000), 511–31, and "(In)digestible Material: Illness and Dialectic in Hegel's *Philosophy of Nature*" (collected in this volume, "Veganism and the Politics of Nostalgia").
27. *Ezyklopädie der philosophischen Wissenschaften im Grundrisse* (1830*), Zweiter Teil: Die Naturphilosophie, Mit den münlichen Zusätzen*, in *Werke*, Vol. 9. The English translation hereafter *PN*, the German *W*.
28. Andrzej Warminski, *Readings in Interpretation: Hölderlin, Hegel, and Heidegger* (Minneapolis: University of Minnesota Press, 1987), 113–62.
29. Balachandra Rajan, *Under Western Eyes: India from Milton to Macaulay* (Durham: Duke UP, 1999), 140, 108.
30. Wilhelm Halbfass, *India and Europe: An Essay in Understanding* (Albany: State University of New York Press, 1988), 105.
31. Cited in Jean W. Sedlar, *India in the Mind of Germany: Schelling, Schopenhauer, and their Times* (Washington: UP of America, 1982), 42.
32. I discuss this rhetoric in "'The Necessary Heritage of Darkness': Tropics of Negativity in Schelling, Derrida, and de Man," *Intersections: Nineteenth-Century Philosophy and Contemporary Theory*, ed. Tilottama Rajan and David L. Clark, 79–146, "Heidegger's Craving: Being-on-Schelling," *Diacritics* 27.3 (1997), 8–33, and "Mourning Becomes Theory: Schelling and the Absent Body of Philosophy," "Schelling and Romanticism," *Romantic Circles Praxis* Series, ed. David Ferris (June 2000); http://www.rc.umd.edu/praxis/schelling/clark/clark.html.

Part II

Aesthetics & Representation

Dark Veganism: The Instrumental Intimacies of Matthew Herbert's *One Pig*

Michael D. Sloane

I put on my headphones and press play. While listening to the first track, "August 2009," on British electronic musician Matthew Herbert's album *One Pig* (2011), I hear the rustling of the wind being picked up by a Sennheiser 418 m/s mic and a Nagra V recorder.[1] Within seconds, I notice the murmur of poultry. The clucking quickly fades to the background. Then, eerily, there is nothing but dead air for forty seconds. After this lull, there is a jarring exhalation of breath that is treated with both delay and reverb. For the next two and half minutes, the track features the uneasy sounds of a sow undergoing labour. As birds chirp in the background, the sow's breathing increases in tempo and moves toward a climax that comprises snorts. Then, the outburst trails off and leads to an almost imperceptible sound, which is shortly explained by Herbert himself on the recording: "That was it; it was being born." The remainder

M.D. Sloane (✉)
Language and Liberal Studies, Fanshawe College, London, ON, Canada
e-mail: sloane.mike@gmail.com

J. Castricano, R.R. Simonsen (eds.), *Critical Perspectives on Veganism*,
The Palgrave Macmillan Animal Ethics Series,
DOI 10.1007/978-3-319-33419-6_6

of "August 2009" consists of snorting and squealing sounds accompanied by pitches generated from a sample of Herbert's video camera shutting off. Commenting on this track, Herbert explains how "it becomes like a dialogue between me and my video recording and the pig itself."[2] As *One Pig* opens with a recording of a recording, it asks us not only to listen to the pig's "temporal afterlife"[3] but also to consider the documentation, production, consumption, and ethics of animal media.

One Pig records and remixes the sounds of a pig's life from beginning to end. Although we do not hear the sound of the pig's death, we do hear samples of its body being prepared and later consumed. When it comes to animal ethics, Herbert's album is both good and bad. (While "good" and "bad" sound uncritical and oversimplified, I am purposefully using these polarizing qualifiers to offer one kind of perspective for how one could initially understand *One Pig*; the following comments are but a foil for my nuanced reading of Herbert's album.) On the one hand, then, the album is good because it gives a voice to and raises awareness of the often-silenced nonhuman animal through a visceral auditory barrage of one pig's life from "birth to plate," to borrow from Accidental Records album's description on their Bandcamp website.[4] On the other hand, though, it is bad because it objectifies an animal life by creating a consumer product out of nonhuman pain. I am just scratching the surface here. Indeed, there are other iterations of *One Pig* that warrant critical attention, too, like Herbert's live show with a "StyHarp" or *One Pig (The Micachu Remix EP)*.[5] I will focus on Herbert's album, however, which is both a productive and problematic cultural work, and it is crucial to consider how and why this is the case. As Jeff R. Warren notes, "musical experience involves encounters with others, and ethical responsibilities arise from these encounters"[6]; for *One Pig*, these "others" are animals. While I will address the animal ethics of *One Pig*, I want to avoid a straightforward discussion of what is ethically right or wrong about Herbert's album, one that is more than likely to reach an impasse. To do this, it is necessary to reframe the concept of ethics and then reconsider it in relation to animals.

In an interview from 1983, Michel Foucault offers a number of remarks that help to recalibrate ethics. "My point is not that everything is bad," Foucault says in an interview, "but that everything is dangerous, which is not exactly the same as bad. If everything is dangerous, then we always have something to do. So my position leads not to apathy, but to a hyper- and pessimistic activism."[7] Foucault's idea that "everything is dangerous" is my starting point. This might run the risk of contributing to a culture of fear, paranoia, or suspicion, but we need to consider the gradations of suffering, harm, or injury. Everything is dangerous to varying degrees. Instead of thinking in terms of diametric opposites like right and wrong, it is useful to start with a more flexible foundation, one that motivates both critical thinking and action, without disregarding lessons learned from the long history of work on animal ethics, which includes anyone from Peter Singer to Carol Adams, or from Jacques Derrida to Cary Wolfe. For Foucault, though, the "ethico-political choice we have to make every day is to determine which is the main danger."[8] The presumption that there is a "main danger" is part of Foucault's "hyper- and pessimistic activism," which ostensibly seems problematic because it might lead to the projection of imaginary or nonexistent threats. Thinking about danger is thinking dangerously. Yet, seeking the "main danger" helps to combat how the workings of capitalism—for instance, alienation from the mode of production and the proliferation of consumerism—often facilitate and encourage a lifestyle centered on ignorance and bliss, comfort, and complacency. However, this pleasurable, painless lifestyle does not always extend to nonhumans. Indeed, capitalism and animals have a long, troubling history,[9] one premised on an economy of exploitation. We know this, yet in a culture of carnism, we often choose to ignore or forget about animal suffering; moreover, it does not help that capitalism continually tries to obscure this reality from us. A pessimistic fixation on danger is a useful tactic for responding to how capitalism produces blind spots. In turn, asking what might be dangerous about one's actions and their implications might help to foreground the lives of animals. Acknowledging the ubiquity of danger when it comes to production and consumption as such will compel us to ask about and for alternatives. While thinking about danger when listening to *One Pig*, I hear what I call "dark veganism."

On 1 November 1944, the word "veganism" was born.[10] Veganism says "no" to the consumption of anything and everything animal related, and it says "yes" to a view of animals as having a right to their own life. Veganism is more than just a diet, lifestyle, or consumer choice; it is also a responsibility to nonviolence and the eradication of animal exploitation.[11] More often than not, individuals who adhere to the status quo and consume animals or animal by-products also subscribe to some form of speciesism, consciously or not. As Gary L. Francione notes, "it is extremely difficult—perhaps impossible—not to be at least indirectly complicit in animal exploitation as consumers."[12] This issue of indirect complicity is important to my discussion of *One Pig* because it helps to identify a problem in the seemingly innocuous documentation and musical production of a pig's life. "Ethical veganism," writes Francione, "is the personal rejection of the commodity status of nonhuman animals, the notion that animals have only external value, and the notion that animals have less moral value than do humans."[13] He explains how this position is an "unequivocal moral baseline of any social and political movement that recognizes that nonhuman animals have inherent or intrinsic moral value and are not resources for human use."[14] Veganism is the foundation for animal ethics. "Happy meat is simply a myth," as Richard P. Haynes puts it.[15] Yet, many cling to this myth. How do we change this? The answer is to stop consuming animals in any which way. "We will never even be able to see the moral problem with animal use as long as we are continuing to use animals," writes Francione.[16] But what about Herbert's use of an animal on *One Pig*? The ethical nature of this album makes something like veganism much more complicated. If anything, *One Pig* is a cultural text that gestures toward a shift in how we can think about a different, darker part of veganism itself.

Recently, veganism has undergone some groundbreaking changes. Take, for instance, Rasmus Rahbek Simonsen's "Queer Vegan Manifesto," which affirms deviation at the table of heteronormativity and exclaims, "share negativity!"[17] As Simonsen reminds us, moreover, "veganism as a discourse is suffused with the tenets of death and suffering"[18]: "Dead animal bodies daily pass by vegan 'lips'—understood as a figure for the threshold of the self—as it is by internalizing the loss of animal lives that a crucial component of vegan ethics and identity is established."[19] Clearly, there is a dark side to veganism, one that is and will be useful to explore at a time when the positive rhetoric of "vegan-friendly" abounds in the mainstream.[20] Today, the rhetoric of vegan friendliness is infectious, which includes anything from menu options at restaurants to grocery store product packaging to Guinness going vegan.[21] While this is without a doubt useful, it also runs the risk of inoculating us against the reality of animal pain and suffering. This is where dark veganism comes in.

What is dark veganism exactly? Dark veganism is not veganism as we know it. It is less an identity and more of a description, process, mode, or concept, an unnamed and unexplored phenomenon that has always existed within veganism. A part of and a perspective on veganism, dark veganism is animal activism in operation through a close attention to scenes and sounds of animal suffering that directly or indirectly work toward achieving nonviolence across and between humans and animals. Dark veganism does not engage with what is good or bad as such. Instead—dwelling on the productivity and unproductivity of danger, disgust, and death—dark veganism critically engages with an aberrant animal ethics while paradoxically shining a light on the shadows of the animal agriculture industry. Dark veganism is a necessary idea for thinking through limit cases and transgressions involving the lives and deaths of animals in one way or another. It is a means that works toward ending animal suffering; it dabbles and delves into what is dangerous when it comes to addressing animal pain and death in a cultural context. What dark veganism helps us to realize and think about are those objects, events, and institutions around us that engage or exhibit the idea of animal rights directly or indirectly, intentionally or unintentionally, but do so in a way that is somewhat spurious and perhaps not as "pure" or "noble" as an animal activist might hope. For example, regardless of his intentions, Damien Hirst's ethically problematic installation *Mother and Child Divided* (1993), which divides a cow and calf in half and preserves them in formaldehyde, compels the viewer to consider the use and consumption of animals via an uncanny, grotesque, and invasive spectacle. For me, rather paradoxically, dark veganism encourages the investigation of texts that troublingly use animals like *Mother and Child Divided* in order to think through and pursue a form of animal activism linked to veganism, especially in circumstances that seem counterintuitive relative to the desire to end animal suffering. Thinking about and exploring the nature and meaning of a darker veganism—one riddled with contradiction and dissonance—might help to acknowledge and recuperate ethically problematic cultural texts, like *One Pig*, and generate new or different ways of seeing, knowing, and doing. We need to make room for thinking about dark veganism to see the light of a world without nonhuman pain. *One Pig* is dark veganism because it picks apart the life and death of an animal track by track in order for us to hear what not to do in our everyday lives.

As Kattie Basnett argues, *One Pig* uses technologies of sound for both the expression of animal authorship and animal suppression; she suggests that recording risks abstracting the animal in ways that lessen the chances of human ethical responses to the pig.[22] Through a discussion of *One Pig*, however, I explore the implications of starting with the dangers of the dark side of veganism in order to take seriously animal ethics via *One Pig* as a conceptual catalyst. My analysis begins with a thorough explanation of the story around Herbert's album. Then, I follow this with an in-depth discussion of a number of what I call "instrumental intimacies"—fleshy spectres, indirect complicity, ethical disgust, and the pain–pleasure problematic—which will help to think through the useful dangers of engaging and representing animal pain and death in the context of dark veganism.

Matthew Herbert

Born in 1972, son of a BBC sound technician, Herbert was actively involved with music at an early age.[23] He started playing violin and piano at age four, joined an orchestra and choir at age seven, and played keyboards in a band at age thirteen.[24] While studying theatre at Exeter University in the early 1990s, Herbert started to experiment with field recordings and found sounds of everyday things like bottles, jars, pots, and chip bags.[25] After developing an interest in contemporary electronic dance and house music, Herbert started to play his first live gigs, which included opening for acts like Meat Beat Manifesto and Radiohead.[26] After graduating in the mid-1990s, Herbert launched his career as a prolific musician, artist, producer, and writer under many pseudonyms such as Best Boy Electric, Doctor Rockit, Herbert, Mr. Vertigo, Mumblin' Jim, Radio Boy, Slojak, The Music Man, and Wishmountain.[27] To date, Herbert has put out just under one hundred releases including the recent album *The Shakes* (2015), which addresses personal issues like raising a child in an unstable world and incorporates samples of used bullets and shells purchased from eBay on his track "Safety."[28] Far from a gimmick, Herbert's use of unique samples is his signature. To name but a few, he has used often-modified samples of things such as laser eye surgery on *Bodily Functions* (2001),[29] a Chieftan MK 10 battle tank crushing a recreation of Nigella Lawson's meal for George Bush and Tony Blair on *Plat Du Jour* (2005),[30] drums recorded in a hot air balloon on *Scale* (2006),[31] and a pro-Muammar Gaddafi plane dropping a bomb on the Libyan town Ra's Lanuf on *The End of Silence* (2013).[32] It is not surprising that Herbert has been described as an "alchemist of avant-garde sounds."[33]

Herbert's motivation for using idiosyncratic samples is political. Initially, Herbert made music for fun; however, now, he is less compelled by this reason. He explains: "At a time when inequality is rising to unprecedented extremes and when the system we have created is designed to destroy rather than nurture, music's propensity to noodle inconclusively can seem unhelpful at best. Who needs diversion when action is required?"[34] While Herbert fairly points out that music "can soothe and reassure," he adamantly pushes this further: "I think music should challenge…I think it should make people hear differently, make people engage with the world differently."[35] "It's very hard to not see music as part of the status quo," he says.[36] Instead of relying on standard electronic musical tools such as drum machines, synthesizers, and presets—technologies of standardization and mass production—Herbert opts for a do-it-yourself approach. Specifically, his modus operandi is clearly outlined in his rather ascetic manifesto called "Personal Contract for the Composition of Music (Incorporating the Manifesto of Mistakes)" (PCCOM), written in 2005. In this document, Herbert offers a template for his own work, which is "not intended to be a definitive formula for writing music." For example, Herbert restricts his use of prefabricated sounds, encourages compositional accidents, and stresses full disclosure of the production process.[37] PCCOM's severe parameters are a part of Herbert's "exciting realization that the artistic agenda in electronic music was there for the taking," as he puts it.[38] With chagrin, Herbert often notes how he is "surprised others didn't do it first." For him, the artistic process involves asking serious questions—"what is this work about?", "why does it exist now?", or "what is the intended effect?"—that do not really appear in the mainstream.[39]

Herbert's thoughtfulness comes with a catch, however. In a podcast, Herbert evokes the pejorative connotations surrounding labels such as "conceptual artist" and "concept album" that have been used to describe his work; finding these labels rather annoying, Herbert responds to his interviewer, Todd L. Burns on *RA Exchange*:

> You talked about "concept"—that's definitely a label I am stuck with. I find it really irritating just because it's just another word for thinking about something you know and it's like a hang-up from the progressive rock period..... We are so used to accepting this idea that music is other, always other; it's an abstraction of an otherness that is hard to pin down and we shouldn't try to pin down and that you are a party pooper if you try to think about it. We are so used to that idea that music is just an ephemeral expression of universal love or some kind of universal language...and I find that really unconvincing in this day and age. And I think it is one of the reasons why music is in collapse.[40]

Here, Herbert takes umbrage with an anti-intellectual sentiment. Moreover, he is irked by the objectification and fetishization of music as a transcendent phenomenon that exceeds any and all sociopolitical contexts. Surely, this idea of music as other is antithetical to the practitioner of field recordings and found sounds. Indeed, Herbert's music emphasizes that no work can rise above the conditions it was created in. For instance, in "The Truncated Life of a Modern Industrialized Chicken" from *Plat Du Jour*, we hear the recordings of thirty thousand broiler chickens, twenty-four thousand one-minute old chicks, forty free-range chickens (one of which is killed), a dozen organic eggs, a Pyrex bowl, and a "cheep" from a chick pitched down to form the bass line.[41] Collecting context to compose, Herbert actively seeks to undo the fantasy of modern music. "The music industry...is complicit...in the creation of a sort of bubble that we live in," says Herbert, "which [involves] living this incredible life of luxury that is based on [the] overexploitation of minimal resources [and] based on [the] exploitation of others."[42] For him, "part of music's responsibility...is to prick the bubble, not to polish the windows."[43] This kind of ideological critique is not unfamiliar, and it is, yes, hopeful, and perhaps even admirable. Yet, what Herbert says and what his music does are two different things.

One Pig

On 21 May 2009, Herbert announced his new project: an album entirely made up of sounds from the life cycle of a pig. Tentatively titled *The Pig*, the album would include recordings of the pig's birth, life, and death. Then, postmortem, Herbert would capture sounds made during the butchery process, culinary preparation, and feast. He claimed that it would "all be recorded and then turned into music."[44] The musician's announcement closed with anticipation: "i [sic] await a call from the farm in kent [sic] to let me know that my pig is on its way. i [sic] will rush there with a camera and a tape recorder."[45] Less than a month later, Herbert posted that he just "missed one" because he was away from home, and he added that, "it's a strange (and brilliant) feeling to not be able to start a new record until a farm animal allows you to."[46] Here, rather sincerely, Herbert defers agency to the pig-to-be, which, in a way, will help to circumvent the issue of responsibility later. Finally, then, Herbert matter-of-factly declared the arrival of the pig: "on Saturday [sic] 15th of August [sic] my pig was born."[47] While collecting field recordings during its life, Herbert offered up a number of intermittent observations. For instance, he noted how his pig was "the one with the dark triangle on the rear right rump where it was bitten by its mother."[48] Later, Herbert reflected on the end of the pig's life relative to industry norms: "it has about 6 weeks left now, and compared to an industrially raised pig it is much smaller as it has had more exercise and less food. [C]ommercial pigs would be killed about now."[49] Then, on 9 February 2010, Herbert hit an obstacle: due to a British law—or, "the uk's [sic] bizarre corporate secrecy imperative"[50]—he could not document the slaughtering of the pig. Herbert was quite frustrated about this complication because recording the pig's death was "a crucial part of the project." While it was the part he was "looking forward to the least," he still felt that it was the "most pertinent" in his "understanding of this life."[51] On 10 February, the pig was killed. Then, on 19 February, Herbert's label, Accidental Records, paid an anonymous farm £100.00 for the pig.[52] Born, killed, purchased: this makes sense within our meat-eating world. What is strange, however, is Herbert's presumption of ownership prior to a monetary transaction. After the pig's birthday, Herbert referred to the pig as "my pig."[53] Here, Herbert's use of the possessive is a symptom of the hegemonic ideology that animals are always already human property, which happens to be codified in our legal system. Even after the pig's death, Herbert wanted to "take ownership" of the slaughter.[54] Then, after all of this transpired, Herbert's project brought on a backlash well before it was even released.

Shortly after the pig's death in February of 2010, a spokesperson for People for the Ethical Treatment of Animals (PETA) issued a statement about Herbert's forthcoming album (recall: *One Pig* was released in October 2011). PETA's statement reads as follows:

> No one with any true talent or creativity hurts animals to attract attention, but we are sorry Matthew Herbert couldn't include the screams of pigs being made into bacon on his record, as they would have instantly turned some people into vegetarians. Pigs are inquisitive, highly intelligent, sentient animals who become frightened when they are sent to slaughterhouses, where they kick and scream and try to escape the knife. They are far more worthy of respect than Matthew Herbert or anyone else who thinks cruelty is entertainment.[55]

Here, PETA attacks Herbert's so-called art by suggesting that an authentic form of "true talent" precludes nonhuman harm. The organization gets caught up in an almost ad hominem argument and misses an opportunity to interrogate the ethics of using nonhumans in art. PETA also implies that Herbert's project greatly disrespects the pig. Although using animals in art is controversial,[56] PETA's claim is problematic for a number of reasons that Herbert himself details.

Baffled by PETA's "utterly absurd…knee-jerk reaction"[57] to an unfinished, unheard album, Herbert's retort starts off by taking issue with their accusation of disrespect. In response to this, he meticulously describes the pig's environment on the family-run farm, which included siblings, straw, a partially open-air sty, and locally sourced cereals for the pig to eat. Describing the pig's life, Herbert notes that it lived twenty-five weeks rather than the usual twenty weeks (based on *One Pig*'s dated track listing, we know that Herbert visited the farm at least nine times). He also adds that he did not kill the pig (I will return to Herbert's clarification of this fact). After pointing out the pig's humane surroundings and remarking on the farmer's care for the "wellbeing of the animals on his small farm,"[58] Herbert explains that the farm is a working one where pigs are raised for meat. As if it were a foregone conclusion, he writes that "the pig was always going to be killed."[59] (I return to this below, but it is worth briefly mentioning that Herbert's presumptuous perspective is troubling because it is mired in a form of fatalism that treats animals teleologically: animals are for eating; therefore, they will be killed.) Then, he frames himself as an objective, nonparticipant observer, someone who "was there to learn, not to preach, prod or adapt the recording to some hidden agenda."[60] His defense turns to a valid critique of how the UK's food system does not allow citizens to witness how animals are kept, killed, or prepared, and as a meat eater, he asserts that it is his responsibility to fully comprehend the consequences of his actions. Indeed, Herbert recognizes a serious problem: "If i'm [sic] not allowed to see or hear the pig die who I may have ended up eating one day, either by my government, or by an over zealous PETA who think that to acknowledge that death in a public sense is tantamount to torture, then we are even further away than I thought from the sustainable, enlightened, engaged society we so desperately need."[61] Herbert is onto something. Interestingly, though, his response to PETA sidesteps the issue of animal rights. (If anything, his remark about how "the pig was always going to be killed" emphasizes that it may as well be impossible to fight against the tyranny of animal agriculture.) Instead, Herbert focuses on an appeal to animal welfare by emphasizing that the pig he recorded had a very comfortable life. Moreover, he avoids fully addressing animal rights by evoking a separate, but related, issue—namely, the way the bureaucracy of the food industry and the state find ways to conceal animal death from those who want to bear witness. While *One Pig* "is not intended as a call to vegetarianism,"[62] as Herbert writes, we should not ignore the issues of animal rights and ethics in and around this album; we do not need to kowtow to his intention. If anything, Herbert's explicit disassociation of vegetarianism is suspect in itself—indeed, why even make this remark? Is this disavowal an anxious symptom concerning an ethical conundrum or impasse? Even if the groundbreaking *One Pig* is working hard to unveil what is "hidden in the absurd corners of our consumer lifestyles," is it ethically dangerous, and if so, how and why?

After the release of *One Pig* on 11 October 2011, Herbert's album was widely taken up by journalists, bloggers, and critics. "The album can make even the most committed carnivore squirm," wrote Ben Sisario in *The New York Times*. *Pitchfork*'s Brandon Soderberg noted how *One Pig* defied expectations by not being "a slab of musical vegan didacticism." And PETA spoke up again, too. This time Jobst Eggert, editor-in-chief of PETA Germany, had a conversation with Herbert, and he admitted to appreciating the concept behind *One Pig*, but ultimately defaulted to circular thinking: "Murder is murder—no matter what."[63] Unsurprisingly, Herbert is asked to explain himself when it comes to the issue of animal respect, which he already did in his initial response to PETA on 4 March 2010; this time, however, he is eager to point out to PETA how *One Pig* does, in fact, open up other important matters:

> For most people, with less extreme views, the careful preservation of the memory of one otherwise anonymous pig through using every part of a body that would normally have been thrown in [a] landfill and forgotten, is an act of respect. I'm not here to simply debate the ethics of eating meat or wearing leather, particularly as we need to see this pig as more than just meat. Wouldn't it be more inclusive, less aggressive to state that as a society we should eat [sic] be eating much less meat, and we should be treating the meat we do eat with much greater respect? At least we could move beyond this point in our discussion and start to talk about the relationship between music and activism, between art and protest, between noise and silence.[64]

Even if this is a deflection, Herbert points to how *One Pig* is enmeshed in a network of concepts and relations. Elsewhere, he emphasizes this constellation of complex connections when he reflects on the album's production: "face to face with the blurring of distinctions between sound and music, life and death, right and wrong, this was the hardest record I had made: an awkward dance with *the other*."[65] Herbert's awkwardness regarding an animal's alterity is a symptom of how he does not or cannot actually confront or reconcile his disregard for an animal's right to life.

Picking up on this bigger, yet blurry picture, Hayden Lorimer includes a discussion of Herbert's album to demonstrate a number of key ideas. Writing on familiar, recent themes such as human–nonhuman relations, nonhuman agency, and posthumanism within the field of human geography, Lorimer refers to *One Pig* as "the animal-art-agriculture-advocacy assemblage."[66] After telling *One Pig*'s story, Lorimer offers a brief reflection of Herbert's "musical portrait of an animal bred, ultimately, for human consumption,"[67] which highlights the significance of animal lives raised for human consumption, encourages the scrutiny of the animal agriculture industry, and reveals the cause and effect of dietary desires.[68] Lorimer offers an excellent overview of *One Pig*'s political dimension, and he takes the time to itemize a litany of sounds. There is, however, a need for even more rigorous scholarship on the instrumental intimacies that we hear on Herbert's album. After all, as the musician himself explains, the "most important message of a lot of my projects is that I would like us to listen to the world a little more carefully."[69]

Fleshy Spectres

One Pig's album art is simple and stunning, and it introduces us to a strange feature of dark veganism on Herbert's record. Artist Lenka Clayton's work exhibits what I call a "fleshy spectre." On the cover, we see the capitalization of the album's title and artist's name in white on a black background. Around the word "PIG," there is a faint, white smudge. The smudge is a figure for and of the pig. This textural detail not only registers a life being erased, but also resembles the pig that you hear on the album. Kattie Basnett offers a strong reading of this image when she writes that the "vague, washed-out image of the pig is so faint and indeterminate that it is only because it is over-lain with the word 'PIG' that we realize that what lies behind the word is, in fact, a pig and not a blackboard eraser's smudge…. [T]he pig's corporeal self is most definitely conveyed on the left of the image by the jutting-out of the pig's snout."[70] Now, the recording of the pig is something other than what Herbert bore witness to during his visits to the family farm. Neither alive nor dead, what you see and hear on *One Pig* is a fleshy spectre, a trace of the undead. "To haunt does not mean to be present," Jacques Derrida reminds us in his text about hauntology. On this, he writes that "it is necessary to introduce haunting into the very construction of a concept. Of every concept, beginning with the concepts of being and time. That is what we would be calling here a hauntology. Ontology opposes it only in a movement of exorcism. Ontology is a conjuration."[71] Overlapping ontology with hauntology, Derrida gestures to how "[t]*here must be* disjunction,"[72] one that resides in being itself, as he indicates with his play on the infinitive "to be." This haunting disjunction of being and time plays itself out aurally on *One Pig*. If, as Mark Fisher argues, "hauntology has an intrinsically sonic dimension,"[73] then *One Pig* is undoubtedly a ghostly soundscape. What haunts Herbert's album is what you never hear: the cessation of the life of the pig. Recall that Herbert was unable to record the slaughter due to a British law. Perhaps the ultimate absent referent,[74] the pig's death on 10 February escapes technological reproduction; however, *One Pig* gestures to this brutal cut with a cut. On the album, the transition from "January" to "February" is quite abrupt. "January" is 4′04″ long and draws its source material from when the pig "is in the trailer waiting to go to the abattoir." "[C]urious still, but certainly unsettled," writes Herbert about "January," "we hear it pace and explore the confines of the trailer alone, the idling of the car engine a constant presence."[75] The last third of this track is chaotic: with a heartthrobbing percussive crescendo, Herbert signals the pig's imminent end and, in turn, imagines and conveys what the pig is feeling, even though he cannot actually know the animal's experience. Following this, Herbert makes his killer cut. Specifically, the *dénouement* is followed by a series of oinks and snorts. At 3′58″, we hear the last snort which is sharply edited: the last snort is interrupted by six seconds of dead silence; then, "February" begins with the sharpening of knives. The implication is fatal. The death of the pig haunts *One Pig* through the sounds of the fleshy spectre, which includes thin-as-skin ambient sounds to meaty beats made from a recording of the pig's head being dropped.[76] *One Pig* is the organization of sounds made by Herbert who uses samples made by a pig that is now dead, whether it is the sound of its first breath in "August 2009" or the sound of the air being beaten out of its lungs in "February 2010."[77] Here, different temporalities and sounds involving an animate and inanimate animal overlap with and complicate one another, generating an uncanny experience. *One Pig* blurs life and death with the sounds of the fleshy spectre haunting the album, which unnerves the listener and initiates some degree of ethical reflection. Indeed, regardless of one's dietary choices, the sound of an undead animal is, to say the least, irksome. Given that *One Pig* dwells on animal death from the moment you see the fast-fading pig-shaped smudge to the moment you hear humans eating "dry-cured streaky bacon,"[78] one is visually and aurally exposed to dark veganism in operation because the fragility of animal life is exhibited in and exploited by *One Pig*, which, in part, points the listener to a nonviolent alternative—namely, abstaining from consuming animals.

Indirect Complicity

On 4 September 2015, I purchased Matthew Herbert's *One Pig* CD from Amazon for $23.79. About a month later, my package arrived. After tearing through the cardboard, I found the shrink-wrapped CD and peeled away the plastic. Then, after examining closely the cover on the front and the tracklist on the back, I gently opened the gatefold to find a centrefold of an unsettling black-and-white photograph of a pig. I was looking at a pig looking at me, and I could not tell if it was alive or dead. Thinking about this still life of a still life now, I cannot help but wonder if I am somehow complicit in endorsing an ethically problematic economic transaction in a very direct way, one that eventually leads back to the loss of an animal life. What about Herbert himself? After all, as noted earlier, his label purchased *One Pig*'s pig and aestheticized this purchase by including a facsimile of the invoice with the CD itself. Given the controversy around *One Pig*, what does Herbert have to say about his engagement with the animal, and how can we understand the situation? To answer these questions, we need to return to Herbert's conversation with PETA:

> "I simply didn't slaughter or cut apart a pig for the sake of art, food or music," says Herbert in response to PETA Germany's editor in chief.[79] The earlier discussed conversation between Herbert and Eggert gets rather intense; at one point, Eggert says the following: "You use big words, but obviously don't see your responsibility for the life of the pig that you slaughtered, cut apart and used for your album. I already mentioned that I appreciate the idea behind 'one pig', but at the same time I am disgusted by it. As an animal rights supporter I believe that your work on this album is truly disrespectful."[80]

After Herbert explicitly denies that he slaughtered a nonhuman animal for *One Pig*, he explains what happened and acknowledges the nature of his participation to a point:

> I simply didn't slaughter or cut apart a pig for the sake of art, food or music. I observed someone else doing it. I did however eat part of the pig and i [sic] remember every mouthful. Your whole point seems to simply end up being reduced to the fact that you are annoyed that I ate the pig. Is that really where this conversation stops? Can you only see this whole project through the prism of strict vegetarianism?[81]

While a concession is made, the language of culpability does not crop up even though Herbert financially facilitated the making of *One Pig*, which cost not only £100, but also a life. Seemingly, Herbert has an ethical blind spot. Elsewhere, however, Herbert admits to an alternative option when responding to a question about the politics of his album. After noting one political aspect having to do with the prohibition of recording the death, Herbert turns to the "moral parts" of *One Pig*:

> The moral part is very different because I could have stepped in and saved this pig but then what would I have done with this pig? It was growing for me, so unless I have a big enough garden and a willingness to lose every single plant in it then I couldn't have brought the pig home. Then there was the moral part of eating it—I didn't feel massively comfortable eating it—and the moral part of the butchery process—recording the bones being sawed through....[82]

Why is it so unfathomable for him to step in? Of course, Herbert was not about to do this in the middle of making *One Pig*; however, the sheer fact that Herbert acknowledges this possibility implies that he is aware of his role in the death of the pig. Strangely, Herbert sees his own blind spot. How are we to understand this inability to fully confront one's indirect complicity in the death of a nonhuman animal? The answer has to do with causality.

Aristotle, Hume, Kant—causation is big philosophy. Whether causality is singular, general, linear, nonlinear, productive, difference making, influence based, or pattern based, there are many, many different ways of thinking about it.[83] Generally speaking, though, causality is widely thought about along the lines of its everyday definitional understanding: "that which produces an effect; that which gives rise to any action, phenomenon, or condition."[84] This is singular causation, seeing y as an effect of x. Because Herbert did not slaughter the pig, he did not cause its death, or so his reasoning goes. Nevertheless, when Herbert denies slaughtering the pig and says, "I observed someone else doing it," he is essentially defaulting to a form of singular causation. Today, then, the common understanding of causality is a given: causation is "now taken to be obvious: one object exerts force over another and makes it change physical position or some of its features."[85] Indeed, even "the theme of causation has largely vanished from philosophy," notes Graham Harman.[86] It is for these reasons that Herbert does not think about indirect complicity, or his role in the death of a pig. If, however, we turn to a different understanding of causation, then things change.

Recently, causation has undergone something like a renaissance, as it undergirds new theories such as actor–network theory, speculative realism, object-oriented ontology, and other offshoots. In these new theoretical frames, we encounter things like "reality itself is not mechanical or linear when it comes to causality"[87] or "[n]othing is an effect of just one agency."[88] Much of this sea change is based on and has to do with what Harman calls "vicarious causation," or how "the relation of objects must always be indirect or *vicarious*, since no object can enter fully into any interaction."[89] Essentially, this means that "things can be in contact with something else without being *fully* in contact with them."[90] While the ethical implications of vicarious causation have not been entirely explored, I would argue that it helps to explain the indirect complicity of Herbert's role in the death of a pig. So, Herbert did not cause the death of a pig; rather, his actions vicariously caused an animal to die, but he is blind—or partially blind—to how this is the case. Granted, Herbert is not Kazuo Shiraga, who hunted, shot, killed, and skinned a boar for his bloody painting *Inoshishi-gari 1* (Wild boar Hunting I) (1963); however, *Wild Boar Hunting I* and *One Pig* vary in degree rather than kind, which means that both are complicit in the death of a nonhuman animal for a variety of aesthetic, ethical, and political reasons. *One Pig* is unsettling, but it importantly works toward dark veganism's ethical disgust, or that which generates a visceral response via a problematic treatment or representation of animality in order to not only register the reality of animal pain but also motivate animal rights advocacy and activism. *One Pig* has a pragmatic power—it got people talking about the treatment of animals. Whether it motivates drastic or gradual change, Herbert's album has made an impact. For instance, although Herbert eats meat, he said the following about *One Pig* in an interview: "I was definitely freaked out when it came back dead, that wasn't a pleasant experience and I haven't eaten pork since I've finished the record."[91] Perhaps *One Pig* has the ability to affect others in a similar way. Ultimately, though, it is up to the perceiver to decide whether *One Pig* is a form of critique or gratuitous violence, and so what really matters is how the album galvanizes one or many to consider—perhaps for the first time—the issue of animal rights, if not veganism. *One Pig* does not draw any hard-and-fast ethical lines; rather, it blurs lines, so much so that it incites, if not requires, more than just armchair activism.

Ethical Disgust

What do we hear on *One Pig*? From birds to pigs and tractors to knives, some of the sounds on Herbert's album are immediately recognizable; however, others require guesswork. Thankfully, Herbert has done some of the work for us, given how his manifesto requires that there must be some form of documentation for each and every source of sound in his work; in fact, he follows his own guideline that a "notation of sounds used [are] to be taken and made public."[92] According to Herbert's sonography for *One Pig*, then, we hear the pig's innards, head, and carcass being dropped; the removal of the pig's kidney, liver, lungs, heart, leg, loin, trotter, and bones during butchery; the pig's blood being poured into a bucket and sloshed around; and a number of cooking and consuming sounds, including a fried pig's tail by English chef Fergus Henderson, who is known for his use of offal.[93] Listening to the dismemberment, preparation, and consumption of the dead is disgusting, to say the least. *One Pig* pushes the affect of disgust even further into the realm of aesthetics, too. William Ian Miller writes that "the disgusting itself has the power to allure."[94] In opposition to our ideas of traditional aesthetics and what we think is beautiful, disgust's "power to allure" points to what Carolyn Korsmeyer calls "aesthetic disgust"; she explains this in the following way:

> By this term [aesthetic disgust] I do not mean disapproval or rejection but rather an emotion appropriately aroused by certain works of art—and by other objects as well—that signals appreciative regard and understanding.... [T]his emotive reaction can assume many forms. Generally speaking, aesthetic disgust is a response that, no matter how unpleasant, can rivet attention to the point where one actually may be said to *savor* the feeling. In virtue of this savoring, this dwelling on the encounter, the emotion constitutes a singular comprehension of the value and significance of its objects.[95]

The affect and aesthetics of disgust come together on *One Pig*. By "rivet[ing] [one's] attention"[96] in different ways, *One Pig* compels the listener to "*savor* the feeling" of what it is like to hear the life of a pig from birth to plate and, in turn, consider the importance of "dwelling on the encounter"[97] with an animal other, which can be a starting point for animal rights activism. This plays itself out by way of *One Pig*'s instrumentation.

Not only is the pig an instrument, but it is used to make one, too. For instance, Stephen Calcutt used the pig's skin to make a drum. On the track entitled "October," we hear what sounds like the beating of the pig drum. This track opens with animals eating followed by *One Pig*'s signature sound of snorts. After, "October" starts to build up a number of textures including "chords…made from a cow in the next-door stables."[98] Over the course of the song, the layers build up and ultimately lead to a melody accompanied by a bright, shimmering effect. Then, at 5′14″, we hear what sounds like hands hitting a bongo, djembe, or conga drum; here, however, I suspect it is the London-based drummer Tom Skinner playing the pig drum, who is, after all, credited on the album for playing drums. We know that a pig drum was used on *One Pig*, but we do not know on which tracks exactly. Regardless, the unmistakable timbre and texture of a hand hitting a drum head leads me to believe that the pig drum appears on "October." If the pig drum was made after the pig's death, then, on "October," we have a provocative juxtaposition between animacy and inanimacy, an instrumental pig and a pig instrument. As far as I can tell, the pig drum only appears once more on *One Pig*: "August 2010" at 3′50″. The stretching of the pig's skin to make a drum head that is then hit on this album symbolically reveals the tension between animal life and death in commodity culture. This tension undergirds a feeling of disgust, one that is resoundingly ethical. This feeling of disgust is ethical because it amplifies a pervasive, yet often forgotten, economy of animal use and exploitation.

The pig has been disgustingly rendered ethical. Nicole Shukin's *Animal Capital* (2009) takes an unwavering materialist approach to animal studies. Writing about "capital's terrestrial costs,"[99] or how "capital becomes animal, and animals become capital,"[100] Shukin foregrounds the phenomenon of rendering. For her, rendering is both the mimetic act of making a copy and the industrial process of boiling down and using animal remains, which is related to what she calls "animal capital" and its mode of production.[101] *One Pig* is the repetition of rendering. Indeed, on the album we have the death, dismemberment, and distillation of pig parts for songs found on mass-reproduced compact discs, let alone infinite MP3s. The pig drum endorses and promotes the use of animals for entertainment. Moreover, the pig's skin is stretched and beaten with each listen, which traumatically reverberates the violence of animal exploitation. The irksome instrumentalization of an animal on *One Pig* is productive, however, thanks to disgust itself.

"Disgust is a feeling *about* something and in response to something, not just raw unattached feeling," writes Miller.[102] For him, disgust must be tied to a type of danger, like "the danger inherent in pollution and contamination" or "the danger of defilement."[103] Disgust is a relational affect that is hierarchically structured, one that is surprisingly open to what is repugnant. While disgust "proclaims the meanness and inferiority of its object" and thus asserts "a claim to superiority," it also "recognizes the vulnerability of that superiority to defiling powers of the low."[104] The power of disgust is its relationality. Miller argues that "disgust do[es] proper moral work,"[105] and it "gives us reasons for withdrawing."[106] Disgust can help us to withdraw from a system of animal capital. Listening to *One Pig* can have this effect. "Disgust is a recognition of danger to our purity," writes Miller.[107] For him, though, there is more to it—namely, experiencing the affect of disgust generates a need or compulsion to purify or cleanse one's self while also equalizing us with the other; Miller explains this in the following way:

> The mere sensation of [disgust] also involves an admission that we did not escape contamination. The experiences of disgust, in other words, does not purify us in the way that experience of anger or indignation can. Disgust signals the need to undertake further labors of purification. It is thus that disgust does not do its moral work[,] so as to allow us unambivalent pleasure in our relative moral superiority to the disgusting other. Disgust admits our own vulnerability and compromise[,] even as it constitutes an assertion of superiority.[108]

If one is disgusted by *One Pig*, then this might make one think twice about the consumption of animals, let alone how they are conceptualized relative to humans. *One Pig* is a dangerous work that indirectly endorses animal exploitation through permissibility, yet it is important to acknowledge how the "idiom of disgust consistently invokes the *sensory* experience of what it feels like to be put in danger by the disgusting."[109] It is unpleasant to hear the throb of the pig drum or the injured, "jawless piglet's voice rising above all the others"[110] on "September." *One Pig*'s disgusting instruments and sounds are working to reveal capitalism's disregard for animal ethics; this disgust works to combat animal rights activists' feelings of despair in the face of the immensity of the animal agricultural industry while also to expose a peculiar pain–pleasure problematic.

Pain–Pleasure Problematic

One Pig's penultimate track, "August 2010," features a strange instrument called the "Organis Draculatus,"[111] or, commonly, a pig-blood organ. Commissioned by Herbert, the Organis Draculatus was made by Henry Dagg, a sound sculptor or "blood instrument maker,"[112] as per his album credit. Explaining this instrument to Sara Mohr-Pietsch, who was visiting the studio, Herbert says the following about Dagg's creation:

> I wanted to make musical instruments from the pig…what was left over… I didn't want anything to go to waste. And one of the things that we had a lot of was blood. It's an antique mahogany drawer; you can still see the lock down the bottom here. He's bored some holes down through the drawer, so you can mount eight, large glass…they look like test tubes…but inside the tube you'll put some pig blood mixed with vinegar or water, and then on top there's some plungers; as you push them down, the blood squeals up through the valve and sort of gurgles and then it's tuned diatonically.[113]

Perhaps even more disgusting than the pig drum, given the reanimation of pig's blood in a sonic register, the Organis Draculatus is another Frankensteinian example of the instrumentalization of an animal; however, the use of it on *One Pig* evokes a pain–pleasure problematic.

On "August 2010," there is a pleasurable earworm derived from a painful event. The track is 6′06″ of catchy beats, hooks, riffs, melodies, and solos. Of course, the highlight of the song, if not the album, is the earworm generated from Dagg's Organis Draculatus, which begins at 2′14″ and ends at 4′40″. From the German *Ohrwurm*, an earworm is part of the "stuck song syndrome."[114] Usually, an earworm involves a "piece of the song that is typically less than or equal to the capacity of auditory short-term ('echoic') memory: about 15 to 30 seconds."[115] In Herbert's "August 2010," the bass line of the earworm starts at 1′52″ and repeats for some time, but falls well within the noted short-term memory's parameters. This bass line comprises six repeated, ascending notes (F, G#, A#, C, C#, and D#), which is a loose counterpoint for the high-pitched pig-blood organ that comes in at 2′14″ with a bend from C# to D#. At its point of entry, the pig-blood organ's melody forms an augmented fifth with the start of the repeated bass line; a semitone away from a perfect fifth, the sound of this dissonant interval is close to, but far enough removed from, a conventional harmony. This sonic surprise grabs the listener's attention and draws him or her into one of the most cohesive moments on *One Pig*. Given that Herbert's avant-garde songs on *One Pig* do not really conform to a traditional verse–chorus–verse structure, the majority of "August 2010" bucks this trend. Although the track avoids a traditional song structure, its repeated bass line, riff-based pig-blood organ solo, and unchanging, motorik beat effects an earworm, a "'musical itch'"[116] that cannot be scratched away.

Even if one gets caught up in the groove for a majority of the song, however, there is a tension that arises at the 5′08″ mark when we hear lip smacking, chewing, swallowing, and grunts of satisfaction as the song fades out. Here, the listener consumes the consumption of that which made *One Pig* possible—namely, the life and death of an animal. This dizzying effect is part of *One Pig*'s tension between aesthetic pleasure and animal pain, which is not necessarily resolved. Rather, this tension subtends *One Pig*, and it is most manifest in "August 2010" as a form of *schadenfreude*. Traditionally, *schadenfreude*, or harm-joy, involves a person experiencing pleasure from another person's pain. To think about *schadenfreude* in the context of *One Pig* might seem strange; however, a human listening to entertaining traces of what will have been an animal suffering resembles the relation between harm and joy tied to *schadenfreude* itself. To be clear, though, there is not necessarily a direct one-to-one event where a listener enjoys listening to a pig squeal in pain; rather, Herbert's aestheticized, remixed samples-turned-songs are provocatively pleasurable, given that they are rooted in a history of an animal's anxiety, stress, fear, and pain. In a weird way, the listener of the catchy, aesthetically appealing "August 2010" derives pleasure out of the pig's pain. One's enjoyment of *One Pig* is contingent on a conscious or unconscious acceptance of animal suffering, which runs the risk of fostering complacency and, thus, complicity. Yet, dark veganism's pain–pleasure problematic perseveres because Herbert's album establishes a relation founded on gradations of danger: if you are enjoying this, then who or what could be suffering, and to what degree? (Here, we could understand this sadistic relation as a symptom of Sigmund Freud's beating fantasy, which has "feelings of pleasure attached to it."[117]) And, importantly, how much suffering can the vegan listener accept? In part, *One Pig* "acknowledge[s] the realities of what it is to eat meat," says Herbert, given how there is, he continues, "a huge disconnect between what we do and the consequences of our actions."[118] Even if, as Herbert notes, *One Pig* "is not intended as a call to vegetarianism," we can and should take seriously how it evokes the stirrings of a dark veganism, which opts to acknowledge danger through an often tenuous, often tenacious relation between pain and pleasure, between us and animals.

Coda

One Pig officially ends with what seems like a sentimental ode to the pig. In "May 2011," returning to the family farm, Herbert sings the following lyrics:

> At this time, this time of year
> I take my happiness and disappear.
> Tell these lungs to breathe again,
> When the sun is back to my refrain.
> A simple life is all we need,
> A love to multiply, magnify, dignify each day.
> And so to rest I put my head,
> Let you occupy my thoughts instead.

In remembrance of the pig, Herbert sings of a utopic retreat to what could be a "simple life," one premised on the reproducibility of love. But then, right after the song ends, we hear something else; Herbert explains what happened:

> there's a brilliant part which is one of my favourite parts of the whole record which is when I'd very earnestly finished singing my song. And the minute I finish the last line the pig starts taking a piss, about two foot [sic] away from me, you can just hear like a big stream of piss and it's like: "that's what I think of you and your record and probably your species as well."[119]

Aside from Herbert's anthropomorphic projection, which can be useful at times because, as Marc Bekoff and Jessica Pierce note, "empathy is empathy,"[120] this anecdote demonstrates that utopia is not bliss, but piss. After all, in reading the lyrics, the singer of "May 2011" opts to let the dead pig occupy his thoughts, thoughts that are not vegan friendly, but darkly vegan. And so, while listening to the blood and guts of *One Pig* is not necessarily "magical," it is undoubtedly "dark," as Herbert puts it.[121] Indeed, now that "music can *be* life and death,"[122] as he says in an interview, it is time to think carefully about what this means today whether it involves a human or nonhuman. Going into the dark can be hazardous, but if we make it out, then we might just be ready for what ethical dangers lie ahead.

Notes

1. Matthew Herbert, in accompanying booklet, *One Pig*. Accidental Records Ltd. AC48CD, 2011, compact disc, 26.
2. Accidentalist, "ONE PIG, by Matthew Herbert – The Story Behind the Album," *YouTube* Video, 10:17, September 12, 2011. https://www.youtube.com/watch?v=GddErv81vOY.
3. Herbert, in accompanying booklet, *One Pig*, 2.
4. See https://accidentalrecords.bandcamp.com/album/one-pig.
5. Ben Sisario explains that Yann Seznec, sound designer, created a "StyHarp" for Herbert's live show, which is a device made to look like a pigsty that uses strings to trigger samples from *One Pig*. For more information on the StyHarp, see Seznec's blog, http://www.yannseznec.com/category/styharp/.
6. Jeff R. Warren, *Music and Ethical Responsibility* (Cambridge, UK: Cambridge University Press, 2014), 1.
7. Michel Foucault, *Ethics: Subjectivity and Truth*, ed. Paul Rainbow, trans. Robert Hurley et al. (New York: The New Press, 1997), 256.
8. Ibid.

9. For more information, see Nicole Shukin's *Animal Capital* (2009), 1–49.
10. For a facsimile of Donald Watson's first issue of *The Vegan News*, visit http://www.ukveggie.com/vegan_news/.
11. Gary L. Francione, *Animals as Persons: Essays on the Abolition of Animal Exploitation* (New York: Columbia University Press, 2008), 16.
12. Ibid.
13. Gary L. Francione, "Animal Welfare, Happy Meat, and Veganism as the Moral Baseline," in *The Philosophy of Food*. Ed. David M. Kaplan (Berkley: University of California Press, 2012), 181–182.
14. Ibid.
15. Haynes, Richard P Haynes, "The Myth of Happy Meat" in *The Philosophy of Food*. Ed. David M. Kaplan (Berkley: University of California Press, 2012), 165.
16. Francione, "Animal Welfare," 182.
17. Rasmus Rahbek Simonsen, "A Queer Vegan Manifesto." *Journal for Critical Animal Studies* 10, no. 3 (2012): 51–80. Accessed May 5, 2015, 57.
18. Simonsen, "A Queer Vegan Manifesto," 67.
19. Simonsen, "A Queer Vegan Manifesto," 71.
20. See, for instance, Russell Simmons' *The Happy Vegan: a Guide to Living a Long, Healthy, and Successful Life* (2015).
21. Denver Nicks, "Guinness is Going Vegan," November 3, 2015, http://time.com/4098272/guinness-is-going-vegan/.
22. Kattie Basnett, "Animal Remainders, Remaining Animals: Cross-Species Collaborative Encounters in Victorian Literature and Culture" (Ph.D. diss., Rice University, 2014), 178.
23. "Matthew Herbert," *Accidental Records*, 2014, http://accidentalrecords.com/portfolio/matthew-herbert/; Dimitri Nasrallah, "Herbert: Pitch Control," *Exclaim.ca*, October 26, 2008, http://exclaim.ca/music/article/herbert-pitch_control.
24. Nasrallah, "Herbert."
25. Ibid.
26. Ibid.
27. See http://www.discogs.com/for more information.
28. By my count on *Discogs.com*, at the time of writing, Herbert has put a total of ninety-five releases, which includes LPs, EPs, singles, and other miscellany; Phill Savidge, "Matthew Herbert Biography," *Matthew Herbert.com*, 2013, http://matthewherbert.com/about-contact/.

29. Savidge, "Matthew Herbert Biography."
30. Matthew Herbert, "Plat du Jour," *Platdujour.co.uk*, September 9, 2005, http://platdujour.co.uk/notes.php?pageNum_displayNotes=4&totalRows_displayNotes=14&theme=11.
31. Savidge, "Matthew Herbert Biography."
32. Tim Jonze, "Matthew Herbert: 'I can make music out of a banana or David Cameron or Belgium,'" *The Guardian*, August 20, 2015, http://www.theguardian.com/music/2015/aug/20/matthew-herbert-the-shakes-interview.
33. "Matthew Herbert." *Accidental Records*.
34. Savidge, "Matthew Herbert Biography."
35. Todd L. Burns and Matthew Herbert, *RA Exchange 32*, podcast audio, "Matthew Herbert," May 6, 2011, http://www.residentadvisor.net/podcast-episode.aspx?exchange=32.
36. Jonze, "Matthew Herbert."
37. Matthew Herbert, "Manifesto," *MatthewHerbert.com*, 2013 [2005, 2011], http://matthewherbert.com/about-contact/manifesto/.
38. Savidge, "Matthew Herbert Biography."
39. Savidge, "Matthew Herbert Biography."
40. Todd L. Burns and Matthew Herbert, *RA Exchange*.
41. Matthew Herbert, "Plat du Jour," *Platdujour.co.uk*, September 9, 2005, http://www.platdujour.co.uk/notes.php?theme=1.
42. Todd L. Burns and Matthew Herbert, *RA Exchange*.
43. Ibid.
44. Herbert, "a project is born," *MatthewHerbert.com,* May 21, 2009, http://matthewherbert.com/a-project-is-born/.
45. Ibid.
46. Matthew Herbert, "missed one," *MatthewHerbert.com*, June 18, 2009, http://matthewherbert.com/missed-one/.
47. Matthew Herbert, "ARRIVED!," *MatthewHerbert.com*, August 17, 2009, http://matthewherbert.com/arrived/.
48. Matthew Herbert, "[Untitled]," *MatthewHerbert.com*, August 20, 2009, http://matthewherbert.com/category/one-pig/page/2/.
49. Matthew Herbert, "pig moved to new sty," *MatthewHerbert.com,* December 19, 2009, http://matthewherbert.com/pig-moved-to-new-sty/.
50. Herbert, in accompanying booklet, *One Pig*, 2.
51. Matthew Herbert, "abattoir," *MatthewHebert.com*, February 9, 2010, http://matthewherbert.com/abattoir/.

52. *One Pig* comes with a facsimile of the invoice sent to Herbert's label, Accidental Records, which is marked "PAID" (dated February 19, 2010). Also, while *One Pig*'s liner notes state that the "farm wish[es] to remain anonymous," an online post from FACT Magazine dated September 1, 2009, reports that on August 15 Herbert's pig was born at Monkshill Farm in Kent (Matthew Herbert's blog corroborates this date, too). Elsewhere online, Monkshill Farm appears in connection with Matthew Herbert. Moreover, the copy of the invoice that comes with *One Pig* does not do the best job of censoring the name of the farm (the word "Monkshill" is discernible because you can see the top of the word even though it is blacked out).
53. Herbert, "ARRIVED!."
54. Matthew Herbert, interview by Matthew Bennett, "Matthew Herbert's One Pig," *www.clashmusic.com*, October 12, 2011, http://www.clashmusic.com/feature/matthew-herberts-one-pig.
55. PETA, "Animal Rights Group PETA Slam Matthew Herbert's 'Pig' Album," *www.gigwise.com*, February 12, 2010, http://www.gigwise.com/news/54640/Animal-Rights-Group-PETA-Slam-Matthew-Herbert's-'Pig'-Album.
56. See Steve Baker's *Postmodern Animal* (2000) or *Artist Animal* (2012).
57. Matthew Herbert, "full response to PETA," *MatthewHerbert.com*, March 4, 2010, http://matthewherbert.com/full-response-to-peta/.
58. Ibid.
59. Ibid.
60. Ibid.
61. Ibid.
62. Herbert, in accompanying booklet, *One Pig*, 2.
63. Herbert, interview by Jobst Eggert. "Matthew Herbert vs. PETA."
64. Ibid.
65. Herbert, in accompanying booklet, *One Pig*, 2.
66. Haydn Lorimer, "Human—non-human." In *Introducing Human Geographies*. Eds. Paul Cloke, Philip Crang, and Mark Goodwin. Third edition (New York: Routledge, 2014), 44.
67. Lorimer, "Human—non-human," 46.
68. Ibid.
69. Ben Sisario, "Raising an Album, From Pigpen to Studio."
70. Basnett, "Animal Remainders, Remaining Animals," 204.
71. Jacques Derrida, *Spectres of Marx: the State of the Debt, the Work of Mourning and the New International*. Trans. Peggy Kamuf (New York: Routledge, 1994), 202.

72. Derrida, *Spectres of Marx*, 42.
73. Mark, Fisher, *Ghosts of My Life: Writings on Depression, Hauntology and Lost Futures* (United Kingdom: Zero Book, 2014), 120.
74. See Carol Adams' *The Sexual Politics of Meat* (2010 [1990]), 66–67.
75. Herbert, in accompanying booklet, *One Pig*, 2.
76. Matthew Herbert, "Sonography," *www.MatthewHerbert.com*, 2013. http://matthewherbert.com/sonography/.
77. Herbert, interview by Matthew Bennett, "Matthew Herbert's One Pig."
78. Herbert, in accompanying booklet, *One Pig*, 4.
79. Herbert, interview by Jobst Eggert.
80. Ibid.
81. Ibid.
82. Herbert, interview by Matthew Bennett, "Matthew Herbert's One Pig."
83. Douglas Kutach, *Causation* (Cambridge, United Kingdom: Polity, 2014), 2.
84. *OED Online*.
85. Graham Harman, "On Vicarious Causation." *Collapse II*. Ed. R. Mackay (Oxford: Urbanomic, 2007), 172.
86. Graham Harman, "Time, Space, Essence, and Eidos: a New Theory of Causation," *Cosmos and History: The Journal of Natural and Social Philosophy* 6, no. 1 (2010), 1.
87. Timothy Morton, *Realist Magic: Objects, Ontology, Causation* (Ann Arbor: Open Humanities Press, 2013), 17.
88. Levi R. Bryant, *Onto-Cartography: an Ontology of Machines and Media* (Edinburg: Edinburg University Press, 2014), 245.
89. Graham Harman, *Guerrilla Metaphysics: Phenomenology and the Carpentry of Things* (Chicago: Open Court, 2005), 174.
90. Harman, Graham. *Circus Philosophicus* (United Kingdom: Zero Books, 2010), 51.
91. Herbert, interview by Matthew Bennett, "Matthew Herbert's One Pig."
92. Herbert, "Manifesto."
93. Herbert, "Sonography."
94. William Ian Miller, *The Anatomy of Disgust* (Cambridge, Massachusetts: Harvard University Press, 1997), 111.

95. Korsmeyer, Carolyn Korsmeyer, *Savoring Disgust: the Foul and the Fair in Aesthetics*. (Oxford: Oxford University Press, 2011), 3.
96. Ibid.
97. Ibid.
98. Herbert, interview by Matthew Bennett.
99. Nicole Shukin, *Animal Capital: Rendering Life in Biopolitical Times* (Minneapolis: University of Minnesota Press, 2009), 14.
100. Shukin, *Animal Capital*, 16.
101. Shukin, *Animal Capital*, 20.
102. Miller, *The Anatomy of Disgust*, 8.
103. Ibid.
104. Miller, *The Anatomy of Disgust*, 9.
105. Miller, *The Anatomy of Disgust*, 203.
106. Miller, *The Anatomy of Disgust*, 204.
107. Ibid.
108. Ibid.
109. Miller, *The Anatomy of Disgust*, 9.
110. Herbert, in accompanying booklet, *One Pig*, 2.
111. Matthew Herbert, interview by Sam Inglis, "Matthew Herbert: Sampling Pig Noises," *www.soundonsound.com*, November 2011. http://www.soundonsound.com/sos/nov11/articles/herbert.htm.
112. Herbert, in accompanying booklet, *One Pig*, 26.
113. Matthew Herbert and Sara Mohr-Pietsch, *Composers' Rooms: No. 6 Matthew Herbert,* podcast audio, May 11, 2014, http://www.bbc.co.uk/programmes/p01yx9c1.
114. Daniel J. Levitin, *This is Your Brain on Music: the Science of Human Obsession* (New York: Dutton, 2006), 151.
115. Ibid.
116. Vadim Prokhorov quoted in Peter Szendy, *Hits Hits: Philosophy in the Jukebox* (New York: Fordham University Press, 2012), 2.
117. Sigmund Freud, "'A Child is Being Beaten' A Contribution to the Study of the Origin of Sexual Perversions" in *The Standard Edition of the Complete Psychological Works of Sigmund Freud, Volume XVII (1917–1919): An Infantile Neurosis and Other Works* (London: Hogarth Press, 1953), 175.
118. Accidentalist, "ONE PIG, by Matthew Herbert – The Story Behind the Album."

119. Herbert, interview by Matthew Bennett.
120. Marc Beckoff and Jessica Pierce, *Wild Justice: the Moral Lives of Animals* (Chicago: The University of Chicago Press, 2009), 41.
121. Herbert, interview by Jobst Eggert. "Matthew Herbert vs. PETA."
122. Matthew Herbert, interview by Mike Doherty. "Q&A: Producer Matthew Herbert on the mortality of his music." *National Post*, June 26, 2013.

References

Accidentalist. 2011. ONE PIG, by Matthew Herbert – The Story Behind the Album. *YouTube* Video, 10:17. September 12. https://www.youtube.com/watch?v=GddErv81vOY.

Basnett, Kattie. 2014. *Animal remainders, remaining animals: Cross-species collaborative encounters in Victorian literature and culture*. PhD dissertation, Rice University. Houston, Texas.

Bekoff, Marc, and Jessica Pierce. 2009. *Wild justice: The moral lives of animals*. Chicago: The University of Chicago Press.

Bryant, Levi R. 2014. *Onto-cartography: An ontology of machines and media*. Edinburg: Edinburg University Press.

Burns, Todd L. and Matthew Herbert. 2011. *RA Exchange 32*, podcast audio. "Matthew Herbert." May 6. http://www.residentadvisor.net/podcast-episode.aspx?exchange=32.

Derrida, Jacques. 1994. *Spectres of Marx: The state of the debt, the work of mourning and the new international*. Trans. Peggy Kamuf. New York: Routledge.

Fisher, Mark. 2014. *Ghosts of my life: Writings on depression, hauntology and lost futures*. Winchester: Zero Book.

Foucault, Michel. 1997. *Ethics: Subjectivity and truth*. Ed. Paul Rainbow. Trans. Robert Hurley et al. New York: The New Press.

Francione, Gary L. 2008. *Animals as persons: Essays on the abolition of animal exploitation*. New York: Columbia University Press.

Francione, Gary L. 2012. Animal welfare, happy meat, and veganism as the moral baseline. In *The philosophy of food*, ed. David M. Kaplan. Berkley: University of California Press.

Freud, Sigmund. 1953.'A child is being beaten' A contribution to the study of the origin of sexual perversions. In *The standard edition of the complete psychological works of Sigmund Freud, Volume XVII (1917–1919): An infantile neurosis and other works*. Trans. James Strachley. London: Hogarth Press.

Harman, Graham. 2005. *Guerrilla metaphysics: Phenomenology and the carpentry of things*. Chicago: Open Court.

Harman, Graham. 2007. On vicarious causation. In *Collapse II*, ed. R. Mackay. Oxford: Urbanomic.

Harman, Graham. 2010a. *Circus philosophicus*. Winchester: Zero Books.

Harman, Graham. 2010b. Time, space, essence, and eidos: A new theory of causation. *Cosmos and History: The Journal of Natural and Social Philosophy* 6(1): 1–17.

Haynes, Richard P. 2012. The myth of happy meat. In *The philosophy of food*, ed. Kaplan David M. Berkley: University of California Press. Print.

Herbert, Matthew. *Matthew Herbert* (blog). http://matthewherbert.com/home/.

Herbert, Matthew. 2005. Pat du Jour. *Platdujour.co.uk*. September 9. http://platdujour.co.uk/notes.php?pageNum_displayNotes=4&totalRows_displayNotes=14&theme=11.

"Matthew Herbert's Pig Is Born!" FACT Magazine. September 1, 2009. http://www.factmag.com/2009/09/01/matthew-herberts-pig-is-born/. Accessed 11 Oct 2015.

Herbert, Matthew. 2011a. In accompanying booklet, *One Pig*. Accidental Records Ltd. AC48CD, 2011, compact disc.

Herbert, Matthew. 2011b. Interviewed by Matthew Bennett. "Matthew Herbert's One Pig." *www.clashmusic.com*. October 12. http://www.clashmusic.com/feature/matthew-herberts-one-pig.

Herbert, Matthew. 2011c. Interview by Jobst Eggert, "Mathew Herbert vs. PETA." *MatthewHerbert.com*. November 3. http://matthewherbert.com/matthew-herbert-vs-peta/. Accessed 11 Oct 2015.

Herbert, Matthew. 2011d. Interview by Sam Inglis. "Matthew Herbert: Sampling Pig Noises," *www.soundonsound.com*, November 2011. http://www.soundonsound.com/sos/nov11/articles/herbert.htm.

Herbert, Matthew. 2011e. *One Pig*. Accidental Records Ltd. AC48CD, 2011, compact disc.

Herbert, Matthew. 2013a. Interview by Mike Doherty. "Q&A: Producer Matthew Herbert on the mortality of his music." *National Post*. June 26. http://news.nationalpost.com/arts/music/qa-producer-matthew-herbert-on-the-mortality-of-his-music. Accessed 11 Oct 2015.

Herbert, Matthew. 2013b. Sonography. *www.MatthewHerbert.com*; http://matthewherbert.com/sonography/.

"Matthew Herbert." 2014. *Accidental records*. http://accidentalrecords.com/portfolio/matthew-herbert/.

Herbert, Matthew and Sara Mohr-Pietsch. 2014. *Composers' Rooms: No. 6 Matthew Herbert*, podcast audio. May 11. http://www.bbc.co.uk/programmes/p01yx9c1.

Herbert, Matthew. 2013 [2005, 2011]. "Manifesto." *MatthewHerbert.com*. http://matthewherbert.com/about-contact/manifesto/.

Jonze, Tim. 2015. Matthew Herbert: 'I can make music out of a banana or David Cameron or Belgium.' *The Guardian*, August 20. http://www.theguardian.com/music/2015/aug/20/matthew-herbert-the-shakes-interview.

Korsmeyer, Carolyn. 2011. *Savoring disgust: The foul and the fair in aesthetics*. Oxford: Oxford University Press.

Kutach, Douglas. 2014. *Causation*. Cambridge: Polity.

Levitin, Daniel J. 2006. *This is your brain on music: The science of human obsession*. New York: Dutton.

Lorimer, Haydn. 2014. Human—non-human. In *Introducing human geographies*, 3rd ed, ed. Paul Cloke, Philip Crang, and Mark Goodwin. New York: Routledge.

Miller, William Ian. 1997. *The anatomy of disgust*. Cambridge, MA: Harvard University Press.

Morton, Timothy. 2013. *Realist magic: Objects, ontology, causation*. Ann Arbor: Open Humanities Press.

Nasrallah, Dimitri. 2008. Herbert: Pitch control. *Exclaim.ca*, October 26. http://exclaim.ca/music/article/herbert-pitch_control.

Nicks, Denver. 2015. Guinness is going vegan. *Time.com*, November 3. http://time.com/4098272/guinness-is-going-vegan/.

PETA. 2010. Animal Rights Group PETA Slam Matthew Herbert's 'Pig' Album. *www.gigwise.com*. February 12. http://www.gigwise.com/news/54640/Animal-Rights-Group-PETA-Slam-Matthew-Herbert's-'Pig'-Album.

Savidge, Phill. 2013. Matthew Herbert Biography. *Matthew Herbert.com*. http://matthewherbert.com/about-contact/.

Shukin, Nicole. 2009. *Animal capital: Rendering life in biopolitical times*. Minneapolis: University of Minnesota Press.

Simonsen, Rasmus Rahbek. 2012. A Queer Vegan Manifesto. *Journal for Critical Animal Studies* 10(3): 51–80. Accessed 5 May 2015.

Sisario, Ben. 2011. Raising an album, from Pigpen to Studio. *The New York Times*, October 17. http://www.nytimes.com/2011/10/18/arts/music/in-one-pig-matthew-herbert-finds-music-in-a-pig.html. Accessed 28 May 2015.

Szendy, Peter. 2012. *Hits: Philosophy in the Jukebox*. New York: Fordham University Press.

Warren, Jeff R. 2014. *Music and ethical responsibility*. Cambridge: Cambridge University Press.

The Compassion Manifesto: An Ethics for Art + Design and Animals

Julie Andreyev

Preface

In January 2015, an undergraduate student—I'll call D—at Emily Carr University of Art + Design (ECUAD) submitted plans for a project to be installed in the campus' Abe Rogatnick Media Gallery. The project was to involve live captive birds. Upon hearing of the proposal, a group of faculty and staff, troubled by the birds' captivity and their potential harm, voiced their concerns. The faculty and staff who oversaw the exhibitions for the gallery unanimously rejected the proposal. *The Compassion Manifesto: An Ethics for Art + Design and Animals* was written as a response to the proposal and the larger context of contemporary art and design practices that involve nonhuman animals. There are tendencies in art and design genres, such as bioart, to exploit living beings in aesthetic experiments aimed at exploring human conditions. *The Compassion*

J. Andreyev (✉)
Emily Carr University of Art + Design, Vancouver, BC, Canada
e-mail: jandreye@ecuad.ca

J. Castricano, R.R. Simonsen (eds.), *Critical Perspectives on Veganism*,
The Palgrave Macmillan Animal Ethics Series,
DOI 10.1007/978-3-319-33419-6_7

Manifesto critiques practices that result in captivity, harm, and death of nonhumans and proposes an ethics of care and compassion as alternatives to anthropocentric methods. *The Compassion Manifesto* was inspired by earlier manifestos, such as by the historical art group the Situationists International[1] and the recent *Animal Manifesto* by Marc Bekoff.[2]

A reading of *The Compassion Manifesto* was performed in the Abe Rogatnick Media Gallery on March 17, 2015, by *Vegan Congress*[3] members Maria Lantin and myself, and fellow faculty member Alexandra Phillips. The reading was produced as a relational art event where the University community, and public, was invited to participate in a free-form discussion. (The concerns raised in the discussion, the developments of D's project, and the resulting initiative by the University are annotated in the Afterword of this chapter.) At the end of the discussion, a signing "ceremony" took place where a printout of *The Manifesto* was signed by Maria Lantin, Alexandra Phillips, Trudy Chalmers, Lucy Chen, Karolle Wall, Carol Gigliottii, Ben Bogart, Greg Snider, and myself.

The original *Compassion Manifesto* was designed to be read aloud as a participatory performance. The version below is modified from the original, with an expansion on key points as was necessary for this text version. In order to retain the performative character of the original, I chose not to include the argument's details in the body of the text, but instead in the endnotes.

Compassion Manifesto: An Ethics for Art + Design and Animals

In a brief address during a Thanksgiving Observance, Buddhist monk Tashi Nyima speaks about the "Brightly lit aisles [that] conceal the horrible darkness where animals are confined, enslaved, tortured and slaughtered for pleasure."[4] He speaks about the need to reduce the suffering of not just humans, but of all sentient beings. Nyima, in his call for compassionate living, quotes the Buddha: "Having abandoned the taking of life, refraining from the taking of life, we dwell without violence, with the knife laid down—scrupulous, full of mercy—trembling with compassion for all sentient beings."

The *Compassion Manifesto: An Ethics for Art + Design and Animals* expands on compassionate living by advocating that ethics for nonhuman others be integrated into cultural practices. Historical and contemporary art and design that involve nonhuman others most often fails to consider what is at stake for those other beings. The majority of these practices construct nonhuman others as objects, not participants, as materials, not lives. The ideological belief in freedom of expression instead justifies the use and abuse of nonhuman others. *The Compassion Manifesto* calls for compassionate thought and action informed by an interdisciplinary investigation into cultural theory, critical animal studies, philosophy, Buddhism, veganism, indigenous cultures, ethics of care studies, biology and cognitive ethology. The knowledge gained from these investigations can contribute toward the enhancement of art and design, integrating more ethically and ecologically sound thinking and making.

Representations of nonhuman others appeared in cultural forms for millennia beginning with the cave paintings of migrating herds of animals. Contemporary culture continues this tradition with depictions of non-humans in movies, animations, nature programs, newspapers, magazines, social media, advertising, and so on. Most often these depictions reinscribe detrimental thinking about animals, serving to define the human while distancing the animal.[5] Much work needs to be done to create more caring representations within culture. *The Compassion Manifesto* focuses on art and design, fields that can lead in developing ethically improved cultural thought and form. But here too, attention and care needs to be developed.

Many art and design practices, historical and contemporary, operate within instrumental frameworks, resulting in harm and death of other beings. A bioart project that combines plants, fish, and computers to explore the ecological relationships between them and experiment with created closed, sustainable energy sources for human benefit does not consider the lives of those nonhuman beings involved.[6] *The Compassion Manifesto* draws attention to these detrimental practices and advocates for inspiration, creativity, and feeling toward more just and caring processes. As writer and *Vegan Congress* member Carol Gigliotti stresses, "[We] believe that animals are sentient, conscious, intelligent, and creative beings who are just as necessary to the world as human animals."[7]

Millions of nonhuman beings suffer physically and psychologically in factory farms and laboratories, in zoos, aquariums, and other spectacles of entertainment. Anthropocentric views are at the root of these abuses that conceptualize human existence as superior to nonhuman existences. Speciesism emerges out of anthropocentrism and formulates systems that exclude most nonhumans from ethical consideration. Speciesism leads to mechanisms and practices that contrive the nonhuman as resource for human exploitation, experimentation, and consumption.[8] In this instrumental framework, nature and nonhumans are seen as means through which the world is produced for human ends.[9]

Western history of thought privileges human reason and language as the yardstick of valuation for all other species. Aristotle, Descartes, Kant, and others laid the foundation for human-centered being, arrogantly declaring the human at the top of an anthropocentrically created "great chain of being."[10] This doctrine creates a culture of consent for imposing violence on so-called lesser animals, and is at work in art and design under the guise of freedom of expression. Aspects of art and design culture perpetuate ideas about unfettered creativity as a holy grail.[11] This (un)creative expression justifies the suffering, even death, of nonconsenting others. The history of art and design is written with the bodies of animals.

sable-hair paint brushes,[12]
silver gelatin prints,[13]
rabbit skin glue,[14]
meat dresses,[15]
meat orgies,[16]
pony skin chaise lounges,[17]
an Eames chair reskinned with elk hide,[18]
a captive deer in a gallery,[19]
a coyote penned-in with a so-called shamanic artist,[20]
an aviary of tightly caged birds,[21]
insect-controlled robots,[22]
livestock slaughter machines,[23]
fruit flies as hazardous chemical detectors,[24]
a miniature stadium of insects, spiders, scorpions forced into coexistence in a bleak hard architectural model,[25]
a glowing genetically modified Bunny,[26]

interactive blenders full of goldfish,[27]
a starving dog on public display.[28]

These projects destroyed lives in the production of culture. They constructed the human form from the remains of the nonhuman. Forces of domination and commodification are at work in art and design that joins with the laboratory.[29] Using mechanisms of oppression, these practices transform living, breathing, sensing beings into material for aesthetic use. *The Compassion Manifesto* asks: who are these animals that suffer and die so that art can be made?[30] It calls on artists and designers to expand their consciousness—to learn about and pay attention to nonhumans and our shared ecological being.

The Compassion Manifesto calls for the consideration of nonhumans as subjects of their own lives. Nonhuman beings have languages,[31] cultures, families, and communities; they are creative and have concerns and projects of their own. Methodologies of neutral objectivity, as upheld by science, are not adequate to understanding and instead cause harm. *The Compassion Manifesto* calls for methods of "biocentric anthropomorphism," to allow nonhuman thoughts, feelings, and states of being to be considered.[32] Nonhuman animals have emotions that are similar to humans, such as sadness, happiness, and empathy.[33] Plants respond to the environment by foraging; they perceive other plant communications, remember stresses from the past, and look to the future.[34] *The Compassion Manifesto* calls for artists and designers to think-like-a-bird, feel-like-a-dog, and attend-to-the-earth-like-a-plant in order to provide for awareness and ethical interactions.

Indigenous forms of relating with nonhumans have much to teach creative practitioners. The sable, rabbit, deer, coyote, bird, spider, fish, and plant are our brothers and sisters. The Earth depends on humans to have good relations with other beings, not to think they own them. *The Compassion Manifesto* calls for openness to indigenous forms of knowledge and awareness, to generate understanding, and promote wide-eyed expanded curiosity so important to creative fields.

The problems characteristic of the anthropocene, such as the loss of biodiversity and the destruction of environmental systems, have spurred artists and designers to respond. Recent art that investigates ecological systems and the lives of nonhumans, known as bioart, has some creative models that allow for human reflection on nonhuman intention,[35] that ethically reveal otherwise hidden forms of being,[36] and that point to shared states of ecological being.[37] However, the majority of bioart is dominated by anthropocentric views where nonhumans are treated as living material to support explorations on the human condition and the human challenges posed by ecological degradation. Biomimetic design methods are inspired by physical forms, organic systems, and the movement of living beings to design robotic and other systems. These methods often depend on laboratory experiments on animals, dead or living. *The Compassion Manifesto* calls for rethinking how we respond to the anthropocene by developing advancements on cultural forms without causing additional harm.

The Compassion Manifesto invites the abandonment of destructive, outmoded, unecological beliefs generated by anthropocentrism. It summons practices that engage two interconnected tasks: resituating the human within the *continuum* of nature and reconsidering nonhumans in *ethical* terms.[38] These tasks begin with the understanding that anthropocentrism affects all life, including human life. Reconsidering reason through the lens of "ecological thought" reveals that Being on Earth is an interconnected web, not a hierarchy.[39] The interconnectedness between all sentient and nonsentient beings forms a relational ethic of entwined existences.[40]

The Compassion Manifesto advocates an expansion of our compassion footprint.[41] It calls for resistance to rationalist culture by reminding us that we are feeling, sensing, creative beings. Emotion and empathy contribute forms of knowledge that can be extended to nonhuman others. The suffering of another, including nonhuman beings, can be felt, and can awaken right attention. As one becomes aware of others and their own states of being, consciousness expands to become more attentive to the world. Attending to the needs of another is ethical. This right attention renders freedom an illusion because in an aware state, the ethical choice is the desired choice.[42] Have you ever been held in the gaze of an animal?[43] Have you ever walked beside, shared experience, cultivated life with another animal? Have you ever felt their intention, curiosity, joy, or sadness? *The Compassion Manifesto* calls for art and design processes that include practicing loving attention and right action directed toward Earth's others.

The Compassion Manifesto questions the binary opposition of human, and all other beings established by anthropocentric thought. The opposition of human and animal as developed by the historical humanist project is a great "self-interested mis-recognition."[44] The category "animal" itself is problematic, as it unifies all other-than-human animals into one kind, apart from the human. Let us be more accurate. There are infinite varieties of being, not only species, but individuals. An ethics of maximum respect[45] allows us to ask: "What are *you* going through?"[46]

The Compassion Manifesto calls for

- self-critical examinations into problematic relationships with other beings and ecologies.
- attending to the continuities between humans, other animals, and plant life;
- attending to relatedness and shared states of being;
- methodologies of openness rather than closedness in relation to others;
- curiosity about minds that take different forms—pheromone-, pollen-, scent-, and sonar-based minds;
- explorations into different knowledges—ocean, sky, and soil wisdom;
- examinations into creativity as it occurs in nonhuman cultures—hives, schools, pods, and flocks;
- expansions on human humility in the face of other beings' agency.

The Compassion Manifesto advances a nonhierarchical, nonbinary consideration of being, a recognition of "We." It is an ethics of "maximum respect"[47] with regards to all of us.

The Compassion Manifesto declares that we "not kill, eat, torture, [or] exploit [others], because they do not want to be so treated, and we know that. If we listen, we can hear them."[48]

Afterword

Informed by vegan ethics, *The Compassion Manifesto* argues against creative practices that use living nonhuman animals in unethical ways and calls for a reconsideration of material sources used in art and design. In contemporary practices, most tools and materials made of nonhuman animal by-products can be avoided. However, this is not the case for many traditional forms of practice. During the discussion following the reading of *The Compassion Manifesto*, concerns about the conflict between vegan ethics and indigenous traditional practices were raised. In local indigenous practices, skins, fur, and feathers are used in the production of drums and other cultural objects. It was argued in the discussion that the practice of hunting and the use of nonhuman animal remains in indigenous traditions is key to the identity of those cultures. A participant was critical that the views in *The Compassion Manifesto* were another form of violence, in this case against indigenous cultures. I suggest that this critique is itself problematic because it assumes a homogenizing view on indigenous cultures, suggesting that, for example, there are no vegan indigenous people. The need to respect indigenous peoples is imperative for a global expansion of ethics. So, how may vegan ethics and indigenous traditions be reconciled?

Alfred Irving Hallowell, in his essay, "Ojibwa Ontology, Behavior and World View,"[49] argues that the Ojibwa people's traditional narratives indicate a worldview that considers other-than-human animate beings as persons. The outward manifestation of a person, as a human or other animal, is incidental—changeability is an inherent capacity of animate beings. Some stories relate how nonhumans may be animals or may be human ancestors in nonhuman form. Dreaming and awake states also form a relational continuum where other-than-humans and humans communicate, and where humans may take on other-than-human forms. Mutual obligation is also present in the Ojibwa worldview where other-than-human ancestors are seen as important contributors to the health of all life. The Ojibwa's is only one example of an indigenous worldview that includes ethics for nonhuman animals and ecological existence.[50] Hallowell also relates Ojibwa hunting practices where considerations are extended to the hunted animal so as not to cause suffering. Can traditions that argue for killing nonhuman animals, and for the use of their remains in cultural forms, be reconciled with vegan ethics?

Mi'kmaq scholar Margaret Robinson argues that vegan ethics are not at odds with the indigenous Mi'kmaq worldview.[51] Using an ecofeminist critique, she suggests two reasons as barriers to indigenous veganism: the belief that meat eating is culturally more authentic and the view that veganism is a form of racial privilege. Hunting in Mi'kmaq culture is seen as a male practice that reinforces views on virility and masculinity, such as in a boy's first hunt as an entry into manhood. Rejecting hunting practices is seen as a rejection of rituals crucial to the formation of male identity. However, she argues that "[m]eat, as a symbol of patriarchy shared with colonizing forces, is arguably more assimilating than practices such as vegetarianism."[52]

Robinson's argument for indigenous veganism is based on two aspects of Mi'kmaq culture: the worldview that includes respectful relating with nonhuman others and the need to consider culture and its living forms. Activities normally performed by Mi'kmaq women, such as gathering fruits, nuts, and vegetables, contribute counternarratives to hunting. Robinson argues that the belief in preserving traditional rituals, such as hunting, can be seen as joining with colonial views that reject contemporary indigeneity: "When Native is defined exclusively as a primordial lifestyle it reflects our intentional extinction as a people."[53] The changing circumstances of indigenous peoples, she argues, must take into account a need for reinterpreting rituals within retained set of values. Traditional values—respect for life and recognition of relationality between humans and nonhuman persons—can be upheld in new rituals. Traditional Mi'kmaq, like Ojibwa, value kinship relations with nonhuman others. Nonhuman animals are seen as persons, and their value is not in their utility to humans, but in their intrinsic essence as living beings. Robinson argues that veganism can provide a sense of belonging for a community that values life in daily practice. Indigenous women can determine authenticity for themselves, rejecting dominant masculine notions of preservation for precolonial pasts.

Later in the spring semester of 2015, one month after *The Compassion Manifesto* was performed as a reading, D reproposed her project to the University, this time to take place in a small secluded room in the sculpture area, a space normally used by sculpture students to install their work, and have it viewed by their classmates and instructor. D provided elaborations on the installation including information on the birds—four pigeons to be "rented" from a "fancy pigeon" breeder who shows, rents, and sells his birds for events such as weddings. The proposed project would contain a bed, and the birds would be allowed to freely move around the room and interact with the bed and other items in the space. D proposed that students and faculty be allowed to enter the room, like into a gallery. At the time, there was no policy in place to address the use of nonhuman animals in creative practices at the University. Each instance was treated on an ad hoc basis. Historically, this lack of policy had generally resulted in abuses with little to no oversight or review processes. What was at stake for the nonhuman animal was not meaningfully considered up until this case. Based on our initial protests, the administration provided D with a set of guidelines that called on the student to

- review the safe practice of using animals in the arts with the instructor;
- follow the British Columbia Society for the Prevention of Cruelty to Animals' (BCSPCA's) position statement on the use of animals in the arts[54];
- use breeder pigeons with written approval from the pigeon owner to use and transport the pigeons for the installation (the University will contact the owner to verify this);
- obtain written approval from the pigeon owner that the pigeons are free of transmissible disease and are regularly checked for health and medical requirements;
- provide direction from the pigeon owner in safe transportation, feeding requirements and recaging once the installation is over;
- provide that the pigeons will only be on-site for a day, from 9 am to 5 pm;
- provide someone in place at all times to ensure the well-being of the pigeons;
- ensure the critique room is secured and that the pigeons cannot escape.

Upon hearing that the project proposal had reemerged, the original group of concerned faculty and staff requested a meeting with D and the administration involved. Prior to the meeting, I contacted the chief science officer at the SPCA to gain information on pigeons and on the pigeon breeder. My concern was that if the University allowed the project to take place, how could we determine distress behavior in the birds or if they were being harmed? The officer suggested that perching objects be installed in the space to allow the birds to rest high above the ground, a normal behavior for pigeons. Distressed behavior may include the birds flying around in an agitated way, attempting to flee the space, or bumping into objects potentially causing harm to themselves. She advised to have a vet on hand to attend to any injuries if necessary. She confirmed that the breeder was known by the SPCA. She lamented that the SPCA was unable to confirm that harmful processes were being used in his business because firsthand accounts were unavailable. She clarified that the SPCA could be called in to the University if distress or harm occurred, but that the organization could not be involved in monitoring for potential harm.

During the meeting with D, the administration, and members of the newly forming Duty of Care committee, it became evident that there were a number of shortfalls in relation to how this case was being handled. It was assumed that the student was able to interpret the guidelines set by the University and to self-evaluate care processes with the birds. There was no meaningful mentorship in place for the student on the ethics of involving nonhuman animals in art practices. There was a presumption that robust ethics of care was being practiced by the pigeon breeder. There was a lack of critical consideration with respect to practices of breeding nonhuman animals for entertainment purposes. Surprisingly, it became evident that D was intending a critical examination of animal exploitation, such as in the food industry, but lacked the guidance to determine an ethical form for the project. D intended the project as a means to reconsider animal being, in a space that provided nondominating forms of relating. D had not considered that the use of birds in captive conditions that pose potential harm was another form of exploitation. D was unaware of the dubious care of the breeder and grappled with the idea that the breeder could be providing harmful conditions to the birds. It was clear to the concerned members of the community that the student was ill-equipped to carry out meaningful ethics of care for the birds. We were unanimous that the proposal not be allowed to move forward.

However, the provost defended D's project based on freedom of expression and potential learning and allowed the installation to go ahead if that was D's wishes. Based on the information from the SPCA Officer, the provost agreed to our suggestions for improved care: not allowing anyone else to enter the room while the birds were there and providing a safe window into the space that the birds would not mistake for an opening. D was genuinely interested in extending care toward the birds and was eager to provide safeguards against harm. The day and night before the project was to be installed, individuals from our group, the provost, and the dean tried to dissuade D from carrying out the project. Despite this, D was undeterred.

In the early morning of the installation, the administration approached a member of our group concerned about how we would respond to the project's approval. We sensed that they feared public controversy for the University if we contacted the press, the Humane Society, or SPCA. As a group, we discussed the complexity of the case and the constellation of potential outcomes. Because of D's seemingly good intentions and openness to suggestions, there was a potential for improved ethics in D's future projects. We did not want to alienate D from this potential. It was clear that the provost, who had been recently appointed, was irked by the lack of in-place policy and procedures for nonhuman animals at the University. Members of our group sensed the potential for continued dialogue about future policy. We decided that the best option was to refrain from involving outside bodies unless we observed harm, and that we would extend additional care to the birds and to D's learning process. We volunteered to monitor the installation and to be on hand to observe the birds for any signs of distress, and to provide support to D if needed. This also allowed us to continue dialogue with D during the exhibition, posing questions and providing information on art, research and ethics. As each of us took shifts, we became affected by the presence of the birds. They interacted affectionately with each other and seemed calm despite being in an unfamiliar space. They slept on the bed or rested on the provided perch. The exhibition proceeded uneventfully until the end of the day when D was returning the birds to their carrier.

Around 4:30pm, the dean's assistant and *Vegan Congress* member Trudy Chalmers, who was monitoring the exhibition, texted me. D was trying to catch the birds with a fishing net. Trudy learned from D that this was the method suggested by the breeder. The birds were clearly agitated. They flew around the room, bumping into the objects and walls. It all happened very quickly. Once back in the carrier, Trudy monitored the birds to see if they had suffered any injuries or prolonged distress. She texted me that the birds seemed to be calming down and were perched on the bar inside the carrier. Trudy and I formulated a report and sent it to our group and to the provost. Alexandra Phillips, ECUAD professor and one of the readers of *The Compassion Manifesto*, who previously had had companion birds, observed that this net method was unnecessary and that birds could be coaxed into a carrier through nonviolent means. Because of the distress caused to the birds, it was my belief that the University and our group had failed to adequately provide an ethics of care for the birds. Given the lack of meaningful process and the lack of information on correct bird handling, we felt that the event warranted further examination. The distress and potential harm caused to the birds, as a result of these deficits, clearly indicated the need for ethical frameworks to be developed at the University.[55]

We let our concerns be known to the administration. The provost held a post-exhibition "debriefing" that consisted of faculty, staff, students—including D—the provost, and other members of the administration, as a means to voice our concerns. The outcome was a recommendation for a working group to address the need for a clear policy on the involvement of nonhuman animals in student projects, university research, and curriculum. This recommendation was subsequently approved, and the Animals, Ethics and Creativity Working Group was formed. Members of the working group include the readers of *The Compassion Manifesto*, D, the dean, and other interested faculty and staff. Alex andra Phillips and myself are the cochairs of the working group.

Admittedly, the complicated and troubling events that followed the reading of *The Manifesto* were not ideal. Decisions were made under duress, favoring long-term potentials over immediate risks. Despite this, I believe that the best possible outcome was realized. The Animals, Ethics and Creativity Working Group has just completed its first year of meetings. Some working group members argue that any involvement of nonhuman animals in creative practices not be tolerated. Others maintain a belief in freedom of expression for the artist or designer. A few argue for the potential of improved relating latent in the human–nonhuman encounter. The working group is now in the process of developing language for the proposed policy and procedures.

The Compassion Manifesto: An Ethics for Art + Design and Animals was initially created as a means to draw attention to the need for ethics for nonhuman animals in creative practices. The current work toward drafting a policy and set of procedures at Emily Carr University of Art + Design is the direct result of vegan ethics practiced in public events by the *Vegan Congress*, such as with the reading of *The Compassion Manifesto*. By making vegan practice more visible in the public sphere through these kinds of projects, the *Vegan Congress* has been able to attract individuals from ECUAD and other universities as well as from the public. The work of the collaboration has helped catalyze community support for improving the lives of animals.

Notes

1. Situationists International wrote the 1966 pamphlet, "On the Poverty of Student Life: A Consideration of Its Economic, Political, Sexual, Psychological and Notably Intellectual Aspects and of a Few Ways to Cure it" as a means to draw attention to oppressive ideologies of the state and institutions such as the university. Ten thousand copies were printed and distributed at the University of Strasbourg. The pamphlet was a key text inspiring the student uprisings in France and Germany in 1968. Knabb, Ken. *The Situationist International Anthology*. Berkeley: Bureau of Public Secrets, 1981.
2. Bekoff, Marc. *The Animal Manifesto: Six Reasons for Expanding Our Compassion Footprint*. Novato, California: New World Library, 2010.
3. The *Vegan Congress* is an activist and relational art and design collective providing events and information about vegan practice to help develop discourse and applied ethics. The *Vegan Congress* consists of like-minded independent researchers and of faculty, staff, and students at universities in Canada. www.vegancongress.org.
4. Nyima, Tashi. "Bright Aisles, Dark Alleys," 2014. Great Middle Way. https://greatmiddleway.wordpress.com. Accessed Oct. 26, 2015.

5. Contributions to cultural theory that extend the discourse on how the nonhuman animal is constructed in popular cultural forms are offered by writers such as John Berger, Jonathan Burt, and John Sorenson. Berger critiques systems of power, such as zoos that contain the animal, and conduct violence for entertainment. Berger John. "Why Look At Animals?" *About Looking*. New York: Random House, 1980. Print. Burt examines how animals are portrayed in film and other cultural products and how these reinscribe detrimental thinking. Burt, Jonathan. "The Illumination of the Animal Kingdom: The Role of Light and Electricity in Animal Representation," in Kalof, Linda and Amy Fitzgerald, (eds.). *The Animals Reader: The Essential Classic and Contemporary Writings*. New York: Berg, 2007. 289–301. Print. Sorenson examines how the animal is represented by media and in-popular culture and how these portrayals serve to define the human and distance the animal. Sorenson, John. *Ape*. London: Reaktion Books, 2009. Print.
6. See, for example, "Biomodd [NYC4]" (2012) by Diego S. Maranan. Artists webpage: http://www.diegomaranan.com/?portfolio=biomodd-nyc4. The project is an installation that combines plants, fish, and computers to explore the ecological relationships between them and to experiment with created closed, sustainable energy sources. The heat generated by the plant tanks is used to heat the fish tanks containing live goldfish, and heat grows algae in the tanks which feed the fish. The fish tanks also cool the computers that are part of the system. On the artists' website and in talks delivered by the artist, there is little mention of the lives of the plants and fish. Instead, the artist focuses on the so-called positive aspects of the system's thermal dynamics.
7. Gigliotti takes issue with Steve Baker's defense of artworks where artists harm and even kill animals in the making of the work. Gigliotti, Carol. Book review on Steve Baker's new book, *Artist | Animal*, in *Humanimalia: a journal of human/animal interface studies,* Volume 6, Number 1 (Fall 2014). Accessed Oct. 26, 2015.
8. An overview of the various shades of anthropocentrism is well articulated in Weitzenfeld and Joy "An Overview of Anthropocentrism, Humanism, and Speciesism in Critical Animal Theory." Nocella, Anthony J. et al. *Defining Critical Animal Studies: An Intersectional Social Justice Approach for Liberation*. New York: Peter Lang Publishing, 2014. Print.
9. Adorno and Horkheimer argue that culture has an instrumentalist view on Being "apprehended under the aspect of manufacture and administration. Everything—even the human individual, not to speak of the animal— is converted into the repeatable, replaceable process, into mere

example for the conceptual models of the system." Adorno, Theodor and Max Horkheimer. *Dialectic of Enlightenment.* London: Verso Editions, 1979. Print. 83.

10. Aristotle, in *The History of Animals*, proposed a fixed category of being as a hierarchy of all animals, with humans at the top and insects at the bottom. This later developed into the metaphor "scala naturae" or "The Great Chain of Being" which continues to inform Western beliefs on how animals are valued in relation to their placement in the hierarchy. Kalof, Linda and Amy Fitzgerald (eds.). *The Animals Reader: The Essential Classic and Contemporary Writings*. New York: Berg, 2007. Print. 5–7.
11. Gigliotti debates with Steve Baker on the ethics of limitless artistic freedom, when artists are involved with other animals. Baker argues "that artists be allowed certain freedoms that scientists should not be allowed," while Gigliotti calls for an examination on the ethics of "of unfettered creativity [as] the holy grail, not only in the arts, but in the sciences and society at large." Gigliotti, Carol. (ed.). *Leonardo's Choice: Genetic Technologies and Animals*. New York: Springer, 2009.
12. Sable hair is a traditional material used in the production of watercolor brushes. Brushes used for oil and acrylic come from other animals, such as pigs and horses. There are new acrylic brushes that do a good a job without killing animals for their hair.
13. Silver gelatin, mostly a historical technique, was used in the production of traditional photography, and is derived from animal by-products.
14. Rabbit skin glue was historically used in the process of preparing canvas as a painting surface.
15. Jana Sterbak's artwork "Vanitas: Flesh Dress for an Albino Anorectic," 1987, was first displayed in the National Gallery of Canada. It is an edition of 2, each composed of 50 pounds of flank steak sewn together into a dress hung on a tailor's form, with a photograph hung nearby of a model wearing the dress (Walker Art Center, www.walkerart.org). Sterbak claims the work is feminist indicating cultural issues on fashion, consumption, and the female body. While the work does indicate the problematics of fashion and women, it unreflectively makes use of *animals'* bodies in its production.
16. Carolee Schneemann's work from 1964, called "Meat Joy," was originally performed by the Judson Church performance group, NYC. The performers interacted with each other and pieces of real meat. Schneemann describes the work as "a celebration of flesh as material: raw fish, chickens, sausages, wet paint, transparent plastic, rope brushes, paper scrap.

It's propulsion is toward the ecstatic—shifting and turning between tenderness, wilderness, precision, abandon: qualities which could at any moment be sensual, comic, joyous, repellent" (Schneemann, www.caroleeschneemann.com) The work is considered to be a seminal performance art piece that experiments with flesh and pleasure. However, the use of dead animals' bodies as materials remains unproblematized in discourse on the performance.

17. The chaise lounge designed by Le Corbusier, considered an icon of modernist furniture, was produced in 1928, and continues to be reproduced today. Originally covered in fabric, later models used pony skin, cow skin, or leather. Vitra Design Museum, www.design-museum.de.
18. Canadian sculptor Brian Jungen recently created a series of works, including "My Decoy," that use icons of modernist furniture design, covered over with real elk hide to transform the original object into shapes recalling indigenous drums. He says the works are inspired by his cultural connection to First Nations' Dane-zaa, heritage of hunting and drum making (Kunst Verein Hannover, http://www.kunstverein-hannover.de). By "colonizing" various modernist forms, such as the Eames chair or car parts, with the hide of an elk that he himself killed, he claims that the gesture performs an indigenization of Western culture. While the work critiques modernist frameworks of oppression indigenous peoples, here, the role of the colonizer is transferred to the artist in oppressing the animal.
19. Artist Mircea Cantor created a video work called "Deeparture," 2005, in which a live deer and a wolf are placed together inside a gallery and video recorded. There is no documented violence between the animals, but the final video relies on the tension created by our preconceptions of the predator prey relationship. Cantor claims "It's the power of humanity, the ability to control. That's why we are above other creatures, because we can control and sublimate the tension" (Ting 2015). Cantor does not acknowledge the problematic ethics involved with placing unconsenting animals in the gallery, and the harm and stress caused to them.
20. Joseph Beuys' work "I Like America and America Likes Me," 1974, is considered to be an iconic "action" work. Beuys spent three days in a room with a captive coyote. Beuys engaged with the coyote and the relationship shifted over the three days from cautious to playful. The discourse on the work contrasts views on the coyote, seen as a powerful god in Native American cultures, but as a pest for agriculturalists. Beuys

claims the work is a metaphor for the damage caused by white Americans to indigenous cultures, and that the action provided for a "healing" process (Tate Museum, www.tate.org.uk). The majority of discourse on this work does not consider the ethics of using captivity for the coyote.

21. "Rara Avis," by Eduardo Kac, 1996, consists of an aviary of live birds installed inside a gallery, with a telerobotic bird providing a webcam view of inside the cage and the live birds to remote viewers. Kac restricts his reflections on the work to formalist observations—mixing virtual and real, online, and in-space participants—but he has little to say about the problematics of involving live birds and what *their* points of view may be. Kac, www.ekac.org.
22. See the work of Garnet Hertz, "Cockroach Controlled Mobile Robot" that combines computer technology onto the bodies of living insects. Hertz claims the work is a reflection on post-humanist ideas, but has nothing to say about the captivity and labor of the insects. Hertz, conceptlab.com.
23. Temple Grandin's design work with industrial farming methods has focused on producing "humane" livestock facilities that she believes eliminates fear and pain from slaughter. The implementation of these designs may have reduced the stress to factory farmed animals. However, these "humane" systems are also means to justify the ongoing slaughter of millions of animals killed per year for consumption and to assuage guilty conscience of producers and consumers. Grandin, Temple and Catherine Johnson. *Animals in Translation: Using the Mysteries of Autism to Decode Animal Behavior.* New York: Harcourt Inc., 2005.
24. Experiments with fruit flies to evaluate whether natural or artificial scent detection can be used to determine hazardous chemicals. Nowotny et al. (eds). "Drosophila olfactory receptors as classifiers for volatiles from disparate real world applications." Bioinspiration & Biomimetics. IOPScience. 14 October, 2014. http://iopscience.iop.org. Accessed Oct. 26, 2015.
25. Huang Yong Ping's artwork "Theatre of the World" contains these living beings forced together into a small space, as an enactment of *Gu*, referred to in the *I Ching* as a magical potion made of five venomous animals. When the work was shown in the Vancouver Art Gallery in 2007, protest ensued and the SPCA forced the closure of the work. Ping objected that the order had "violently interfered with the rights of an artwork to be freely exhibited in an art museum." Phillips critiques the artist's use of

animals as "the colonial Other for the Empire of Man." Phillips, www.alexphillips.ca.

26. Artist Eduardo Kac created a transgenic project called *GFP Bunny* (GFP referring to green florescent protein). The rabbit was genetically modified to include a gene from a jellyfish that is naturally florescent green. He writes, "This must be done with great care, with acknowledgment of the complex issues thus raised and, above all, with a commitment to respect, nurture, and love the life thus created." The controversy is his apparent ethics outlined in his writing that is contradicted by his practice. Kac, www.ekac.org.
27. *Helena* by Marcus Evaristti is a participatory art project that displays blenders filled with water and live goldfish. Participants in the gallery were allowed to turn on the blenders. Evaristti, www.evaristti.com.
28. Guillermo Vargas, in his piece *Eres Lo Que Lees* (You Are What You Read), included an emaciated dog tied to a wall in the Codice Gallery in Manuagua, Nicaragua. In the display, the dog was without food or water. Visitors to the gallery seemed to ignore the plight of the dog. There was protest on blogs and news outlets, and conflicting stories about whether the artist and gallery workers allowed the dog to starve and die, or whether the dog survived. The artist refused to clarify. The artist claimed that he used the dog in the artwork to make a statement about an immigrant who was killed by two dogs. Gigliotti, Carol. "Heartburn: Indigestion, Contention and Animals in Contemporary Art" in *Antennae: The Journal of Nature and Visual Culture,* Issue 14, 2011. www.antennae.org.uk.
29. Linda Birke critiques laboratory experiments with animals and makes connections between science's objective method and cruelty. Many bioart practices consider the studio as a laboratory where organisms are manipulated and experimented with. Birke, Linda. "Into the Laboratory." Kalof, Linda and Amy Fitzgerald, (eds.). *The Animals Reader: The Essential Classic and Contemporary Writings.* New York: Berg, 2007. Print. 323–335.
30. Modified from Lori Gruen in "Dismantling Oppression: An Analysis of the Connection Between Women and Animals." The full quote is, "Our responsibility for our own actions has been mediated. Who are these animals who suffer and die so that I can eat pot roast? I do not deprive them of movement and comfort; I do not take their young from them; I do not have to look into their eyes as I cut their throats." Gaard, Greta.

(Ed.) *Ecofeminism: Women, Animals, Nature.* Philadelphia: Temple University, 1993. 79.

31. Recent studies by biologists and cognitive ethologists call into question the denial of language to nonhuman animals. These researchers argue against the anthropocentric impulse to force human language on other animals as a way to test intelligence. Cognitive ethology proposes the more difficult task of decoding nonhuman languages, a move toward understanding the animal on their own terms. Slobodchikoff has recently determined prairie dog vocal language that contains signifiers for predator types, whether airborne or approaching by land, their rate of approach, what color they are, how large they are, and so on. He has compiled numerous studies on nonhuman languages. Slobodchikoff, Con. *Chasing Doctor Dolittle: Learning the Language of Animals.* New York: St. Martin's Press, 2012. Print.
32. Bekoff, Marc. *The Emotional Lives of Animals.* Novano, California: New World Library, 2007. Print. 123.
33. ibid.
34. Marder, Michael. "Plant Intelligence and Attention." *Plant Signaling & Behavior* 8:5, e23902; May 2013. PDF. Web. http://www.michaelmarder.org. Accessed August 28, 2015.
35. Elizabeth Demaray's project "The IndaPlant Project: An Act of Trans-Species Giving," 2014, draws awareness to the needs of plants. Plant photo- and hydrotropism is made visible using sensors and robotic technologies created in collaboration with engineers, biologists, and computer scientists at Rutgers University. The sculptural forms in the project are called floraborgs, describing the plant–robot combinations, where each plant lives in a specially outfitted pot atop robot programmed to read the health of the plant and to respond accordingly with assistance. The robots are powered by solar panels, and they move around the lab using sonar sensors. The robots take on the phototropic ability of the plants by moving through the indoor space of the university, locating well-lit spots as the light shifts throughout the day. Sensors respond to the plant's need for water, and the robot signals and invites human passersby to water the plants. The project acts as a catalyst to encourage conditions for considering nonhuman intentionality. Demaray, http://elizbethdemaray.org.
36. See Nathalie Jeremijenko's "Amphibious Architecture," created with architect David Benjamin and installed in the New York Harbor in 2009. Jeremijenko observes that the below-water environment sur-

rounding New York City is largely unconsidered by the community. She critiques some communities' "do-not-disturb" ethic toward the Hudson River by proposing positive reciprocal engagement with the sea life. In "Amphibious Architecture," 16 slender buoys fitted with sensors and LED lights flash above the water when fish are nearby. Human participants can send a text message to the fish and receive back a text in the form of a chatty informational response poetically imagined from a fish's point of view. For example "Hey there! There are 11 of us, and it's pretty nice down here. I mean, Dissolved oxygen is higher last week" (Weiner 2013). Jeremijenko creates spaces for humans and nonhuman to connect. This work allows for the otherwise hidden world of fish to be revealed through current technologies, indicating the shared ecology of the Hudson River. Jeremijenko, http://www.environmentalhealthclinic.net.

37. See Camilla Nelson's practice that includes poetry walks, called "Grass Roots," where she leads a group on investigations into the nonhuman living beings in her neighborhood. She directs attention to those urban objects and beings not normally considered on walking tours. The walks blend information on the architecture with information on insects and plants, combined with discrete paper slips of micro-poetry previously inserted into the site that participants may come across along the way. The project draws attention to urban space and its relational environment with nonhuman neighbors. Nelson, www.singingapplepress.com.
38. Val Plumwood calls for two interconnected tasks toward a positive restructuring of human relationships with nature and other than human animals. She argues for resituating humans in ecological terms, focusing on continuities and relatedness with nonhuman others. She argues for considering nonhumans in ethical terms through a critical evaluation of "Otherising" of the nonhuman world that creates destructive hierarchical views, and through decentering the human centeredness of language used in ethics. Plumwood, Val. *Environmental Culture: The Ecological Crisis of Reason.* New York: Routledge, 2002.
39. Timothy Morton proposes the philosophy of "ecological thought" that views the interconnectedness of all things, sentient and nonsenient, and the enmeshed nature of all beings, each influencing the other's world. Morton 2010.
40. Graham Parkes writes about the Mahayana Buddhist promise of salvation to all sentient beings, based on the belief in the "dependent co-arising." The philosophy is expanded on in the Tang dynasty (618–907) by philosopher Zhanran, from the Tiantai School, who wrote that "even non-

sentient beings have Buddha-nature." The philosophy of dependent co-arising of all sentient and nonsentient beings was transmitted to Japan by the monk Saicho, and it later became incorporated into Zen Buddhism. Parkes, Graham. "The awareness of rock: East-Asian understandings and implications," Skrbina, David (Ed.). *Mind that Abides: Panpsychism in the New Millennium.* Philadelphia: John Benjamins Publishing Company, 2009.

41. Marc Bekoff uses chapters to define reasons why expanding our compassion footprint is good, both ethically and ecologically. Bekoff, Marc. *The Animal Manifesto: Six Reasons for Expanding Our Compassion Footprint.* Novato, California: New World Library, 2010.
42. Iris Murdoch proposes a moral philosophy that has as a central task to defeat "the fat relentless ego" and its attendant obsession with individual freedoms. She argues that action is normally associated with ideas of freedom. But, she argues, right attention as moral effort renders freedom as an illusion; the ethical choice is always chosen in an aware condition. Murdoch, Iris. *The Sovereignty of Good.* New York: Routledge Classics, 2001. Print. 36).
43. Jacques Derrida refers to the "epoch" of Aristotle, Descartes, Kant, Lacan, Levinas, and other philosophers that created an immense "disavowal" of the animal. Derrida suggests that these philosophers "made of the animal a *theorem*, something seen and not seeing." Derrida argues that the "disavowal" is a refusal of the subjectivity, agency, and creativity of the animal. Throughout the text, Derrida relates himself standing naked and being observed—seen—by his cat, a proposition of the cat's own subjectivity and agency. Derrida asks, "Who therefore?" is the cat who [*chooses*] to address him? Derrida, Jacques, *The Animal that Therefore I am (More to Follow).* New York: Fordham University Press, 2008. Print. 10–14.
44. Derrida, Jacques and Elizabeth Roudinesco. *Violence Against Animals. For What Tomorrow:...A Dialogue.* Stanford: Stanford University Press, 2004. Print. 63.
45. Derrida argues for an ethics of maximum respect, or hyperbolic ethics, where one continually and relentlessly examines one's intentions for instrumental motives. He proposes that using this kind of critical self-reflective process, ethics can be extended to animals. Derrida, Jacques, *The Animal that Therefore I am (More to Follow).* New York: Fordham University Press, 2008.
46. Josephine Donovan, in her chapter "Animal Rights and Feminist Theory," refers to an ethic of humility developed by Sara Ruddick and originating

in writings by Iris Murdoch and Simone Weil. The ethic proposes an attitude of 'attentive love' in relation to the other, using the practice of asking, "What are you going through?" Gaard, Greta. (Ed.) *Ecofeminism: Women, Animals, Nature*. Philadelphia: Temple University, 1993. Print. 183.

47. See Calarco on Derrida's ethics of maximum respect. Calarco, Matthew. *Zoographies: The Question of the Animal from Heidegger to Derrida*. New York: Columbia University Press, 2008. Print., 103–149.
48. From Josephine Donovan: "We should not kill, eat, torture, and exploit animals because they do not want to be so treated, and we know that. If we listen, we can hear them." Gaard, Greta. (Ed.) *Ecofeminism: Women, Animals, Nature*. Philadelphia: Temple University, 1993. Print. 185.
49. Hallowell Alfred Irving. "Ojibwa Ontology, Behavior and World View". 1960. PDF. This essay was passed on to me by Mimi Gellman, an Ojibwa scholar and faculty at Emily Carr University of Art + Design, who suggested it as a good indicator of Ojibwa worldview on nonhuman animals.
50. Also see: Kohn, Eduardo. *How Forests Think: Toward an Anthropology beyond the Human*. Berkeley: University of California, 2014. Laws, Rita. "Native Americans and Vegetarianism." *International Vegetarian Union*. http://www.ivu.org/history/native_americans.html. Accessed Dec 22, 2015.
51. Robinson Margaret. "Veganism And Mi'kmaq Legends: Feminist Natives Do Eat Tofu." PDF. www.margaretrobinson.com. Accessed December 21, 2015.
52. ibid.
53. ibid.
54. The British Columbia Society for the Prevention of Cruelty to Animals, in 2008, released their *Position Statement on Animals Used for Clothing, Fashion and Art*: "The BC SPCA is opposed to the infliction of pain or suffering upon, or the killing of any animal, explicitly for clothing or any aesthetic purpose. This position applies, but is not limited to, the killing or use of animals for their fur and the use of animals for artistic display. The BC SPCA accepts the use of animals for clothing or aesthetic purpose only when the methods used to raise the animals meet the *Five Freedoms* and only if the harvest of the fibre or product: is a by-product of food production (e.g., leather); (e.g., wool) or does not necessitate the killing of the animal. The BC SPCA's *Five Freedoms* describe conditions that must be fulfilled in order to prevent the suffering of domesticated animals in human care. We acknowledge that absolute provision of these

freedoms may not be possible, but we expect all animal guardians to strive to provide them. The BC SPCA's Five Freedoms are: Freedom from hunger and thirst; Freedom from pain, injury and disease; Freedom from distress; Freedom from discomfort; Freedom to express behaviours that promote well-being." PDF http://www.spca.bc.ca/assets/documents/welfare/position-statements/animals-used-for-clothing.pdf.

55. In the sciences, where experimentation on nonhuman animals takes place on large scales, such as at the University of British Columbia near Emily Carr University, policy and procedures—albeit problematic ones—have been created to reduce harm. I am not advocating for these practices. I strongly oppose the use of nonhumans in the search for new medicines, genetic research, the cosmetic industry, or other systems of violence. However, the point is that the arts lag behind the sciences in the failure to acknowledge the potential for harm of nonhumans in arts practice and the need for policy and procedures.

References

Adorno, Theodor, and Max Horkheimer. 1979. *Dialectic of enlightenment*. London: Verso Editions. Print.

Bekoff, Marc. 2007. *The emotional lives of animals*. Novano: New World Library. Print.

Bekoff, Marc. 2010. *The animal manifesto: Six reasons for expanding our compassion footprint*. Novato: New World Library. Print.

Berger, John. 1980. Why look at animals? In *About looking*. New York: Random House. Print.

British Columbia Society for the Provention of Cruelty to Animals. *Position statement animals used for clothing, fashion and art*. PDF. http://www.spca.bc.ca/assets/documents/welfare/position-statements/animals-used-for-clothing.pdf.

Burt, Jonathan. 2001. The illumination of the animal kingdom: The role of light and electricity in animal representation. Kalof, Linda and Amy Fitzgerald, (eds.). *The Animals Reader: The Essential Classic and Contemporary Writings*. New York: Berg, 2007. Print. 289–301.

Calarco, Matthew. 2008. *Zoographies: The question of the animal from Heidegger to Derrida*. New York: Columbia University Press. Print.

Demaray, Elizabeth. Artist's website: http://elizbethdemaray.org. Accessed 26 Oct 2015.

Derrida, Jacques. 2008. *The animal that therefore I am (more to follow)*. New York: Fordham University Press. Print.

Derrida, Jacques, and Elizabeth Roudinesco. 2004. *Violence against animals. For what tomorrow:... a dialogue*. Stanford: Stanford University Press. Print.

Evaristti, Marcus. Artist's website. www.evaristti.com. Accessed 26 Oct 2015.

Gaard, Greta (ed.). 1993. *Ecofeminism: Women, animals, nature*. Philadelphia: Temple University. Print.

Gigliotti, Carol (ed.). 2009. *Leonardo's choice: Genetic technologies and animals*. New York: Springer. Print.

Gigliotti, Carol. 2011. Heartburn: Indigestion, contention and animals in contemporary art. *Antennae: The Journal of Nature and Visual Culture*, (14). www.antennae.org.uk

Gigliotti, Carol. 2014. Book review on Steve Baker's new book, *Artist | Animal. Humanimalia: a Journal of Human/Animal Interface Studies* 6(1). Accessed 26 Oct 2015.

Grandin, Temple, and Catherine Johnson. 2005. *Animals in translation: Using the mysteries of autism to decode animal behavior*. New York: Harcourt Inc.. Print.

Hallowell, Alfred Irving. 1960. Ojibwa Ontology, Behavior and Wold View. Diamond, Stanley (Ed.). Culture in History: Essays in Honor of Paul Radin. New York: Columbia University Press. PDF.

Haraway, Donna. 2003. *The companion species manifesto: Dogs, people, and significant others*. Chicago: Prickly Paradigm Press. Print.

Hertz, Garnet. 2015. Artist's website, conceptlab.com. Accessed 26 Oct 2015.

Jeremijenko, Natalie. 2015. Artist's website. http://www.environmentalhealthclinic.net. Accessed 12 Dec 2015.

Kac, Eduardo. 2015. Artist's website, www.ekac.org. Accessed 26 Oct 2015.

Kalof, Linda, and Amy Fitzgerald (eds.). 2007 *The animals reader: The essential classic and contemporary writings*. New York: Berg. Print.

Knabb, Ken. 1981. *The situationist international anthology*. Berkeley: Bureau of Public Secrets. Print.

Kohn, Eduardo. 2014. *How forests think: Toward an anthropology beyond the human*. Berkeley: University of California. Print.

Kunst Verein Hannover. 2015. Brian Jungen 20.4.–16.6.2013. http://www.kunstverein-hannover.de. Accessed 26 Oct 2015.

Laws, Rita. 2015. Native Americans and Vegetarianism. *International Vegetarian Union*. http://www.ivu.org/history/native_americans.html. Accessed 22 Dec 2015.

Marder, Michael. 2013. Plant intelligence and attention. *Plant Signaling & Behavior* 8:5, e23902; May 2013. PDF. Web. http://www.michaelmarder.org. Accessed 28 Aug 2015.

Morton, Timothy. 2010. *The ecological thought*. Cambridge, MA: Harvard University Press. Print.

Murdoch, Iris. 2001. *The sovereignty of good*. New York: Routledge Classics. Print.

Nelson, Camilla. 2015. Artist's website. www.singingapplepress.com. Accessed 26 Oct 2015.

Nocella, Anthony J., et al. 2014. *Defining critical animal studies: An intersectional social justice approach for liberation*. New York: Peter Lang Publishing. Print.

Nowotny et al. (eds.). 2014. Drosophila olfactory receptors as classifiers for volatiles from disparate real world applications. Bioinspiration & Biomimetics. IOPScience. 14 October 2014. http://iopscience.iop.org. Accessed 26 Oct 2015.

Nyima, Tashi. 2014. Bright Aisles, Dark Alleys. Great Middle Way. https://greatmiddleway.wordpress.com. Accessed 26 Oct 2015.

Parkes, Graham. 2009. The awareness of rock: East-Asian understandings and implications. In *Mind that abides: Panpsychism in the new millennium*, ed. David Skrbina. Philadelphia: John Benjamins Publishing Company. PDF.

Phillips, Alexandra. 2015. A curtain closes on theatre of the world. www.alex-phillips.ca. Accessed 26 Oct 2015.

Plumwood, Val. 2002. *Environmental culture: The ecological crisis of reason*. New York: Routledge. Print.

Robinson, Margaret. 2015. Veganism and Mi'kmaq legends: Feminist natives do eat Tofu. PDF. www.margaretrobinson.com. Accessed 21 Dec 2015.

Schneemann, Carolee. 2015. Artist's website, www.caroleeschneemann.com. Accessed 26 Oct 2015.

Simon Fraaser University. 2013. Feeling our way: New pathways in environmental education workshop. 2 Oct 2013, Vancouver.

Slobodchikoff, Con. 2012. *Chasing doctor Dolittle: Learning the language of animals*. New York: St. Martin's Press. Print.

Sorenson, John. 2009. *Ape*. London: Reaktion Books. Print.

Tate Museum, collection. www.tate.org.uk. Accessed 26 Oct 2015.

Ting, Selina. 2015. Interview: Mircea Cantor. In *initiArt Magazine*. http://www.initiartmagazine.com. Accessed 26 Oct 2015.

Turner, Michael. 2015. Brian Jungen shows continuity & contrast in German survey. Review, Canadian art. www.canadianart.ca. Accessed 26 Oct 2015.

Vegan Congress. www.vegancongress.org. Web.

Vitra Design Museum, Weil am Rhein, Germany. www.design-museum.de. Accessed 26 Oct 2015.

Walker Art Center, collection. www.walkerart.org. Accessed 26 Oct 2015.

Weiner, Jonah. 2013. The artist who talks with the fishes. *The New York Times Magazine*, June 26. http://www.nytimes.com/2013/06/30/magazine/the-artist-who-talks-with-the-fishes.html. Accessed 12 Dec 2015.

Lives of their Own: Animal Death and Animal Flesh in J.M. Coetzee's Writings

Parag Kumar Deka

The Question of Meat Eating in Coetzee

Any attempt at looking for overt, clear-cut themes of veganism in J.M. Coetzee's works might prove not only frustrating but also futile, as despite his scrupulous vegetarianism,[1] in his fiction, Coetzee approaches the issue of eating animals only in an oblique way. In his 1995 essay "Meat Country," published in the special food issue of the *Granta*, Coetzee rather intriguingly writes, "The question of whether we should eat meat is not a serious question … [The craving for meat] is a given, it is the human condition."[2] But, at the same time, many of Coetzee's novels present us with characters in the process of reconsidering their perception of animals as well as their food habits. Coetzee's most celebrated work in this respect is

P.K. Deka (✉)
English, University of Gauhati, Jalukbari, Guwahati, India
e-mail: paragk.deka@gmail.com

J. Castricano, R.R. Simonsen (eds.), *Critical Perspectives on Veganism*,
The Palgrave Macmillan Animal Ethics Series,
DOI 10.1007/978-3-319-33419-6_8

undoubtedly his fictionalized 1998 Tanner Lectures on Human Values, published in 1999 under the title *The Lives of Animals*. This work, along with his highly acclaimed Booker prize winning novel of the same year, *Disgrace*, has "become crucial to the new animal studies."[3] However, even in *The Lives of Animals*, Coetzee's fiercely vegan protagonist, Elizabeth Costello, whose veganism is a constant source of anxiety and unease for the other characters, tells her audience,[4] "I have never been much interested in proscriptions, dietary or otherwise. Proscriptions, laws. I am more interested in what lies behind them."[5] Like his own character, in "Meat Country," Coetzee too expresses his own lack of "interest in making converts" to his own "diet without flesh."[6]

Critics have tried to explain away Coetzee's observation about the "question" of meat eating in the Granta essay either as an example of his "sardonic … irony," as Jennifer Schuessler puts it in a *New York Times* article on November 13, 2009, or as a sign of his early vacillation about meat eating. Karen Dawn and Peter Singer, for instance, say in reference to this comment, "An examination of Coetzee's interviews suggests that during the decade that followed the publication of 'Meat Country,' he may have become less attached to the idea that eating meat is a fundamental part of our nature."[7] In the very next sentence, however, the authors mention that "in a 2004 interview in the Australian weekly magazine *The Bulletin*, Coetzee says he gave up meat thirty years earlier."[8] This indicates that unlike their conjecture, Coetzee's comment in "Meat Country" can hardly be attributed to any indecisiveness on his part about the naturalness of meat eating. Rather, as I would argue, these observations underline the strong precedence of the questions of life, death and suffering of animals for Coetzee over the narrow concern of eating their flesh. As we shall see, Coetzee's particular mode of vegetarianism is rooted in his convictions about the joys and ecstasies of animal being and the intrinsic value of individual animal lives.

Similarly, in order to understand Coetzee's ideas about what Melanie Joy calls carnism, and about disgust and various taboos related to meat eating, we need to recognize the particular emphasis he puts on human sympathy, the ability to adopt the perspective of the Other. It is only after considering all these aspects that a clearer picture of Coetzee's unique vegan stance begins to emerge.

The Ecstatic Lives of Animals

In *The Lives of Animals* (henceforth, *The Lives*), Coetzee's vegan protagonist Elizabeth Costello fiercely attacks the Cartesian view of animals as soulless automata, the idea that "an animal is no more than the mechanism that constitutes it."[9] Descartes's famous dictum *cogito ergo sum*, Costello says, relegates non-thinking animals to a subordinate order of beings and thereby justifies their exploitation and oppression by human beings. In a radical departure from the Western rationalist tradition, she declares:

> To thinking, cogitation, I oppose fullness, embodiedness, the sensation of being—not a consciousness of yourself as a kind of ghostly reasoning machine thinking thoughts, but on the contrary the sensation—a heavily affective sensation—of being a body with limbs that have extension in space, of being alive to the world.[10]

This fullness of being is diametrically opposed to the Cartesian version of the mind–body dualism, where the disembodied mind gets precedence over the physical, material body. Whereas in the Cartesian mode of being there is a sharp break between the thinking mind and the breathing body, in Costello's idea of *fullness*, the mind and the body are fused into one indistinguishable whole. Responding to philosopher Thomas Nagel's observation that "human beings are restricted to the resources of [their] own mind[s]" when trying to "know what it is like for a *bat* to be a bat,"[11] Costello says:

> What it is like to be a bat? Before we can answer such a question, Nagel suggests, we need to be able to experience bat-life through the sense-modalities of a bat. But he is wrong; or at least he is sending us down a false trail. To be a living bat is to be full of being; being fully a bat is like being fully human, which is also to be full of being. Bat-being in the first case, human-being in the second, maybe; but those are secondary considerations. To be full of being is to live as a body-soul. One name for the experience of full being is *joy*.[12]

For Costello, as well as for Coetzee, it is only by forgoing the dualistic mode of existence that human beings can achieve this fullness of being. But animals, by virtue of their native non-dualistic, non-rational perception, are the most natural practitioners of this mode of being. In a letter to Paul Auster dated 31 May 2011, Coetzee writes: "I've always found it interesting that whereas we human beings think of our bodies as having parts—arms, legs and so forth—animals don't. In fact, I doubt that animals think of themselves as 'having' bodies at all. They just are their bodies."[13] This position about the nature of animal being has two important implications. Firstly, due to their organic perception of self, animals live more intensely and continuously joyful lives than human beings; and, secondly, due to their emphasis on the superiority of reason over sensation, and the resulting fragmentation of and alienation from their selves, most human beings cannot attain this purity of being.[14] This, in turn, implies that an overemphasis on rationality underplays the joy of living as an embodied being, which Costello, as well as Coetzee, identifies as the primary mode of animal existence—and this justifies the killing of animals.[15]

In sharp contrast to this "organic" view of animals, many of Coetzee's characters can be seen constantly struggling with their sense of self, fractured by Cartesian duality. In Coetzee's very first novel *Dusklands*, which comprises two thematically linked narratives, Eugene Dawn, the narrator of the first story, "The Vietnam Project," says that the American "Chieu Hoi (surrender/reconciliation) programming" in Vietnam is "wholly Cartesian": "the voice which our broadcasting projects into Vietnamese homes … is the voice of the doubting self, the voice of René Descartes driving his wedge between the self in the world and the self that contemplates that self."[16] Dawn is a prototype of the Cartesian fragmented self, who constantly talks of the "enemy in [his] body … throw[ing] up walls against the forays of [his] brain," and his struggle against "the pressure of [his] enemy body."[17] Likewise, early in Coetzee's second novel, *In the Heart of the Country*, the narrator-protagonist Magda exclaims with delight at having stumbled upon a way of disposing her father's corpse: "How clearly my mind works, like the mind of a machine!"[18] And later, she reflects: "Perhaps I never had an *animal integrity*, or lost it before I was six, perhaps by the age of six I was already *a corporeal machine* trotting around the yard."[19] It is this condition of "animal integrity" that Coetzee advances as an alternative to the Cartesian view of the divided self, of the body as a biological automaton. Though Magda can never experience this integrity herself, she too, like Elizabeth Costello, ascribes this state of being to nonhuman animals:

> The flies, which ought to be in transports of joy, sound merely cross … Why are they not singing? But perhaps what I take for petulance is the sound of insect ecstasy. Perhaps their lives from cradle to grave, so to speak, are one long ecstasy, which I mistake. Perhaps the lives of animals too are one long ecstasy interrupted only at the moment when they know with full knowledge that the knife has found their secret and they will never again see the goodly sun which even at this instant goes black before them.[20]

The only possibility for human beings to attain this animal integrity, as Magda realizes, is to "give [one]self more to sensation."[21] "I am faced with a choice," she reflects, "that flies do not have to make."[22]

Similarly, in his foreword to Jonathan Balcombe's book, Coetzee commends the author for "the humanity with which [he] approaches the lives of animals."[23] More pertinently:

> [Balcombe] takes seriously—not just as rhetoric—William Blake's immortal question, "How do you know but that every bird that cleaves the aerial way is not an immense world of delight closed to your senses five?" … Far from being absorbed in a grim battle to survive, he contends, animals actually enjoy life minute by minute, day by day.[24]

Through this emphasis on the joyfulness of individual animal lives, Coetzee not only presents a corrective to the rationalist philosophy of animal ethics but also challenges the modern ecological philosophy that places the survival of the species over the survival and well-being of individual nonhuman animals.

Coetzee's Alternative Ecology

In "Meat Country," discussing the killing of rare animals and its emotional effect on people, Coetzee poses a quandary: "The death of the bear, the deaths of the flamingos, disturb us as the death of the beef ox does not. Why?" The general answer, Coetzee points out, is that this is "because there are so many more oxen than bears on earth."[25] According to this position, which Coetzee calls the "species argument," it is permissible for human beings to kill a particular animal as long as that animal is not threatened with extinction *as a species*. This position automatically assumes that the "life of the species is of a higher order than the life of the individual."[26] Later, in *The Lives*, Coetzee has his character Elizabeth Costello articulate the same argument, this time in a more detailed and forceful manner:

> In the ecological vision, the salmon and the river-weeds and the water-insects interact in a great, complex dance with the earth and the weather. The whole is greater than the sum of the parts. In the dance, each organism has a role: it is these multiple roles, rather than the particular beings who play them, that participate in the dance. As for actual role-players, as long as they are self-renewing, as long as they keep coming forward, we need pay them no heed.[27]

Costello calls this version of ecological philosophy Platonic due to its emphasis on the abstract idea of the species and the abstract systems of interaction between the species and the environment rather than on the individual creatures themselves; and also because, according to this vision, the individual organisms are viewed as no more than embodied manifestations of the interactions between different abstract forces. Whereas human beings look at nonhumans as representatives of particular species, the natural world is inhabited by uniquely individual living beings. The idea of the species is simply an abstraction of these individual beings. But the contemporary ecological vision takes the living nonhumans themselves to be embodiments of the abstract idea of the species rather than seeing the concept of the species as a human construct. The greatest and devastating irony of this vision, Costello tells us, is that while it asks human beings to live in harmony with other living creatures, it "justifies itself by appealing to an idea, an idea of a higher order than any living creature … an idea which no creature except Man is capable of comprehending."[28] But for Costello, what invalidates this rationalist vision of ecology is each and every organism's struggle for its own life: "Every living creature fights for its own, individual life, refuses, by fighting, to accede to the idea that the salmon or the gnat is of a lower order of importance than the idea of the salmon or the idea of the gnat."[29]

It is only in the light of this anti-Platonic ecological vision and his notion of animal being that we can appreciate Coetzee's views on eating and killing of animals. In the various animal slaughter scenes in Coetzee's novels, the animals' desperate fight for their lives demonstrates the suffering and cruelty inherent in their killing. In *Life & Times of Michael K*, for instance, when a famished K tries to strangle a goat with his bare hands, he feels the goat "kicking like a fish in the mud to regain its footing … He could feel the goat's hindquarters heaving beneath him; it bleated again and again in terror; its body jerked in spasms … The hindquarters thrashed."[30] This description prefigures Elizabeth Costello's observation in *The Lives*:

> Anyone who says that life matters less to animals than it does to us has not held in his hands an animal fighting for its life. The whole of the being of the animal is thrown into that fight, without reserve. When you say that the fight lacks a dimension of intellectual or imaginative horror, I agree. It is not the mode of being of animals to have an intellectual horror: their whole being is in the living flesh.[31]

Likewise, when in *Age of Iron* the narrator-protagonist Elizabeth Curren describes three men killing chicken in a farm enclosure, the focus is equally on the desperate fight of the chicken for their lives and on their suffering as well as the killers' total and mechanical disregard for it.

> He, William, Florence's husband, had a job and the job could not be interrupted. His job was to pounce on a chicken, swing it upside down, grip *the struggling body* between his knees, twist a wire band around its legs, and pass it on to a second, younger man, who would hang it, *squawking and flapping*, on a hook on a clattering overhead conveyor that took it deeper into the shed where a third man in oilskins splashed with blood *gripped its head, drew it taut, and cut it through with a knife* so small it seemed part of his hand, tossing the head in the same movement into a bin full of other dead heads.[32]

The cruelty in these descriptions derives not only from the animals' suffering or painful death but from their death in itself, from their being deprived from the "goodly" life. As Elizabeth Curren observes later on: "Such a good thing, life! Such a wonderful idea for God to have had! The best idea there had ever been. A gift, the most generous of all gifts, renewing itself endlessly through the generations."[33]

This emphasis on each and every creature's fight for his or her individual life frees the question of animal ethics, particularly the debate about the value of animal life vis-à-vis human life, of its inherent anthropocentrism. Elizabeth Costello captures this position succinctly when she says of "the Dulgannon mudflats of [her] childhood and of the frogs who live there … : In my account … the life cycle of the frog may sound allegorical, but to the frogs themselves it is no allegory, it is the thing itself, the only thing."[34]

Among all of Coetzee's characters, perhaps it is Michal K, who perceptually comes closest to this anti-Platonic ecology. After escaping from the Kenilworth Relocation Camp, as he "walked among the rocks peering into the tidal pools … [K] saw snails and anemones living lives of their own."[35] K's sensitivity towards the individual lives of the most "insignificant" of creatures, however, is not incidental. K is the only character in all of Coetzee's novels to rid himself completely of the mind–body dualism.[36] At one point in the novel, while K is in a physical state akin to hibernation in animals, "it came home to him that he might die, he or his body, it was the same thing."[37] Later, we can see that this unity of the body and soul leads K to the joyfulness of pure being that Elizabeth Costello (and Magda) talks about:

> After the hardships of the mountains and the camp there was nothing but bone and muscle on his body. His clothes, tattered already, hung on him without shape. Yet as he moved about his field he felt *a deep joy in his physical being*. His step was so light that he barely touched the earth. It seemed possible to fly; *it seemed possible to be both body and spirit.*[38]

Though Magda, unlike Michael K, can never achieve the "animal integrity" and the joyfulness she so much hankers for, she is able to arrive at the following conclusion: "Perhaps if I talked less and gave myself more to sensation I would know more of ecstasy."[39] This, along with her supposition about the ecstatic inner lives of animals, displays Magda's undeveloped capacity for imaginative identification with the Other. For Coetzee, this imagination has a very important role to play in man–animal relations.

The Faculty of Sympathy

Despite the intrinsic value of each and every creature's lives, when it comes to questions of meat eating and veganism, humans become central to the debate. As Coetzee himself has said about the animal welfare movement, "This enterprise is a curious one in one respect: that the fellow beings on whose behalf we are acting are unaware of what we are up to and, if we succeed, are unlikely to thank us. There is even a sense in which they do not know what is wrong."[40] It is exactly this unawareness of the animals about the broader condition of their lives that makes the animal welfare movement "a human enterprise from beginning to end."[41] However, despite his emphasis on the essential value of individual animal lives, despite his insistence that "the most important of all rights is the right to life," Coetzee maintains:

> Strictly speaking, my interest is not in the legal rights for animals but in a change of heart towards animals … I cannot foresee a day when domesticated animals will be granted … [the] right [to life] in law. If you concede that the animal rights movement can never succeed in this primary goal, then it seems that the best we can achieve is to show to as many people as we can what the spiritual and psychic cost is of continuing to treat animals as we do, and thus perhaps to change their hearts.[42]

For Coetzee, this change of heart is to be achieved by appealing to human sympathy, the ability to imaginatively identify ourselves with other sentient beings. Stressing the necessity of sympathy in our dealings with the Other constitutes one of the central concerns of Coetzee's novels. As Martin Woessner notes, "What Coetzee's novels do not do is tell us how to live. They give us stories and vocabularies that might help us to expand the boundaries of our sympathies."[43] In fact, for Coetzee, this capacity plays a very crucial role in the entire system of animal slaughter and meat eating, exerting influences beyond the understanding of the conscious mind. "Sympathy," Elizabeth Costello says, "allows us to share at times the being of another … [to] think [our] way into the existence of a bat or a chimpanzee or an oyster, any being with whom [we] share the substrate of life."[44] The horror of the Holocaust, Costello says, was not that the victims of the concentrations camps were treated like pests *despite* their shared humanity with the killers, but that "the killers refused to think themselves into the place of their victims, as did everyone else … In other words, they closed their hearts."[45] In our day-to-day dealings with animals, in our refusal to see their suffering and death, we too, Costello maintains, "in a huge communal effort … close our hearts."[46]

However, according to Coetzee, this denial of sympathy is not a result of an inherently cruel human nature. In fact, he asserts:

> We are not by nature cruel. In order to be cruel we have to close our hearts to the suffering of the other. It is not inherently easier to close off our sympathies as we wring the neck of the chicken we are going to eat than it is to close off our sympathies to the man we send to the electric chair … but we have evolved psychic, social and philosophical mechanisms to cope with killing poultry that, for complex reasons, we use to allow ourselves to kill human beings only in time of war.[47]

But these psychic and philosophical mechanisms are never entirely successful in blocking off our sympathies. As Coetzee suggests, it is not possible to close our hearts completely to the suffering of other sentient beings, be they humans or nonhuman animals. The closely guarded secrets about the actual treatment of the slaughterhouse and farm animals, the fiercely promoted myths about the idyllic living conditions of the farm animals,[48] the various carnistic systems evolved and erected in order to keep us in the dark about the suffering and killing of these animals, and the steady movement of the European table culture "towards greater discretion, greater delicacy regarding the unpleasant off-stage business of the slaughterhouse and kitchen"[49]—all these point towards the fact that "somehow the imagination knows what the other's pain is like, even the ant's pain."[50] It is this knowledge, Coetzee suggests, that is the source of all the food taboos regarding animal flesh:

> The bans spelled out with such maniacal exactitude [in the Book of Leviticus] are all on animal flesh. There are no proscriptions on plant foods. The branch of human knowledge that tells which kinds of flesh may be eaten and which are to be avoided seems to be separate from the branch that tells which kinds of flesh may be eaten and which are unclean … The standard for allowing unfamiliar vegetable matter into the body seems to be of a quite different order from the standard for unfamiliar flesh. In the first case, the criterion is taste alone: if it tastes good, I will eat it. In the second, a deep-seated resistance has to be overcome, a resistance which is intimately related to taboo and the horror to which food taboos give expression.[51]

In fact, this knowledge about the pain of the other and the resulting sense of pity are very strong:

> In the visceral imagination there appears to be some mistrust of the alive/dead distinction itself, some reluctance to accept what is dead is henceforth and forever devoid of life. At its deepest level, this mistrust expresses itself as a fear that forbidden flesh—flesh that has not been properly killed and ritually pronounced dead—will continue to live some kind of malign life in one's belly—that it will be, as Leviticus calls it, an abomination inside one. Hence the intimate relations, in so many religions, between priests and butchers, and the requirement for a priestly presence in the slaughter-house. Hence too, perhaps the custom of praying before eating: an effort to placate the angry spirit of the sacrificed beast.[52]

In *The Lives*, at the dinner after the first talk, "The Philosophers and the Animals," all the theories about the various taboos associated with flesh food, forwarded by the other diners (composed entirely of rationalist advocates of meat eating), focus only on issues of sanitation as the sole criterion for eating or not eating certain types of animals. Thus, one of the guests says: "It all has to do with cleanness and uncleanness … clean and unclean animals." To this, Norma, Costello's fiercely rationalist daughter-in-law, rejoins: "There are specific kinds of animal that we don't eat. Surely *those* are the unclean ones, not animals in general."[53] This entire debate appears very ironic in the light of Coetzee's observation in "Meat Country" regarding the nature of this taboo:

> Even in the case of so-called clean meat, like beef, the same people who eat the muscle flesh of cattle are revolted at the thought of eating their eyes, their brains, their testicles, their lungs. They would vomit if they had to drink blood. Why? The question is pointless: distaste for certain body parts, and particularly for body fluids in their fluid state, belongs to the penumbra of taboo, well outside the realm of rational explanation.[54]

Elizabeth Costello is voicing a similar thought when she says that "the ban on certain animals—pigs and so forth—is quite arbitrary … There is no logic in a taboo."[55] In this respect, the tendency of these characters to seek for a rational explanation of flesh taboos on the basis of the alleged uncleanness and hence the essential difference of animals from human beings is an extension of their tendency to rationalize the killing and exploitation of animals for human need and greed.

There are subtle references to this idea of there being "some mistrust of the alive/dead distinction itself" regarding animal flesh in Coetzee's fictional works as well. His most recent novel, *The Childhood of Jesus*, which portrays a fictional city called Novilla, a place curiously devoid of any flesh food,[56] offers such an example. In that novel, Simon forbids his ward David, a child, to eat sausages, as "they put pig meat in sausages … and pigs aren't clean animals. They don't eat grass like sheep and cows."[57] Then he goes on to elaborate:

> It's a matter of hygiene. Ethical hygiene. If you eat pig you become like a pig. In part. Not wholly. You partake of the pig … It is called consubstantiation. Why else do you think there are cannibals? A cannibal is a person who takes consubstantiation seriously. If we eat another person, we embody that person. That is what cannibals believe.[58]

This idea of embodying animals through the consumption of their flesh not only betrays the mistrust discussed above but, at the same time, contrasts ironically with Elizabeth Costello's assertion that it is possible to "embody animals—by the process called poetic invention that mingles breath and sense in a way that no one has explained and no one ever will."[59] In this context, it is significant that the narrative of *The Childhood of Jesus* places a strong emphasis on David, the child's "lively imagination,"[60] and thus offers it as a counterpoint to the rationalism of the grown-ups.

A New Hope

In Coetzee's novels, children display an imaginativeness and ability to sympathize with the death and suffering of nonhuman animals unlike most of the adults. In *The Lives*, where all the grown-ups are totally untouched by Elizabeth Costello's passionate talks about the death and suffering of animals, and even actively try to deflect her appeal to their sympathy with their rational arguments, it is only her grandchildren who show any signs of being moved, that too by simple "stories" of "poor little veal calves and what the bad men do to them."[61] To their mother's great dismay and irritation, the children "pick at their food and ask, 'Mom, is this veal?'"[62] Likewise, in *The Childhood of Jesus*, after knowing that sausages contain pig meat, David refuses to eat sausages, and also fish, especially "if it has eyes."[63] Talking about the future of animal welfare organizations and the animal rights movement as a whole, Coetzee says in his 2007 address to Voiceless, the Australian non-profit animal protection group:

> In this respect, children provide the brightest hope. Children have tender hearts, that is to say, children have hearts that have not yet been hardened by years of cruel and unnatural battering. Given half a chance, children see through the lies with which advertisers bombard them … It takes but one glance into a slaughterhouse to turn a child into a lifelong vegetarian.[64]

When Elizabeth Curren, the narrator of *Age of Iron*, laments the hearts and minds of South African children under apartheid "turning to stone,"[65] she is referring to the depletion of sympathy, the capacity of imaginative identification with other fellow beings:

> Children scorning childhood, the time of wonder, the growing-time of the soul. Their souls, their organs of wonder, stunted, petrified. And on the other side of the great divide, their white cousins soul-stunted too, spinning

> themselves tighter and tighter, into their sleepy cocoons. Swimming lessons, riding lessons, ballet lessons; cricket on the lawn; lives passed within walled gardens guarded by bulldogs; children of paradise, blond innocent, shining with angelic light, soft as *putti*. Their residence the limbo of the unborn, their innocence the innocence of bee grubs, plump and white, drenched in honey, absorbing sweetness through their soft skins. Slumberous their souls, bliss-filled, abstracted.[66]

Significantly, this passage refers not only to the stifling of the capacity of sympathy, but by using the word "abstracted" in a clearly negative context, it also hints at the increasing rationalization and Cartesian fragmentation of the child mind. This abstraction is to be understood in opposition to the idea of "fullness, embodiedness, the sensation of being" forwarded by Elizabeth Costello, which "contrasts starkly with Descartes's key state, which has an empty feel to it."[67] Moreover, this also refers to Descartes's notion that "the mind works less perfectly in an infant than in an adult."[68] Descartes, in fact, stated in his letter to one Hyperaspistes, in August 1641, that "when thriving in an *adult and healthy body* the mind enjoys some liberty to think of other things than those presented to the senses, we know there is not the same liberty in those who are sick or asleep or *very young; and the younger they are, the less liberty they have*."[69] "It seems most reasonable to think," he said, "that a mind newly united to an infant's body is wholly occupied in perceiving in a confused way or feeling the ideas of pain, pleasure, heat, cold and other similar ideas which arise from its union and, as it were, intermingling with the body."[70] Thus, the fact that children are more given to sensation than to cogitation makes them, like animals, fuller, purer beings than the grown-ups whose sense of self is fraught with dualism and abstraction. This is completely opposed to the Cartesian notion that the rational adult is more fully human than the non-rational, sensual child. Coetzee alludes to this idea when he has Magda reflect that probably by the age of six she had lost her animal integrity and become a corporal machine.

However, as Coetzee's own characters demonstrate, it is possible even for grown-ups to attain this sympathetic attitude towards other human and nonhuman beings. Coetzee's second Booker Prize winning novel, *Disgrace*, which along with *The Lives* has been primarily responsible for the "animal/ethical turn" in Coetzee scholarship, is marked by this strong emphasis on sympathy as an important means of overcoming the inter-personal and inter-species barrier. David Lurie, the protagonist of the novel—who had initially believed that "his temperament is not going to change, he is too old for that,"[71] and who had commented to Bev Shaw, "Do I like animals? I eat them, so I suppose I must like them, some parts of them"[72]—is moved at the slaughter of the two Persian sheep brought for Petrus's feast. Holding a plate containing two mutton chops cut out of the Persians, Lurie thinks: "I am going to eat this ... I am going to eat it and ask forgiveness afterwards."[73] Towards the end of the novel, as Lurie helps Shaw euthanize the unwanted dogs at the clinic, we are told, "He and Bev do not speak. He has learned by now, from her, to concentrate all his attention on the animal they are killing, giving it what he no longer has difficulty in calling by its proper name: love."[74] This is the same Lurie who had earlier sarcastically referred to the cultivation of love and sympathy as "sensitivity training."[75]

Martin Woessner points out that it is love that allows Lurie to see the world from the perspective of the Other.[76] As Iris Murdoch says, "love is the extremely difficult realisation that something other than oneself is real. Love ... is the discovery of reality."[77] For Lurie, this realization occurs towards the very end of the novel and involves acknowledging the existence and intrinsic value of the individual lives of animals, even assuming the perspective of the dogs about to be euthanized, and living their mortal fear and anxiety, even if for a very brief moment.[78] It is through love and sympathy that Lurie is finally able to surmount the species barrier between him and the dogs.

Coetzee's comment in "Meat Country" that the question of eating meat is not a serious one, thus, has to be understood in the context of his overall outlook. It is not possible to have a clearer picture of his position about meat eating without taking into account his emphasis on the scope of human imagination (sympathy) and its implicit and potential role in veganism and animal ethics, as well as his insistence on the essential value of individual animal (and human) life. Also, this position has to be seen with reference to Coetzee's idea about the nature of being and his critique of Western rationalism. A closer look at Coetzee's fictional and non-fictional writings show that rather than being concerned with social and cultural dietary norms and questions of eating or not eating animal flesh *per se,* he is more committed to investigating and unearthing structures of human thought and discourse that justifies and normalizes the oppression and killing of human and nonhuman Others.

Notes

1. In an interview with the Swedish paper *Djurens Rätt* (Animal Rights), Coetzee says, "I am a vegetarian. I find the thought of stuffing fragments of corpses down my throat quite repulsive, and am amazed that so many people do it everyday" (Coetzee 2004e, para 11).
2. J.M. Coetzee, "Meat Country," *Granta* 52 (1995), 46.
3. Marianne Dekovan, "Going to the Dogs in *Disgrace*," *ELH* 76, no. 4 (2009), 871.
4. Coetzee presented his Tanner Lectures in the form of a fictional narrative about Elizabeth Costello, an ageing Australian novelist delivering a two-part talk at the fictional Appleton College. These talks, respectively titled "The Philosophers and the Animals" and "The Poets and the Animals," together form the body of *The Lives of Animals*. Later, this text was incorporated into Coetzee's 2003 novel *Elizabeth Costello*.
5. J.M. Coetzee et al., *The Lives of Animals* (New Jersey: Princeton University Press, 1999), 37.
6. Coetzee, "Meat Country," 44.
7. Karen Dawn and Peter Singer, "Converging Convictions: Coetzee and His Characters on Animals," in *J.M. Coetzee and Ethics: Philosophical Perspectives on Literature*, ed. Anton Leist and Peter Singer (New York: Columbia University Press, 2010), 113.
8. Ibid.
9. Coetzee et al., *The Lives*, 33.
10. Ibid.
11. Thomas Nagel, "What Is It Like to Be a Bat?" *The Philosophical Review* 83, no. 4 (1974), 439.
12. Coetzee et al., *The Lives*, 31, original emphasis.
13. J.M. Coetzee and Paul Auster, *Here and Now: Letters 2008–2011* (London: Faber and Faber, 2013), 242. In his novel *Diary of a Bad Year*, Coetzee's protagonist J.C. writes in an essay titled "On the Body": "We speak of *the dog with the sore foot* or *the bird with the broken wing*. But the

dog does not think of itself in those terms, or the bird. To the dog, when it tries to walk, there is simply *I am pain*, to the bird, when it launches itself into flight, simply *I cannot*" (original emphasis, 59).

14. In *Summertime*, the third instalment of Coetzee's fictionalized memoirs, when a biographer of one late J.M. Coetzee, the deceased author of the novels *Dusklands* and *In the Heart of the Country* among others, interviews one of Mr. Coetzee's former colleagues and a temporary love interest about Mr. Coetzee's personal philosophy, he is answered: "He thought of Africans as embodied, in a way that had been lost long ago in Europe … In Africa, he used to say, body and soul were indistinguishable, *the body was the soul*" (emphasis added, 231).
15. Here, it should be noted that most of the leading vegetarians and animal ethicists make reason the primary criterion in their advocacy of the rights of animals (see Bailey 2005). Peter Singer, for instance, while discussing the ethics of killing nonhuman animals, says, "What we are really asking is whether any nonhuman animals are rational and self-conscious beings, aware of themselves as distinct entities with a past and a future" (Singer 1999, 110–111). Reacting to such views, Coetzee observes in his Foreword to Jonathan Balcombe's *Second Nature: The Inner Lives of Animals*:

    > Ever since Aristotle's time we have made the possession of intelligence—intelligence of the kind that enables one to construct intricate machines or ingenious philosophical theories—the crucial test, the test that distinguishes higher from lower, man from beast. Yet why should the crucial test not be a quite different one: for instance, the possession of a faculty that enables a being to find its way home over a long distance? Is the explanation perhaps that the latter is one that *Homo sapiens* would find it hard to pass? (xi–xii)

    Joan Dunayer asks in a similar vein, "Why equate so-called human characteristics with superiority? Because we possess them? … Nonhuman animals have their own ways of knowing" (Dunayer 2001, 19–20).

16. J.M. Coetzee, *Dusklands* (London: Vintage, 2004), 20.
17. Ibid., 6, 8.
18. J.M. Coetzee, *In the Heart of the Country* (London: Vintage, 2004), 16.
19. Ibid., 44, emphasis added.

20. Ibid., 85.
21. Ibid.
22. Ibid.
23. J.M. Coetzee, Foreword to *Second Nature: The Inner Lives of Animals* by Jonathan Balcombe (New York: Palgrave Macmillan, 2010), x.
24. Ibid., xi.
25. Coetzee, "Meat Country," 45.
26. Ibid.
27. Coetzee et al., *The Lives*, 53–54. Joan Dunayer makes the same point when she says, "each sentient being is physically and mentally unique, but speciesist language negates nonhuman individuality" (Dunayer 2001, 6). "The term *specimen*," she says, "turns a unique individual into a species representative. If their species faces extinction, nonhuman individuals are called 'endangered' even if they are personally safe" (ibid). This position of the contemporary ecological vision, where the species finds primacy over the individual nonhumans, is linguistically attested by the fact that "people speak of all members of a nonhuman group" by their species: "they refer to cheetahs as 'the cheetah' and bees as 'the bee'" (ibid).
28. Coetzee et al., *The Lives*, 54.
29. Ibid. It should be noted that the title of Coetzee's Tanner Lectures itself, namely, *The Lives of Animals*, captures this emphasis on the individual lives of animals.
30. J.M. Coetzee, *Life & Times of Michael K* (London: Vintage, 2004), 21.
31. Coetzee et al., *The Lives*, 65.
32. J.M. Coetzee, *Age of Iron* (London: Vintage, 2010), 41–42, emphasis added.
33. Ibid., 109.
34. J.M. Coetzee, *Elizabeth Costello* (London: Vintage, 2004), 217.
35. Coetzee, *Life & Times*, 177.
36. For a detailed analysis of Coetzee's critique of mind–body dualism in *Life & Times of Michael K*, see Deka 2014.
37. Coetzee, *Life & Times*, 69.
38. Ibid., 101–102, emphasis added.
39. Coetzee, *In the Heart*, 85.
40. J.M. Coetzee, "Voiceless: I Feel Therefore I Am," *Hugo Weaving: Random Scribblings*, para 14, http://hugo.random-scribblings.net/?p=7203.
41. Ibid.
42. J.M. Coetzee, "Animals, Humans, Cruelty and Literature: A Rare Interview with J.M. Coetzee," interview by Henrik Engström, *Satya*

Magazine, May 2004, para 11, http://www.satyamag.com/may04/coetzee.html.

43. Martin Woessner, "Coetzee's Critique of Reason," in *J.M. Coetzee and Ethics: Philosophical Perspectives on Literature*," eds. Anton Leist and Peter Singer (New York: Columbia University Press, 2010), 240.
44. Coetzee et al., *The Lives*, 34–35. This notion of "sharing the being of another," no matter however momentarily, is a very strong theme in Tolstoy as well, whom Coetzee identifies as one of the major inspirations behind the "crankhood [that] developed in the England of the 1980s, a creed of … avoidance of alcohol and animal flesh" (Coetzee 1995, 43). In Tolstoy, the necessity for this imaginative identification with other living beings is highlighted, among his other writings, in the story "Esarhaddon, the King of Assyria," where the king momentarily shares the subjectivity not only of his enemy King Lailie but also of a mother-donkey and experiences the "glad feeling of simultaneous life in himself and in his offspring" (Tolstoy 2001, 745). Coetzee's views on the inherently sympathetic nature of human beings have much in common with Tolstoy's views on the same.
45. Coetzee et al., *The Lives*, 34.
46. Ibid., 35. In his essay "The First Step," which discusses vegetarianism and dietary excesses, and which contains some of the most graphic scenes of animal slaughter in his entire oeuvre, Tolstoy writes: "This is dreadful! Not the suffering and death of the animals, but that man suppresses in himself, unnecessarily, the highest spiritual capacity—that of sympathy and pity toward living creatures like himself—and by violating his own feelings becomes cruel. And how deeply seated in the human heart is the injunction not to take life!" (Tolstoy 2015, sec. ix). Coetzee's position, however, is different from Tolstoy's in the sense that for him, the suffering and death of animals is equally important as man's suppression of sympathy in himself.
47. Coetzee, "Animals, Humans, Cruelty," para 10.
48. "The happy chooks that are transformed painlessly into succulent nuggets, the smiling moo-cow that donates to us the bounty of her milk. …" (Coetzee 2007, para 12).
49. Coetzee, "Meat Country," 48.
50. Ibid., 49.
51. Ibid., 49–50. Melanie Joy, in her groundbreaking work on carnism, makes the same point when she says that "taboos regarding the consumption of meat are far more common than those regarding any other food … meat is almost always the object of taboo" (Joy 2010, p. 13).

52. Ibid., 50.
53. Coetzee et al., *The Lives*, 40.
54. Coetzee, "Meat Country," 50. In Coetzee's *The Childhood of Jesus*, Simon's stunned reaction to the prospect of having to eat rats due to the unavailability of any other kinds of meat in Novilla is one example of such irrational disgust for certain types of meat. When Simon asks Alvaro, who had alerted him to the possibility of procuring and eating rat meat, whether he too eats rats, Alvaro replies: "No, I wouldn't dream of it. But you asked me where you could get meat, and that is all I can suggest" (45).
55. Coetzee et al., *The Lives*, 41.
56. At one point of the novel, fed up with his exclusive diet of bread and fruits, Simon rails in his mind: "The food we eat, our dreary diet of bread, lacks substance—lacks the substantiality of animal flesh, with all the gravity of bloodletting and sacrifice behind it" (77).
57. J.M. Coetzee, *The Childhood of Jesus* (London: Vintage, 2013), 202.
58. Ibid., 203.
59. Coetzee et al., *The Lives*, 53.
60. Coetzee, *The Childhood*, 265, 296, 301, 325.
61. Coetzee et al., *The Lives*, 68.
62. Ibid.
63. Coetzee, *The Childhood*, 314.
64. Coetzee, "Voiceless," para 12.
65. Coetzee, *Age of Iron*, 50.
66. Ibid., 7.
67. Coetzee et al., *The Lives*, 33.
68. Anthony Krupp, *Reason's Children: Childhood in Early Modern Philosophy* (New Jersey: Associate University Press, 2009), 38.
69. Rene Descartes, *The Correspondence*, vol. 3 *The Philosophical Writings of Descartes*, trans. John Cottingham, Robert Soothoff, Dugald Murdoch and Anthony Kenny (New York: Cambridge University Press, 1991), 190.
70. Ibid.
71. J.M. Coetzee, *Disgrace* (London: Vintage, 2000), 2.
72. Ibid., 81.
73. Ibid., 131. Similarly, Lurie's daughter Lucy "refuses to touch meat" (121) after her rape and the shooting of her dogs by the rapists.
74. Ibid., 219.
75. Ibid., 43.
76. Woessner, "Coetzee's Critique of Reason," 238.

77. Iris Murdoch, "The Sublime and the Good," *The Chicago Review* 13, no. 3 (1959), 51.
78. Marianne Dekovan discusses how "it is after he agrees to volunteer at the animal shelter that a dog, for the first time, acquires individuality to [Lurie]" (Dekovan 2009, 856). This recognition of the individual identities of animals is in stark contrast to Lurie's earlier belief that "animals [don't] have properly individual lives. Which among them get to live, which get to die, is not, as far as I am concerned, worth agonizing over" (Coetzee 2000, 126–7). Also, this is a significant step towards his sympathetic identification with animals. It is only after he acknowledges the individual identities of various animals that Lurie's carnistic defences finally begin to shake.

References

Bailey, Cathryn. 2005. On the backs of animals: The valorization of reason in contemporary animal ethics. *Ethics and the Environment* 10(1): 1–17, Spring. http://www.jstor.org/stable/40339093.

Coetzee, J.M. 1995. Meat country. *Granta* 52(Winter): 43–52.

Coetzee, J.M., Marjorie Garber, Peter Singer, Wendy Doniger and Barbara Smuts. 1999. *The lives of animals*, ed. Amy Gutmann. New Jersey: Princeton University Press.

Coetzee, J.M. 2000. *Disgrace*. London: Vintage.

Coetzee, J.M. 2004a. *Elizabeth Costello*. London: Vintage.

Coetzee, J.M. 2004b. *In the heart of the country*. London: Vintage.

Coetzee, J.M. 2004c. *Life & times of Michael K*. London: Vintage.

Coetzee, J.M. 2004d. *Dusklands*. London: Vintage.

Coetzee, J.M. 2004e. Animals, humans, cruelty and literature: A rare interview with J.M. Coetzee by Henrik Engström. *Satya Magazine*, May 2004. http://www.satyamag.com/may04/coetzee.html

Coetzee, J.M. 2007. Voiceless: I feel, therefore I am. In *Hugo Weaving: Random scribblings*. http://hugo.random-scribblings.net/?p=7203

Coetzee, J.M. 2008. *Diary of a bad year*. London: Vintage.

Coetzee, J.M. 2010a. *Summertime*. London: Vintage.

Coetzee, J.M. 2010b. *Age of iron*. London: Penguin.

Coetzee, J.M. 2010c. Foreword to *The second nature: The inner lives of animals*, by Jonathan Balcombe.

Coetzee, J.M. 2013. *The childhood of Jesus*. London: Vintage.
Coetzee, J.M., and Paul Auster. 2013. *Here and now: Letters 2008–2011*. London: Faber and Faber.
Dawn, Karen, and Peter Singer. 2010. Converging convictions: Coetzee and his characters on animals. In *J.M. Coetzee and ethics: Philosophical perspectives on literature*, ed. Anton Leist and Peter Singer, 109–118. New York: Columbia University Press.
Deka, Parag Kumar. 2014. *The being in the body: The body of the protagonist in J.M. Coetzee's* Life & Times of Michael K. M.Phil dissertation, University of Gauhati.
Dekovan, Marianne. 2009. Going to the dogs in *disgrace*. *ELH* 76(4): 847–875, Winter. www.jstor.org/stable/pdf/27742965.
Descartes, Rene. 1991. *The correspondence*. Vol. 3 of *the philosophical writings of descartes*. Trans. John Cottingham, Robert Stoothoff, Dugald Murdoch and Anthony Kenny. New York: Cambridge University Press.
Dunayer, Joan. 2001. *Animal equality: Language and liberation*. Derwood: Ryce Publishing.
Joy, Melanie. 2010. *Why we love dogs, eat pigs and wear cows: An introduction to carnism*. San Francisco: Conari Press.
Krupp, Anthony. 2009. *Reason's children: Childhood in early modern philosophy*. Bucknell studies in eighteenth century literature and culture. New Jersey: Associate University Press.
Murdoch, Iris. 1959. The sublime and the good. *The Chicago Review* 13(3): 42–55, Autumn. http://www.jstor.org/stable/25293537.
Nagel, Thomas. 1974. What is it like to be a bat? *The Philosophical Review* 83(4): 435–450. http://www.jstor.org/stable/2183914.
Singer, Peter. 1999. *Practical ethics*, 2nd ed. Cambridge, UK: Cambridge University Press.
Tolstoy, Leo. 2001. Esarhaddon, the king of Assyria. In *Collected shorter fiction, Volume 2*. Trans. Louise and Aylmer Maud and Nigel J. Cooper. New York: Everyman's Library/Alfred A. Knopf.
Tolstoy, Leo. 2015. The first step. *International vegetarian union—Official site*. http://www.ivu.org/history/tolstoy/the_%20first_step.html
Woessner, Martin. 2010. Coetzee's critique of reason. In *J.M. Coetzee and ethics: Philosophical perspectives on literature*, ed. Anton Leist and Peter Singer, 109–118. New York: Columbia University Press.

The Vegan and the Sovereign

Joshua Schuster

In Philip K. Dick's *Do Androids Dream of Electric Sheep?* (first published in 1968), everyone is a vegan. Yet, for some odd reason, nobody talks about it. People do, however, talk a lot about animal ethics and extinction. The novel is set in a time some decades after World War Terminus, a nuclear war that irradiated the planet, leading to a great dying off of animals (plants are not mentioned much—this is not a realist novel). The novel is famous for its humanoid androids, who were built off planet but have managed to make their way to Earth illegally and must be destroyed before they settle in as human imposters. But in a parallel plot, the Earth's ecology has been so devastated that there are extremely few animals left, and this has several consequences: no one ever eats meat; empathy toward animals is the utmost social value; there is a booming trade in mechanical animals; and owning animals—whether living or robotic—has become the highest status symbol.

J. Schuster (✉)
English, Western University, London, ON, Canada
e-mail: jschust@uwo.ca

J. Castricano, R.R. Simonsen (eds.), *Critical Perspectives on Veganism*,
The Palgrave Macmillan Animal Ethics Series,
DOI 10.1007/978-3-319-33419-6_9

Evidently, this novel envisions a kind of veganism that not many contemporary vegans would endorse. Animal liberation is not paramount, even though the living animals that survive receive the highest care. Early in the novel, we hear one neighbor remind another, "You know how people are about not taking care of an animal; they consider it immoral and anti-empathic. I mean, technically it's not a crime like it was right after [World War Terminus], but the feeling's still there."[1] In a combination of enlightened moral thinking and dark, noir-like anxieties over the status of species, the novel shows us what veganism without animal liberation, and without many animals, might look like. This tumultuous vision is not our world, yet there are identifiable trends that exist today that point to some aspects of what Dick describes. Most relevant here is how the novel offers provocative perceptions of vegan theory and practice. The human characters in the novel do not dare eat or kill an animal, yet they do not avowedly declare themselves vegan or vegetarian. This world does not present wholly consistent animal ethics, since the desire to own animals and display them domestically as objects of conspicuous consumption drives many of the human characters, including Rick Deckard, the protagonist and android bounty hunter. Deckard goes about his job of destroying humanoid androids by first testing suspects for empathy toward animals. If the suspect shows any delay in instantly expressing empathy for animals, the android has been found out, so anti-empathic tendencies toward animals, at least for machines, remains a capital crime.

The novel shows us a veganism that seems to have spontaneously arisen after mass extinction and environmental catastrophe. We find a world where eating animals or treating them violently is taboo, yet there is no account of principles of animal activism, and the remaining human characters are haunted by the absence of animals and the lure of machinic ersatz animal companions. After seeing an owl that Deckard believes might be real (it turns out to be a fake), Deckard has a flashback: "he remembered how in his childhood it had been discovered that species upon species had become extinct and how the 'papes had reported it each day—foxes one morning, badgers the next, until people had stopped reading the perpetual animal obits" (42). The extreme rarity of live animals here makes veganism both inevitable and perhaps irrelevant, given the lack of options. Yet, the humans remaining on Earth—many other humans have fled to other planets—clearly empathize with animals and see the world through the prism of shared human–nonhuman animal vulnerabilities. All the humans left behind on Earth remain there for economic reasons or because they are classified as physically deficient due to biological disability, sterility, or "degeneracy." The novel suggests that widespread human disability and a general post-traumatic condition, in coordination with the scarcity of animals, has led to the sympathetic alliance of humans and animals, an "inoperative community"[2] in a world where everyone is wounded. The arc of the novel, however, complicates this alliance as humans become increasingly emotionally bonded with mechanical animals and humanoid cyborgs, whose flaws and dejected status make them seem like they belong on Earth.

In this climate of veganism, extinction, pollution, nuclear fallout, highly advanced machine artificial intelligence, and the extreme commoditization of animals, animal care becomes the measure of one's ontological and legal status on Earth. This novel is often read as an allegory for humans entering a post-human phase,[3] but in subtle ways it also offers an allegory of the contradictions and compromised positions that inform animal ethics in a disintegrating world that is chock-full with doubles and ambiguous species. This essay will provide a discussion of Dick's book in the context of recent debates on how veganism engages with a number of contemporary philosophical, ethical, and ecological crises in its embrace of both rational commitment to principled living and openness to unknown and unknowable others. I take inspiration from the world of Deckard and its troubled vegans to look closer at some of veganism's own troubles today. I relate Dick's treatment of the ambiguities of animal care to Derrida's ambiguous statements that connect the deconstruction of animality and sovereignty together. Finally, I take a brief look at how vegan abolitionism today might connect to a wider consideration of what abolition could entail when ecology and human sovereignty are in question.

From the outset of Dick's story, we find humans who have the utmost care for animal survival and recognize fellow humans through a shared desire to tend, and own, animals. Androids come equipped with some affective and moral capabilities, but are unable or unwilling to wholly empathize and care for animals. However, Deckard becomes aware that some robots may indeed come to care for their pets. Deckard, who covets both mechanical and living animals, falls for a female android. All of these entanglements threaten to blur the distinctions in this "cybernetic triad," as Dominic Pettman calls it,[4] that intertwines human, animal, and cyborg. Deckard determines who is a cyborg by administering the Voigt-Kampff Empathy Test, which tests primarily for evidence of dietary veganism and protective animal care. Here I want to examine this exam in a little more detail and use it as a curious case for reflecting on the tension inherent in vegan concepts in the novel and in the world of today. Subjects of the test are presented with brief scenarios of moral quandaries involving mostly nonhuman animals, and only rarely are questioned about caring for humans. Presumably, if androids cannot treat animals well, they would not care about humans either and cannot be permitted to remain on Earth. However, very few people today would make the same judgment on fellow humans, given how animals are treated in our world. Furthermore, according to the results of this empathy test, very few humans today would qualify as human. The Voigt-Kampff test measures the reaction time of one's involuntary moral sentiments—blushing, capillary action, and pupil dilation—so any hesitation in showing the proper moral outrage regarding the mistreatment of animals gives the subject up. In one of the few overt references to theorizing vegetarianism, Deckard states, "Empathy, he once decided, must be limited to herbivores or anyhow omnivores who could depart from a meat diet. Because, ultimately, the empathic gift blurred the boundaries between hunter and victim, between the successful and the defeated" (31). Androids, who presumably do not need to eat, are seen as potentially ruthless predators who have no feelings for their prey and "no ability to feel empathic joy for another life form's success or grief at defeat" (32). Their machinic AI has made them too intelligent, as well as too sovereign and too independent of living beings. Here we see some wires being crossed with the norms of contemporary society. From at least Descartes to the present, humans have justified their exceptional status over other animals by stating that superior intelligence combined with the most sophisticated moral sentiments has allowed humans to elevate themselves as sovereign and having dominion over all other animals. Yet, in Dick's novel, humans do not have a monopoly on superior intelligence, and they justify their dominant position over androids by evoking their refusal to prey on others. The humans in the novel proclaim their exceptional status to androids by tying their moral status to the precariousness of all animal lives, including other humans.

Deckard's first use of the Voigt-Kampff test is on a young woman named Rachel, whom he suspects is an android. He reads the scenarios rapidly: "You are given a calf-skin wallet on your birthday" (48). Rachel: "I wouldn't accept it." Next one: "You have a little boy and he shows you his butterfly collection, including his killing jar." Rachel: "I'd take him to the doctor" (49). Next one: "You're sitting watching TV… and suddenly you discover a wasp crawling on your wrist." Rachel: "I'd kill it" (49). This is an odd question since empathy for the insect can be countered by the threat of pain the wasp can cause, thus this scenario cannot be reduced to an obvious moral outcome. But Deckard quickly continues:

> "Now consider this. You're reading a novel written in the old days before the war. The characters are visiting Fisherman's Wharf in San Francisco. They become hungry and enter a seafood restaurant. One of them orders lobster, and the chef drops the lobster into the tub of boiling water while the characters watch."
>
> "Oh god," Rachael said. "That's awful! Did they really do that? It's depraved. You mean a *live* lobster?" (49–50).

Deckard detects a hint of simulation by Rachel in this last scenario. The interrogator is looking for split-second involuntary bodily responses, which would be considered programmed or instinctive in human nature, yet all the scenarios in the test are culturally and historically specific. Empathic capabilities toward animal others are certainly embedded biologically in human affective systems, but they are also culturally conditioned and can vary by situation and from person to person. The test is supposed to catch how humans are biologically programmed for empathy, but the cultural and historical variability of responses to the exam shows that humans may have multiple and conflicting responses to animal welfare based on situational and personal circumstances. Is Deckard then testing for something supposedly biologically invariable and essentially machinic—involuntary motor responses—or something culturally mediated? Which response is the identifiably human–animal response? Furthermore, there exists in the novel an "empathy box" that stimulates moral feelings for humans who might be experiencing a waning of affect, a common symptom of post-traumatic stress disorder, depression, boredom, or even everyday postmodern life, as Fredric Jameson once argued.[5] Empathy can be programmed or deprogrammed and manipulated externally by machine or internally through self-control. In the novel, the tremendous empathy toward animals is conditioned by the devastation they endured, and few people would ever encounter live animals to prove their vegan sensibilities. It is not clear if Rachel is programmed to have vegan sentiments, an interesting programming problem, or programmed to anticipate and assume the attitudes of other humans around her. She says she would kill the wasp on her hand, but that too is an instinctual reaction, and fear of being stung would override most empathic sentiments in humans as well. Consequently, that particular question seems a bit faulty, as the scenarios are designed to find inconsistencies in what should be obvious moral setups. Yet, humans are not empathically or morally perfect, and sensitivity to moral ambiguity is also an ethical trait. The novel's *speculative veganism* then shows us a way to combine empathic devotion to animal lives and a version of vegan abolitionism, while portraying how ethical ambiguities and anxieties toward animals can remain and even intensify.

After several iterations of the test, Deckard is asked if he has ever been tested himself (Deckard insists he has). But throughout the novel, the accuser gets accused, and the investigator is part of the investigation. Deckard's insecure ontological status in the cybernetic triad runs throughout the plot, which leaves open the possibility that Deckard is an android. Deckard is empathic toward animal others, even mechanical ones, yet cold in his duty to destroy androids. Deckard's shifting between care and calculation, profiteering and generosity, hunter and hunted, becomes an allegory for the insecure status of humans in times of widespread cybernetic intelligence and environmental destruction. The Voigt-Kampff test seeks a stable and scientifically verifiable subject—a scientifically verifiable vegan—but the test itself is enmeshed in unstable binaries and ambiguous hermeneutics. As the human and machine distinctions fall apart in the novel, so do other binaries such as intention and non-intention, law and lawlessness, sovereign and subjection, affect and affectless, and life and nonlife. In the midst of all this, everyone is drawn into a veganism that itself is starting to show how its instabilities and inconsistencies might still lay the foundation for a better world.

Vegan Sovereignties

The gap between Dick's world and ours today is not large. We live in times of mass extinction.[6] In addition to the outright disappearance of species, across the planet the vast majority of wild animals are shrinking in population numbers and in habitat range. Many animals number around just 10% of their historical population sizes and occupy a comparably diminished territory.[7] Animal life today is lived under these heavy constraints, and species struggle even to maintain themselves in vastly diminished numbers. Many wild animals are corralled into biodiversity hotspots, which can be well protected, but create a kind of rich and poor distinction for animals outside these zones. At the same time, there are billions of animals being raised for food consumption. These animals effectively have no habitat and exist wholly in a closed physical and economic system. Finally there are also a small number of animals that have learned to live with humans in urban or rural landscapes, finding themselves tolerated, often without predators, but susceptible to all the vagaries of human development and patience. Given the extremely disparate lives of animals in these different zones, one hesitates to make any claims for what animality means in general. Compounded with this ongoing revision of the status of animal lives, we also live in times of massive human influence over weather, geological forces, and natural resources. Humans number over seven billion, and by sheer biomass under our control (including our domesticated animals), we hold an immense biological sovereignty compared to other vertebrates. Yet, these are also uncertain times for determining what is human, as machines become increasingly autonomous and intelligent, and new paradigms of symbiosis, nonhuman agency, and models of distributed cognition become prominent. Furthermore, these are also uncertain times for determining what is animal, as biotechnologies develop more ways to make and remake life, while the nascent field of plant studies challenges the supposedly unmistakable differences between plant and animal.[8] This is the Anthropocene—a world where humans are sovereign, especially with regard to animals and Earth's resources, yet the means and determinations of such sovereignty have displaced and disrupted human authority as well. Humans are increasingly in control of their environs, yet the environs increasingly threaten human control.

Veganism takes on new meaning and new tensions in the context of rising extinction rates and the Anthropocene. Transposing some of what we learned from Dick's book into a different context, I will argue that veganism today presents simultaneously a critique of human sovereignty and an unstable hermeneutics of life in the Anthropocene. Dick's fictitious and futurist exam for determining animal ethics and veganism is used by Deckard to definitively determine who is human, yet the exam, the examiner, and the reliability of any kind of ethical test are repeatedly questioned in the course of the novel.

Most vegans will recognize with some knowing discomfort that contemporary veganism is full of tendencies to test oneself and others for vegan credentials. One can hardly find a vegan who has not submitted herself or another person to such a testing.[9] Dick's book exposes this "test drive" (to borrow Avital Ronell's phrase[10]) between vegans and non-vegans. But instead of approaching the test as a means for definitive verification of the identity of the subject, Dick's test rather reveals the continual need for vegans to engage in questioning others and self-questioning in a world of hermeneutical instabilities and uncanny kinships. To be a vegan means taking a clear and committed stance against eating and instrumentalizing animals; however, the vegan is not a stable subject secured by a fixed discourse. To be a vegan means drawing lines by being committed to animal well-being, but also troubling the drawing of lines by querying the need for stable identities, definitive categories of selfhood, and sovereign assertions that rebuff all critique.

Human–nonhuman animal relations pass through multiple scenes of sovereignty and subjection in the Anthropocene. There are already powerful arguments for vegetarianism, veganism, and animal ethics based on utilitarianism (Peter Singer), capabilities (Martha Nussbaum), rights and duties (Steven Wise), and moral absolutism (Gary Francione).

However, many of these arguments seem to assume that anthropocentric sovereign power simply will dissolve in the face of solid reasoning and expanded moral sympathy, and that an extended analysis of how sovereign power is tied to the rational will and the liberal view of the subject is not needed.[11] Moreover, many of these philosophical arguments assume that the rationality of philosophy must line up with the rationality of the sovereign subject who has full control over his or her determinations. But the case for a reasonable account of veganism does not need to be tied only to a stable view of personhood committed to the rational will. Indeed, veganism in the time of the Anthropocene must question how human rationality and self-sovereignty have effectuated both enlightenment freedoms and ecological disturbances, culminating in an Earth perilously controlled by humans.

Animal Sovereigns

To be a vegan is to call for another world where one stands with animals while disrupting the current order of power, sovereignty, and authority that is built on the exploitations of animals and Earth's others. In order to thicken an analysis of how veganism involves negotiating commitments to animal ethics and to self-examination, I want to turn to Jacques Derrida's last lectures where he sought to tie the understanding of animality and sovereignty together. Derrida's work has already been central to establishing a theoretical zest to contemporary animal studies, although his work has garnered a mix reception for those most interested in veganism. Gary Steiner in particular has harshly assessed Derrida's contributions to theorizing justice for animals because Steiner finds that Derrida never offers a definitive answer for how humans should treat animals. Derrida has written copiously on the necessity of grappling with ethics as a continuous openness to alterity, transformation, and unexpected relations with others, be they human or nonhuman. However, Steiner finds that Derrida ultimately offers no way of determining "that we can never be sure that killing and eating an animal… is unjust—*not in one single situation.*"[12] Steiner is aware that there is a philosophical reason that Derrida did not make such blanket pronouncements. Ethics, in Derrida's view, cannot be something that one calculates in advance as a rational sovereign subject. Rather, ethics happens when one cannot calculate, when self-sovereignty is in question, when one is open to difference and the unknown outcomes of ongoing relations with others who have their own interests. For Derrida, one cannot make blanket assertions of ethical absolutism, for ethics must always be open to reinterpretation and response to the call of the other. There are no absolutes because only when questioning begins can there be the possibility for ethics (recall here Dick's world of vegan questioning). There is no stable sovereign declaration of universal principles because ethics ensues from the very questioning of sovereign individuality and its basis in secured and steady definitions of selfhood.

Yet, however much Derrida's work is claimed or critiqued by theorists of animal rights and animal ontologies, we know that Derrida himself never felt the need to establish a vegan or vegetarian position as an intellectual concern, let alone a life practice. In several influential writings and discussions of eating and animal ethics, Derrida did launch a powerful critique of carnivorism and the failure of philosophy to attend to what is happening to the industrialization of animals for food. Yet, one wonders about the theoretical reasons Derrida used to avoid a philosophical analysis of vegetarianism and to avoid avowing any personal position that clearly declared his own eating practice. Looking more broadly at his philosophy, there is a recurrent methodological double take in Derrida's work as he documents how a philosopher takes a position on a philosophical question and subsequently shows how that position becomes bound with what it initially refused. Every philosophical decision inscribes one into a metaphysical system of oppositional presuppositions as well as a countersystem of critique and deconstruction by unraveling those oppositions. Critics of Derrida, such as Steiner, claim that Derrida takes very few clear-cut positions (be they ontological, political, or ethical) and prefers instead to undermine or destabilize the history of position takings in philosophy. However, it can be shown in a much lengthier analysis that the stance of refuting positions is certainly already a stance toward truth telling as a display of differences and double binds rather than absolutizing one's position via abstract reason, axiom, matheme, or making a sharp, irrevocable cut into the real.

For Derrida, the philosopher's leap into any decision is both a necessary act and a risk, one that always involves unknown and unknowable consequences. Each decision is a sovereign assertion of an individual's will that also exposes the individual to situations that are continually undergoing change. Every decision is inflected by "madness"[13] in that it cannot wholly proceed by reason or method as the decision inscribes the subject into a realm where others act and react to these constantly shifting positions. Yet, what Derrida further points to is how taking a leap philosophically can also be a way to open up new paths and concepts for thought and action that had not appeared in advance. The mad decision, initially an act of sovereign will, yet which exceeds any sovereign control, might also open oneself up to otherness and new relations that one would not encounter by standing on the sidelines. One might further argue that there is a kind of madness embedded in every decision to eat, a decision that is both necessary and yet never without its unforeseeable issues. Still, Derrida himself never made a "mad" leap into analyzing philosophical veganism, aside from some intriguing remarks in the interview "Eating Well" and in conversations with Elizabeth Roudinesco. In the latter, Derrida states unequivocally "I do not believe in the existence of the non-carnivore."[14] Here the herbivore, let alone the vegan, cannot even be named without using the negative. It is as if the carnivore speaks for all kinds of eating, all differences in food consumption and incorporation of the other. Derrida's own personal beliefs and eating habits here are not as important as showing how vegan philosophies might be implicit or explicit in philosophical attempts to situate animal life, consumption, and what is now called "food sovereignty," the association of food with state security and national identity. Derrida never analyzed the long history of arguments for veganism in Western and non-Western philosophy as counter-positions to the many instances of philosophical sovereignty over animals, even when doing so could have further put "sovereignties in question," as Derrida phrases it.[15]

The same vigilance over positions yet hesitation to be inscribed in one appears throughout Derrida's two years of lectures published as *The Beast and the Sovereign*, delivered at the École des hautes études en sciences sociales from fall 2001 to spring 2003, the last formal lectures he gave at his home university. The main thread of these lectures follows through an analysis of how human sovereignty, especially in the figure of the king, has consistently been cast as beastly, wolf or lionlike, and "zoological" when using mere brute force. At the same time, the human is said to be sovereign to the animal due to all kinds of declared exclusive qualities, from reason, to speech, to the capacity to negate the self's immediate concerns for future prospects. Early in these lectures, Derrida says that it will not suffice simply to take the position that sovereignty is, at heart, "merely disguised manifestations of animal force."[16] Sovereignty is not wholly the domain of the human, nor is the animal what the human must overcome to be sovereign. Instead, Derrida will follow how sovereignty circulates between and among humans and animals, remaining ambiguous and embedded in fluctuating power relations. Hence, Derrida does not affirm in advance any sovereign declarations for or against the politics and ethics of animals.

> The only rule that for the moment I believe we should give ourselves in this seminar is no more to rely on commonly accredited oppositional limits between what is called nature and culture, nature/law, *physis/nomos*, God, man, and animal or concerning what is "proper to man" than to muddle everything and rush, by analogism, toward resemblances and identities. Every time one puts an oppositional limit in question, far from concluding there is identity, we must on the contrary multiply attention to differences, refine the analysis in a structure[d] field (*Beast* 15–16).

The "rule" or position here is not to leap into "oppositional limits" between binary terms and favoring one side over another, nor try to collapse them into a "muddle." One must instead analyze the multiplicity of differences within a structured set of positions. One must refuse the sovereign decision that asserts strictly either unbridgeable difference or complete identity across human–animal relations. But it also takes an act of sovereignty to stand in watch over these positions.

Derrida continues in this vein by insisting there is no single position on human sovereignty over animals or within the "kingdom" of animality; rather, there is a "haunting of the sovereign by the beast and the beast by the sovereign" (*Beast* 18). Think again of Dick's character Deckard, who claims a certain animality and sovereignty yet becomes involved in the undoing of the certainty of these categories. Entwined with this double haunting, Derrida points to the double nature of eating and being eaten between humans and animals: "You have no doubt already noticed the recurrence of the lexicon of devourment ('devour,' 'devouring'): the beast is on this account devouring, and man devours the beast" (*Beast* 23). And here Derrida extends the motif: "Might sovereignty be devouring? Might its force, its power, its greatest force, its absolute potency be, in essence and always in the last instance, a power of devourment (mouth, teeth, tongue, violent rush to bite, engulf, swallow the other, to take the other into oneself too, to kill it or mourn it)?" (*Beast* 23). Just right here one might expect an extended rethinking of eating and not eating, of questioning the animal as always already devourable. Here one might open the question of sovereignty by asking about animals who are not always in an eat-or-be-eaten relationship to oneself. Here one can put the idea of sovereignty "in the last instance" to critique and turn to new pathways for thinking the range of possible nonconsuming relations among animals. As Derrida already insisted, one must multiply figures and gestures, thereby dislodging supposed last instances. Furthermore, before one devours, there must be a leap, a decision to eat. But what about the metaphor of the leap—is it the animal gesture *par excellence*, such that only humans are capable of not leaping, of asserting rational choice? Is every leap mad, before and beyond reason, and therefore zoological rather than purely logical? The vegan decision not to devour, not to leap, turns on a sovereign refusal of sovereignty. There does not seem to be any reason not to raise such vegan possibilities right here, but Derrida quickly moves into other matters, leaping elsewhere to a discussion of humans as political animals. Right here, there is a sovereign neglect to put sovereignty in question.

Again, it is not a matter of wanting to see Derrida himself avow veganism. At issue here is the missed opportunity to force the dialectical repercussions of confronting sovereignty with another sovereign decision (refusing meat) that is not reducible to sovereignty's "greatest force." Veganism is not an antidote to all the egregious work human sovereignty has wrought—that is way too much to ask. But veganism, as an ethical, political, and philosophical position does not take for granted that "multiplying the attention to differences" that both link and separate the carnivorous beast and the carnivorous sovereign constitute a thorough analysis of differences within sovereign power. Veganism hews closer to the "politics of friendship" that ensues from cross- and counter-sovereign alliances between humans and animals that are instable and under continuous pressure in a world divided between vastly depleted numbers of wild animals and massively populated domesticated animals. Because we can be sovereign, we can question our sovereignty; we can act in ways both rational and open to reassessment and responsibility toward forms of alterity not determined in advance. Veganism is never wholly an abandoning of human sovereignty, nor does it need to assume that universalizing liberal humanist sovereign traits and extending these to all animals, were this possible, would be case closed. The vegan does not think that power and violence will go away in a fully vegan world—but there is no reason to relent on a desire for utopian ways of living together either. Indeed, to be vegan is to continually throw oneself into the midst of power dynamics to diagnose, critique, and effectuate livable justice across species lines. To be vegan is to confront the hierarchical, guarded, and often abject zones of animal life spread across the planet that have been formed by the extension of human self-interest and self-sovereignty around the globe. In the midst of this zoning of animality, the vegan embodies a position that can be described as *sovereign unsovereign*. Veganism is an attempt to achieve collective species justice in a way that leaves open and probably unanswerable questions that are caught up in moral and situational ambiguities. One version of this position is found in the character of Deckard in Dick's book, who obsesses over animals and tests others for veganism but ultimately admits to himself that he is not sure what an animal is. By the end of the novel, Deckard ponders over a mechanical toad he found that he initially thought was real, and states in a sadder and wiser tone, "[E]lectric things have their lives, too. Paltry as those lives are" (241). While we do not live in Deckard's of postapocalyptic world, the rise of extinction rates and the accelerating developments of AI today bring us closer each day. What vegans today share with Dick's vegans and Derrida's "sovereignties in question" is the insistence on making questioning life into a way of life. One becomes vegan not to presume to have all the answers, nor to be able to tell others how to act with animals in all circumstances, but because one wants to make living with animals into a lifelong mutual inquiry. To be vegan is to subject oneself to a test similar to the Voigt-Kampff test, and not necessarily know if one passes–yet also to choose to combine this uncertainty and drive for justice across species lines as the groundwork for a better world.

Sovereign Abolitions

"Man needs sovereignty more than bread," declares Georges Bataille.[17] This sounds like another argument for anthropocentric dominion, but Bataille's work as a whole presents an intriguing case for redefining what sovereignty means. Building on the ambiguities of sovereign life presented in Dick's novel and Derrida's philosophy, I want to offer a brief argument for exploiting Bataille's thought to tie veganism and an Anthropocene critique together. Sensing that philosophy had exhausted itself with claims to access the totality of knowledge and achieve a stability between reason and science, Bataille began in the 1940s to develop what he would eventually call "the unfinished system of nonknowledge." To push philosophy to its utmost limits, Bataille proposed not another claim for a universal system but rather that philosophy become an exercise of thought in the act of expiating itself. This would be Bataille's definition, then, of sovereign nonknowledge: a thought and life that does not submit to any project, anything useful, or anything productive for capital. Instead, Bataille argued that humans should pursue intensity for its own sake by avowing enjoyment, sadness, poetry, sex, self-experimentation, and death. Bataille thought that human self-sovereignty that expiates itself in embodied intensities had a parallel in the way animal existence is "like water in water."[18] Animals are immersed in their moment and surroundings and exist wholly in a state of immanence to their immediate conditions. Bataille certainly romantically overstates the notion that animals are captivated by the intensity of their immediate situation and have no sense of time past or future. Also, there is no question of Bataille arguing for veganism or animal welfare, and he is known for claiming that the sacrifice of animals has historically been a way for humans to tie death to the intensity of the sacred. However, what I would like to borrow from Bataille is the notion that sovereignty can be revalued as a refusal of servility to practical systems, stabilizing norms, and capitalism in an

Anthropocene age. Speaking in the wake of the devastation of World War II, Bataille asserted that he saw the "planet congested by death and wealth."[19] For Bataille, nonknowledge and sovereign expiation rejected the wealth system and the biopolitics of control over life and death for an alternative practice of the self that "entangle[d] morality with intensity."[20]

I want to pirate Bataille's notion of undone sovereignty and put it in dialogue with vegan abolition arguments to raise some further provocations that tie animality and sovereignty together. Vegan abolitionism is a sovereign call to declare that animals shall never be eaten by humans or harmed for the sake of human convenience and consumption. The refusal to eat animals is a form of nonknowledge; there is no human need to know animals as food. With this declaration of abolitionist sovereignty and nonknowledge in mind, I want to remark on another recent use of the term "abolition" in the context of planetary activism. If we want any chance for a planet not to be overheated by greenhouse gases, especially from carbon emissions, we need to leave the vast majority of remaining oil resources in the ground. A global movement to "leave oil in the soil" is calling for a sovereign declaration on a planetary scale to say that oil in the ground will no longer be useful or valued. A recent article in *The Nation* by Christopher Hayes called this movement "the new abolitionism" and estimated that the value of all the fossil fuels in the ground to be around $10 trillion.[21] Hayes argued that the loss of this oil and coal wealth has a historical comparison to the end of the US Civil War, when slaveholders overnight lost the value of keeping humans as chattel. In a global, sovereign declaration of vegan abolitionism, a similar figure of animal capital would be wiped out overnight. Vegan abolitionism also asserts that ignoring animal suffering for human convenience is always morally wrong, just as slavery is always morally wrong. One can also claim that valuing animals as capital, as money, and as exchangeable good is always wrong. Animals are not walking dollar signs, just as humans are not walking dollar signs. A similar logic can also apply even to oil and coal, which if burned would ultimately decimate the planet we share with plants and animals. And if we care about the future of life on this planet, the capitalization and consumption of oil and coal has to be abolished by sovereign decree as well, as an act of fiat asserted by the "kingly" human species. It seems only an act of sovereign abolition can truly change the lives of animals and the ends of the Earth.

Sovereigns decide on coinage, on what can or cannot count as money. The same sovereign power that decides on what constitutes money is at work in deciding what constitutes permissible violence and permissible suffering. Yet, sovereigns can also abolish value. Vegans, as I argued earlier, are sovereign unsovereign. The vegan is still modeled on the liberal subject who is self-sovereign, not owned by anyone else, but also who refuses the crown of anthropocentrism by recognizing that humans do not run the planet and are not just responsible to themselves. In a world driven by the coupling of biopolitics with sovereign decisions over life and death, value and worthlessness, veganism presents a counterpower based on multilayered and shifting relations with animals. Such relationality involves listening, uncertainty, esteeming multiple and asymmetrical engagements with animals, and trying to find the right questions to ask animals (a challenge raised by the philosopher of ethology Vinciane Despret,[22] and which dovetails in intriguing ways with Dick's and Derrida's animal questionings). To be vegan is to make a rational critique of animal abuse, yet it is also to recognize the unstable relational ontologies between humans and animals that need not be reducible either to comprehensive sameness or complete difference. As discussed earlier, Deckard's world is one example of being vegan and negotiating a faltering human sovereignty in times of extinction. Dick's vision of speculative veganism finds characters dealing with a variety of animal attachments and disaffections in a world drawn up not by abolitionism but by destructive force that left all life precarious in its wake. Indeed, taking another page from Dick's book, I would argue that being vegan is similar in feel to being in a science fiction story. Being a vegan means living in a partially alternate world that has a science fiction feel because it involves continual cognitive estrangement from social norms. Vegans must find a way to form a speculative life that bridges this world with a future world of animal justice. The vegan cannot feel sovereign in a world dominated by human sovereignty. Instead, veganism ultimately redefines sovereignty as living one's ideals, committing one's life to one's ideas, and involving animals in these ideals as well.

Notes

1. Philip K. Dick, *Do Androids Dream of Electric Sheep?* (New York: Del Ray, 1975), 13. Page numbers for the novel will be provided in text.
2. The phrase is from Jean-Luc Nancy to describe a situation in which a community is not bound together by an identity, but by mutual exposure and by being "without essence" together. Jean-Luc Nancy, *The Inoperative Community*, tr. Peter Connor (Minneapolis: U of Minnesota P, 1991).
3. For a post-human reading of the novel, see Jill Galvan, "Entering the Post-Human Collective in Philip K. Dick's *Do Androids Dream of Electric Sheep?*," *Science Fiction Studies* 24.3 (1997): 413–429. More recently, critics have become keen to discuss the complicated status of animals and food in the novel. See, for example, Sherryl Vint, "Speciesism and Species Being in Do Androids Dream of Electric Sheep?," *Mosaic: A Journal for the Interdisciplinary Study of Literature* 40.1 (2007 Mar): 111–126; Josh Toth, "Do Androids Eat Electric Sheep? Egotism, Empathy, and the Ethics of Eating in the Work of Philip K. Dick," *LIT: Literature Interpretation Theory* 24 (2013 Jan-Mar): 65–85; David Huebert, "Species Panic: Human Continuums, Trans Andys, and Cyberotic Triangles in *Do Androids Dream of Electric Sheep?*," *TSQ: Transgender Studies Quarterly* 2.2 (2015): 244–260.
4. Dominic Pettman, *Human Error: Species-Being and Machines* (Minneapolis: U of Minnesota P, 2011), 11.
5. Fredric Jameson, *Postmodernism, or, The Cultural Logic of Late Capitalism* (Durham: Duke UP, 1992).
6. For discussions of mass extinction today, see Richard Leakey and Roger Lewin, *The Sixth Extinction: Patterns of Life and the Future of Humankind* (New York: Anchor Books, 1996); Elizabeth Kolbert, *The Sixth Extinction: An Unnatural History* (New York: Henry Holt, 2014).
7. J. B. MacKinnon, *The Once and Future World: Nature as It Was, as It Is, as It Could Be* (Toronto: Random House Canada, 2013), 38.
8. See Michael Marder, *Plant-Thinking: A Philosophy of Vegetal Life* (New York: Columbia UP, 2013); Jeffrey Nealon, *Plant Theory: Biopower and Vegetable Life* (Stanford: Stanford UP, 2015).
9. Dominique Lestel, in *Apologie du carnivore*, stereotypically characterizes the vegan as pushy, threatening, and naively dogmatic due to this tendency to question others about their eating practices. Lestel argues that

veganism is no help at all in undoing deeply entrenched attitudes of human dominion. According to Lestel, the vegan must defend his or her own position and assert dominion over other humans and their eating habits, and this self-satisfied assertion eventually becomes more important than any advocacy on behalf of animals. Dominique Lestel, *Apologie du carnivore* (Paris: Fayard, 2011).

10. Avital Ronell, *The Test Drive* (Carbondale: U of Illinois P, 2005).
11. Matthew Scully's *Dominion* remains an unusual work in the field because he avowedly claims that the Biblical decree granting humans dominion over animals justifies veganism and human sovereignty at the same time. Scully does not seek to question human dominion, but rather mobilize it as the guideline for veganism that leaves Scully's own conservative political vision intact and triumphant. Scully is unwilling to problematize the philosophical and political force of human sovereignty and consider veganism as connected to left-identified social justice and antiauthoritarian movements. Matthew Scully, *Dominion: The Power of Man, the Suffering of Animals, and the Call to Mercy* (New York: St. Martin's Press, 2002).
12. Gary Steiner, *Animals and the Limits of Postmodernism* (New York: Columbia UP, 2013), 81.
13. A concise formulation of this madness of decision appears in Derrida's essay "Force of Law": "The moment of *decision as such*, what must be just, *must* always remain a finite moment of urgency and precipitation; it must not be the consequence or the effect of this theoretical or historical knowledge, of this reflection or this deliberation, since the decision always marks the interruption of the juridico-, ethico-, or politico-cognitive deliberation that precedes it, and that *must* precede it. The instant of a decision is a madness, says Kierkegaard." Jacques Derrida, *Acts of Religion*, tr. Gil Anidjar (New York: Routledge, 2002), 255.
14. Jacques Derrida and Elizabeth Roudinesco, *For What Tomorrow… A Dialogue*, tr. Jeff Fort (Stanford: Stanford UP, 2004), 68. See also Jacques Derrida, "Eating Well," *Who Comes After the Subject?* eds. Eduardo Cadava, Peter Connor, Jean-Luc Nancy (New York: Routledge, 1991), 96–119.
15. Jacques Derrida, *Sovereignties in Question: The Poetics of Paul Celan*, eds. Thomas Dutoit and Outi Pasanen (New York: Fordham UP, 2005).
16. Jacques Derrida, *The Beast and the Sovereign: Volume I*, tr. Geoffrey Bennington (Chicago: U of Chicago P, 2009), 14. Cited hereafter as *Beast*.

17. Georges Bataille, *Theory of Religion*, tr. Robert Hurley (New York: Zone Books, 1992), 161.
18. Bataille, *Theory of Religion*, 25.
19. Georges Bataille, *The Unfinished System of Nonknowledge*, tr. Michelle Kendall and Stuart Kendall (Minneapolis: U of Minnesota P, 2001), 221.
20. Bataille, *The Unfinished System of Nonknowledge*, 116.
21. Christopher Hayes, "The New Abolitionism," *The Nation* (April 22, 2014). http://www.thenation.com/article/new-abolitionism/.
22. Vincianne Despret, *What Would Animals Say If We Asked the Right Questions*? tr. Brett Buchanan (Minneapolis: U of Minnesota P, 2016).

References

Bataille, Georges. 1992. *Theory of religion*. Trans. Robert Hurley. New York: Zone Books.

Bataille, Georges. 2001. *The unfinished system of nonknowledge*. Trans. Michelle Kendall and Stuart Kendall. Minneapolis: University of Minnesota Press.

Derrida, Jacques. 1991. Eating well. In *Who comes after the subject?* ed. Eduardo Cadava, Peter Connor, and Jean-Luc Nancy, 96–119. New York: Routledge.

Derrida, Jacques. 2002. *Acts of religion*. Trans. Gil Anidjar. New York: Routledge.

Derrida, Jacques. 2005. In *Sovereignties in question: The poetics of Paul Celan*, ed. Thomas Dutoit and Outi Pasanen. New York: Fordham University Press.

Derrida, Jacques. 2009. *The beast and the sovereign: Volume I*. Trans. Geoffrey Bennington. Chicago: University of Chicago Press.

Derrida, Jacques, and Elizabeth Roudinesco. 2004. *For what tomorrow… A dialogue*. Trans. Jeff Fort. Stanford: Stanford University Press.

Despret, Vincianne. 2016. *What would animals say if we asked the right questions*? Trans. Brett Buchanan. Minneapolis: University of Minnesota Press.

Dick, Philip K. 1975. *Do androids dream of electric sheep?* New York: Del Ray.

Dominic, Pettman. 2011. *Human error: Species-being and machines*. Minneapolis: University of Minnesota Press.

Galvan, Jill. 1997. Entering the post-human collective in Philip K. Dick's *Do androids dream of electric sheep? Science Fiction Studies* 24(3): 413–429.

Hayes, Christopher. 2014. The new abolitionism. *The Nation,* April 22. http://www.thenation.com/article/new-abolitionism/

Huebert, David. 2015. Species panic: Human continuums, Trans Andys, and Cyberotic Triangles in *Do androids dream of electric sheep?. TSQ: Transgender Studies Quarterly* 2(2): 244–260.

Jameson, Fredric. 1992. *Postmodernism, or, the cultural logic of late capitalism.* Durham: Duke University Press.

Kolbert, Elizabeth. 2014. *The sixth extinction: An unnatural history*. New York: Henry Holt.

Leakey, Richard, and Roger Lewin. 1996. *The sixth extinction: Patterns of life and the future of humankind.* New York: Anchor Books.

Lestel, Dominique. 2011. *Apologie du carnivore.* Paris: Fayard.

MacKinnon, J.B. 2013. *The once and future world: Nature as it was, as it is, as it could be.* Toronto: Random House Canada.

Marder, Michael. 2013. *Plant-thinking: A philosophy of vegetal life.* New York: Columbia University Press.

Nancy, Jean-Luc. 1991. *The inoperative community.* Trans. Peter Connor. Minneapolis: University of Minnesota Press.

Nealon, Jeffrey. 2015. *Plant theory: Biopower and vegetable life.* Stanford: Stanford University Press.

Ronell, Avital. 2005. *The test drive.* Carbondale: University of Illinois Press.

Scully, Matthew. 2002. *Dominion: The power of man, the suffering of animals, and the call to mercy.* New York: St. Martin's Press.

Steiner, Gary. 2013. *Animals and the limits of postmodernism.* New York: Columbia University Press.

Toth, Josh. 2013. Do androids eat electric sheep? Egotism, empathy, and the ethics of eating in the work of Philip K. Dick. *LIT: Literature Interpretation Theory* 24: 65–85.

Vint, Sherryl. 2007. Speciesism and species being in do androids dream of electric sheep? *Mosaic: A Journal for the Interdisciplinary Study of Literature* 40(1): 111–126.

Part III

Food, Memory, Histories

"Are Vegetarians Good Fighters?": World War I and the Rise of Meatless Patriotism

Adam D. Shprintzen

As the USA approached possible involvement in World War I, debates arose in the popular press about American war readiness. While some writers worried about the size and scale of American military capabilities, others debated the state of American masculinity and strength in the face of a potential mass mobilization. Perhaps surprisingly, one group that focused on issues of masculinity, strength, and violence were vegetarians. In the pages of one vegetarian publication, the issue was answered directly. The editorial staff of *Physical Culture Magazine* responded to a letter from a reader asking if vegetarians were prepared to serve their country during its time of need. The magazine's answer made an explicit connection between meatless living and the strength needed to win a war. Responding to the onset of war, the magazine emphasized the need for what an editorial labeled "men in brain and body." The editorial explained that "the manhood" of the USA had been collectively called to reckoning

A.D. Shprintzen (✉)
History, Marywood University, Scranton, PA, USA
e-mail: shprintzen@marywood.edu

J. Castricano, R.R. Simonsen (eds.), *Critical Perspectives on Veganism*,
The Palgrave Macmillan Animal Ethics Series,
DOI 10.1007/978-3-319-33419-6_10

and that the fate of the war would be decided by those "nations that can furnish the best, the most capable, the most courageous men" of strength and health and this group quite clearly included vegetarians.[1]

An editorial in the same issue of the magazine made the connection between diet and warfare even more explicit, by asking the question, "Are Vegetarians Good Fighters?" Predictably, the article fell squarely on the side of vegetarians as assets to the country's fighting efforts. The article's author explained that vegetarians were not any less patriotic than all other Americans and no less apt to "come to the defense of their country than any other group." The editorial evoked the magazine's longstanding support of Theodore Roosevelt in its rationale for supporting vegetarians as fighters, explaining that, "Anyone who knows anything about the bull moose knows that there is no more dangerous animal in the world. He is a vegetarian."[2]

As the USA entered World War I, a seemingly unlikely group of vegetarians sought to gain legitimacy by connecting dietary choice with combat readiness. On the one hand, the fact that vegetarians positioned their diet as a means to be better, more patriotic citizens seems somewhat illogical given the non-violence attached to the diet. However, when one considers that the vegetarian movement in the USA spent the early decades of the twentieth century embracing normative values and culture (such as consumerism and personal advancement), it is perhaps unsurprising that many vegetarians used the war years as a way to further legitimize their cause.[3]

While *Physical Culture* raised the issue of dietary choice and fighting fitness early on, the question remained vitally important throughout the war in a direct way that affected civilians on the home front. In October 1917, the recently formed United States Food Administration under the direction of Herbert Hoover began encouraging American citizens to take part in its "Meatless Tuesdays" program, as well as pledge to have one meatless meal a day in order to save meat to ship to allied troops abroad. The idea was a means to prove the administration's overriding slogan and notion that "Food Will Win the War." The campaign was aimed primarily at women, exploiting the growing home economics movement and connecting it with a notion of domestic civic duty to survive a worldwide crisis.[4]

Meatless and wheatless days were necessary because of reduced productivity and agricultural harvest in Europe as well as the interruption of long distance trade routes. President Woodrow Wilson's resolution explaining the meatless program was bathed in the language of patriotism, duty, and honor. Wilson described dietary restriction as being "one of the most pressing obligations of the war," and necessary because it served "the national interest." The President called on "every loyal American" to follow the guidelines of the Food Administration and emphasized the role of women in ensuring that households follow the new rules. Dietary sacrifice was, as explained by Wilson, citizens' "national service."[5] World War I represented a shift for American vegetarians' relationship with warfare. This change was significant, as it challenged the pacifistic underpinnings of the movement and had implications that lasted long after an armistice was signed to end the fighting of the so-called Great War.

Meat abstention, if only once a week, was ostensibly presented to the American public as a patriotic act of self-control. The program targeted private citizens as well as businesses, as restaurants and large hotels were mandated to follow the meatless day each week.[6] The campaign did not advocate for complete meat abstention or even vegetarianism in name, defining meatless food as being "without any cattle, hog, or sheep products."[7] Chicken, fish, seafood, and other poultries were acceptable alternatives as there was no potential shortage in the amounts shipped abroad. However, vegetarians staked a claim to these developments as well and keenly latched onto a language of patriotic meat abstention. The group was careful to not present the diet as an act of self-sacrifice, but rather one rich in flavors, gustatory pleasures, and general health benefits to all consumers. And despite the fact that meat abstention was presented as an act of self-sacrifice to patriotic Americans by the government, vegetarians themselves recrafted the narrative to utilize the government's temporary meatless endorsement in order to gain popular appeal and legitimacy. Part of vegetarians' desire to reframe the debate was driven by an understanding that the hypermasculine and hyperviolent war years would call into question many of the ideological underpinnings of the movement at the precise time it had gained some popular recognition in America. As a result, the organized movement was focused on finding a way to connect with developments occurring on a national scale.

The Food Administration utilized an intense public relations campaign to encourage Americans to give up meat at least once a week. A massive propaganda operation was launched under Hoover's watchful eye, employing 1500 public employees who produced 43,000 posters and 2000 press releases urging Americans to observe Meatless Tuesdays.[8] Dietary choice was connected with individual duty and given governmental endorsement as having significant social and cultural power. Pamphlets, posters, newspaper advertisements, and cookbooks were all published encouraging Americans to do their part in winning the war.

The United States Food Administration itself, in its advice on how citizens could best survive one meatless day per week, provided information influenced by the new vegetarianism. The government did not have any objections to the use of poultry, fish, or other less necessary animal products as proteins. However, the government endorsed methods first developed and embraced by American vegetarians in its advice on how to substitute for flesh foods during the war. The Food Administration advised Americans to utilize nuts, peas, beans, and other legumes as substitutes for beef, bacon, mutton, and pork.[9] Vegetarians had made similar suggestions for more than two decades, emphasizing nuts and legumes as healthy delicacies that could be easily manipulated to replace flesh foods.[10]

The Council of National Defense—a policy group formed by President Wilson that included the Secretary of War, the Secretary of the Navy, the Secretary of the Interior, the Secretary of Agriculture, the Secretary of Commerce, and the Secretary of Labor—was formulated in order to coordinate the use of industrial production and goods for the war effort. The group worked in conjunction with an advisory commission composed of business and industrial leaders in order to best plan for the war effort. On a local level, the commission functioned through the coordinated efforts of smaller councils and woman's committees that worked to coordinate female volunteers and advise individuals on how to best serve the war effort at home.[11] As part of their duties, the woman's committees produced guides and cookbooks to distribute to local households on how to best prepare meatless meals.

The St. Louis County Unit of the woman's committee published its *Win the War* cookbook in 1918, providing home epicures with meatless menus that utilized the vegetarian methodology of mock meat recipes.

The cookbook advised home cooks on how to best prepare vegetable soup without the benefit of beef or animal bones, while also giving instruction on how to prepare soybean as well as rice and peanut croquettes, cooked in a thick heavy cream sauce. The embrace of dairy products reflected a prevailing confidence that a dairy shortage was not a possibility during the war years. And the use of dairy represented continuity for American vegetarians that dated back to the nineteenth century.[12] Similar recipes could be found in vegetarian cookbooks in the early years of the twentieth century. A recipe for mock mincemeat called for green tomatoes, apples, raisins, allspice, cloves, cinnamon, and nutmeg. The result was so tasty that it was tested on a group of 500 soldiers on a transport who deemed the pie to be "perfect" while clamoring for extra helpings.[13] The recipe could be found in vegetarian cookbooks as early as the late nineteenth century and throughout the early years of the twentieth.[14] The dish was even utilized by vegetarians as early as the second meeting of the American Vegetarian Society in the fall of 1851, though conceived of as a fruit mince pie rather than as a meat substitute.[15]

The woman's committee of Cleveland also offered meatless advice to local women, explicitly emphasizing the duty of meatless living in the title of its *Patriotic Cookbook*. The book explained that home cooks were being called to service for the country, begging readers to do what they could in the hour of "extreme peril to the democratic people of the world." A variety of vegetarian recipes were provided, including green peppers and tomatoes stuffed with cowpeas instead of meat, a pea loaf made of cooked green peas, breadcrumbs, and eggs, and a peanut loaf that resembled many of J.H. Kellogg's meat substitutes developed at the Battle Creek Sanitarium.[16] Pittsburgh's woman's committee put together a wartime cookbook as well, including recipes for mock sausage made from lima beans and a variety of dried spices.[17]

Vegetarians themselves saw an opportunity to grab the banner of meatless patriotism by writing cookbooks that advocated for vegetarianism as both a patriotic cause and a cause onto itself. Eugene Christian, a vegetarian for 21 years and raw food advocate who wrote extensively on the issue of dietary habits, authored a book filled with meatless and wheatless recipes and menus. Christian explained that humanity did its best thinking during "abnormal periods" where habitual acts were questioned and often overturned, and the war provided precisely that type of opportunity for

Americans and their diets. The self-proclaimed "food scientist" explained that the Meatless Tuesdays program was proof of this observation, as the war caused the government to advocate for the cause of less meat as a patriotic duty. Vegetarianism, Christian explained, fueled both the greatest strength and athletic feats, and lowered incidences of infections and cancer. Recipes, including nut and vegetable roasts, were similar to those found in vegetarian restaurants throughout the USA.[18]

Private, non-vegetarian individuals also wrote manuals advising on how to best live a meatless lifestyle once a week, supplementing materials printed by the American government and vegetarians themselves. The contents of these manuals illustrated the concrete effects that vegetarian ideals, identity, and cuisine had on American culture by the start of American involvement in World War I. Alice Bradley, a teacher at Fannie Farmer's famed Miss Farmer's School of Cookery in Boston, published her own meatless and wheatless tract in order to "satisfy the appetites of the family, meet their requirements for nutrition and follow the suggestions of the Food Administration to conserve meat."[19] Bradley explained that meat was expensive and if used at all during the war should be limited to just once a day, on non-meatless days. However, Bradley's implication was clear; during the war it would be most beneficial for individuals and soldiers alike if households cut all meat from their diet. A meat substitutes section closely mimicked those found in vegetarian cookbooks of the early twentieth century, placing vegetable-based quasi-meats as replacements for flesh foods rather than as foods of their own merit. The book included recipes for a soybean loaf, rice and peas, vegetable roast and a rice nut loaf that was already popular with vegetarians beginning in the early twentieth century. Tellingly, the shift toward meatless meals was not targeted at the soldiers themselves, which in some ways would have been a seemingly logical solution to a potential meat shortage. The message was clear: even though the government was advocating meat abstention at home, it was simultaneously endorsing the idea that the war could only be won by beef-fed men fighting abroad. Vegetarian cookbooks provided dietary advice aimed at women, further solidifying and reifying gender roles on the home front.

Ironically, the view pushed by vegetarians reflected though recalibrated the prevailing, normative dietary advice of the time period that emphasized the important role of the domestic sciences and women as the moral guardian of the household, providing sustaining, nourishing, and tasteful meals. The new movement was supported by a proliferation of literature and cookbooks that helped reify gender roles and shape the middle class in America in the early years of the twentieth century.[20] At the same time, Americans by the outbreak of the war were consuming more meat thanks to the simultaneous increase in beef production from the Midwest and the nationalizing of markets thanks to refrigerated car routes found across the country. As explained by one historian of meat consumption in America, by the time of World War I, America had developed into a "beef-eating nation."[21] In the pages of meatless advice manuals, vegetarians sought to exploit America's fascination with the domestic sciences while positioning meatless meals as having similar gustatory and body building properties as meat.

The cookbook *Wheatless and Meatless Days* made the connection between vegetarian victuals and American nationalism explicit through nomenclature, giving meat substitutes patriotic sounding titles. The authors—home economics teachers from a San Diego high school—explained that conservation of food was part of the "battle array" in empowering housewives to help win the war. The "practical self-denial" of meatless days helped ensure success both at home and on the battlefield, "strengthening the arms and hearts" of all Americans. Most importantly, substituting for meat was a sacrifice possible to all who "follow the flag" with sincerity in their hearts.[22]

A section of the book explicitly labeled foods as "meat substitutes," in the same manner as previous years' vegetarian cooking guides. Recipes for common meat substitutes were provided, including a peanut loaf as well as a bean and nut loaf. Less common meat substitutes including a mock crabmeat made of stale bread, mustard, flour, and eggs expanded home cooks' meatless repertoires. However, many of the recipes went further in their appropriation of vegetarian foods for the war effort. Recipes were given explicitly patriotic names, implying that the act of cooking and consuming were rife with political meaning. Combining and baking walnuts, rice, breadcrumbs, cheese, and Worcestershire created a "Liberty Loaf," a meatless meal associated with democratic freedom. Baked beans cooked with breadcrumbs, eggs, ketchup, onion, and a mustard sauce produced a "Navy Loaf" with a "Gunner Sauce" that could make any midshipmen proud.[23] Another cookbook introduced the American public to "Liberty Meat," made of cooked cornmeal, walnuts, and peanut butter fried in oil.[24] Utilizing such names made the point explicit; food had political, social, patriotic, and cultural meaning, a notion that American vegetarians had expressed for nearly 100 years.

During the war years debates over the power of food language were not isolated to the newly rechristened patriotic meatless fare. Food language and choices took on added nationalistic flavor. Sauerkraut, because of its national roots, was renamed "liberty cabbage" in a fit of jingoistic, anti-German sentiment.[25] A similar change occurred with the identity of the hamburger, often referred to instead as a "liberty steak" or "liberty sandwich."[26] Just as vegetarians had done with the marketing of meat substitutes, zealous American nationalists manipulated the nature of food through the power of language. By renaming these products consumers actively manipulated the very identity of the food they ate, disconnecting hamburgers and sauerkraut from their roots that were deemed as socially and politically problematic. Modern society, with its emphasis on empowerment of the self, even allowed individuals to redefine the nature of food products. Food was no longer characterized by its chemical properties, ingredients, or national origin. Rather, it was defined through the preferences of individuals, often guided by marketing, the press, governmental propaganda, or even xenophobic fear. Vegetarians utilized similar methods when advocating for their cause during the war.

The popular press picked up on the association between patriotism and temporary, once-a-week meat abstention, emphasizing that dietary choice helped support American war efforts. On October 13, 1917, a *Chicago Tribune* editorial argued in favor of meatless days by pointing out that throughout history "great wars were won by soldiers who never tasted meat." The author noted that the armies of Cyrus the Great, the ancient Romans, and Napoleon thrived through the sustenance of meatless diets.[27] Two weeks later, on October 29, the paper exhorted its readers to "shun meat tomorrow" to prove their patriotism.[28] The *Tribune*'s domestic culinary guide Jane Eddington—who popularized meat substitutes in the pages of the newspaper as early as 1910—provided readers with regular meatless menus. While the government's meatless advice often included fish and poultry, Eddington advised readers that "the truest meatless meals are those in which a dish of peas or beans is chief." [29]

Legumes were already a preferred meat substitute for vegetarians as early as the turn of the century. Eddington's meatless menus and recipes geared toward patriots continued until March 1918.[30]

The *Tribune* made an explicit connection between Meatless Tuesdays and vegetarianism by suggesting the use of meat substitutes as patriotic and delicious. One article reported that at a downtown luncheon to raise funds for the local woman's committee of the Council of National Defense, a hungry diner experienced a "palpitating moment" as she "drove the carving fork into the breast of what looked like a perfectly good barnyard specimen of roast turkey." However, this was no fowl. Instead, it was a meatless turkey roast made of lima beans, peas, Brazil nuts, eggs, flour, onion, and celery. The mock turkey was described as "absolutely marvelous" and "proved to be of the most delectable flavor, the texture tender and juicy." The roast so accurately approximated a real turkey that it even included imitation white and dark meat, the darker version including grated, rolled wheat to give the food its hue.[31]

The *Tribune* reported on one Great Lakes region restaurant that had gone almost completely flesh free, save for a little bit of lake trout served with lunch. Readers who were afraid that diners might have starved were asked to peruse the day's delectable menu of scrambled eggs and hominy grits for breakfast, bean soup and potatoes for lunch, and baked beans with hot slaw for dinner.[32] Meat consumption was associated with treasonous gluttony, including one restauranteur whose insistence on serving steak caused what the paper described as a "rumpus." The restaurant's owner, Carl Witte, was actually a German-born national, whose menu continued to include dishes such as filet mignon, mutton chops, pork shank, and spareribs. The implication was clear: serving meat on meatless days was an inherently anti-American act.[33]

In contrast, the owner of a restaurant who survived a beating when he refused to serve nine patrons a meat-based meal on a Tuesday was labeled a "patriotic restaurant man" for his decision to stick by his culinary principles.[34] The paper also suggested that Marion F. Sturgis—a housewife who wrote to the paper about her patriotic domesticity—deserved a medal for her ability to make flavorful meatless pies with green tomatoes instead of meat.[35] In another instance, the paper went as far as suggesting that President Wilson sign an executive order calling for all Americans to observe at least one meatless day a week.[36]

The *New York Times* similarly described the meatless program, emphasizing the patriotism attached to weekly meat abstention, labeling the idea a "patriotic rule" soon after its enactment.[37] The *Times* labeled households that did not adhere to Meatless Tuesdays as being "pro-German," proven by families' lack of desire to sacrifice for a greater good. Those who ignored the call for a meatless day who were not treasonous were either "food slackers" or "delinquent Americans," driven by pure selfishness.[38] Even the employed domestic help of New York's elite pledged to do their part and observe meatless days. Henry Physick, J.P. Morgan's butler and a founder of the Butler's Committee in New York, organized to spread the notion of sacrifice throughout the city's domestic help. Physick explained that he was saving food "for our own boys at the front," believing that it was incumbent upon the city's domestics to work for such a goal. Physick pointed out that even if domestics' employers signed pledges to abstain from meat, it would do little good if their employees did not adhere to this policy. The butler viewed his work as thus being "patriotic" and an "honor."[39]

The *Times* reported on the efforts of posh Manhattan hotels to execute meatless Christmas dinners in 1917, despite the fact that the holiday did not even fall on a Tuesday. While some hotels utilized less popular meats such as turtle and guinea hen, others served a full Christmas dinner without an ounce of flesh. The Park Avenue Hotel on 32nd Street and the west side of Fourth Avenue served a Christmas dinner that was "simple…plain, wholesome food" that had been entirely "Hooverized." Instead of flesh, a mock turkey roast known as a "Vermont Turkey" was served along with cream of tomato soup, mashed potatoes, mashed turnips, and a salad of romaine lettuce.[40]

Previous to the Progressive Era vegetarians and their food were mocked by normative society, even demeaned as being anti-American in their supposed ability to build weak individuals. As America entered a world war, vegetarian cuisine was being described as strong and patriotic, some dishes even tied to branches of the military in their name. The path of development that vegetarianism followed from the radical antebellum era politics of groups like the abolitionist-tinged American Vegetarian Society to the twentieth century, muscular vegetarianism of the Progressive Era logically led to the eventual correlation between vegetarianism and military victory.[41]

While the government's program did not explicitly advocate for vegetarianism as an ideologically or ethically driven practice, it did emphasize the use of meat substitutes that vegetarians helped conceive of and popularize. Vegetarians had consumed these meat analogues for nearly two decades and made non-vegetarians aware of these fleshless products. Perhaps, in no small way, vegetarians helped socially prepare Americans and their palates for their brief and limited dalliance with meatless living during the war. The push for Meatless Tuesdays and meatless meals was driven by a logistical need to export meat overseas. However, it was also highly influenced by vegetarian ideals of the time. Vegetarians themselves recalibrated these messages and used the war effort to advocate their dietary cause despite the seemingly disjointed ideals.

Meat substitutes similar to those served in innumerable vegetarian restaurants throughout the USA and prescribed in vegetarian cookbooks and the press were associated with selfless sacrifices that could affect events on battlefields in Europe. Americans were told that their choices in avoiding meat, even just one day a week, had the power to even ultimately win the war. Vegetarianism and vegetarian foods were once associated explicitly with physical and mental weakness by normative American culture in the first half of the nineteenth century.[42] The government's marketing of Meatless Tuesdays and the fact that significant portions of the public responded positively illustrate that by 1917 vegetarianism—at least as a diet—was viewed largely in an opposite, positive way.[43]

However, these changes highlight serious tensions, internal debates, and even contradictions that defined the vegetarian movement as it entered the modern age. While the embrace of Meatless Tuesdays represented a moment in time, it was reflective of larger desires of the movement during the era: commercialization, social cache, and even patriotism. The shift clearly paid dividends for the vegetarian movement in terms of its popular appeal. However, the nature of the group changed significantly from its pacifistic roots, even going so far as embracing vegetarianism as a way to support the war cause. Could vegetarianism—a diet based in avoiding death and bloodshed—reasonably be used to support the carnage wrought by war?

One group of vegetarians outside of the mainstream of the movement answered this important question about the possibility of vegetarianism supporting violence with a definitive no. As movement vegetarians debated about how to best support the war effort—either through vegetarian fighting men or through meatless meals—for the first time, some Americans attempted to utilize their diet as grounds to serve as conscientious objectors. A group known as the "Federation of Humano-Vegetarians" requested recognition to avoid war service on the grounds that "vegetarians, reaffirming our faith in the Universal Kinship of the 'Animal Kingdom' and the 'Brotherhood of Man,' adhere in our allegiance to the elementary human commandment, 'Thou Shalt Not Kill.' "[44] Appeals for recognition, however, did not seem to take hold with the US government. One member of the Federation of Humano-Vegetarians by the name of Jacob W. Rose arrived at Camp Meade in Maryland on November 4, 1917, and immediately began a hunger strike to protest his enlistment. After 25 days, Rose was sent to the base hospital for treatment from the ill effects sustained throughout his protests.[45]

The differences in response between movement vegetarians and fringe groups like the Humano-Vegetarians point toward an inherent tension within American vegetarian practices at the time. Ideologies can be morphed and utilized by a variety of groups to advocate for a wide spectrum of causes. Remarkably, vegetarianism's dominant narrative during the war years was the utilization of meatless dietetics in order to support violence. Those who fell from this narrative—such as Jacob Rose—received scant attention from the movement. The Humano-Vegetarians, for example, were not mentioned at all in *The Vegetarian Magazine*, the official publication of the Vegetarian Society of America.

Movement vegetarians entered into the years of the first world intent on growing the diet's popularity and in the process ignored and distanced themselves from the actions of vegetarian pacifists and conscientious objectors. The result was a more popular and culturally lauded form of vegetarianism. However, ideologically, vegetarianism had turned its back on a long-standing, basic principle in its American history. Pacifists no longer, movement vegetarians received praise during the war years. One must wonder though, at what ideological cost?

Debates surrounding the relationship between meat abstention and warfare have continued well past the World War I years and into the present era, as recently as the American invasion of Iraq. For some American soldiers fighting in Iraq, the question focused less on the ethical implications of war and more on logistical concerns of the availability of meatless meals. As explained by 14-year vegetarian, US Army Staff Sgt. Liza Reiter, while stationed at Al Asad Air Base in Iraq in 2013, trips to the mess hall would sometimes induce "tears" as she took away an empty plate after noticing only meat-filled meals. Reiter fell into a trap of eating junk food, French fries, and grilled cheese sandwiches due to the lack of options, and reported a resulting increase in mood swings and exhaustion as a result.

Interestingly, the lack of vegetarian options reflected the *longue durée* of the Iraq War. As a combat soldier, Reiter previously had access to Meals, Ready to Eat (MREs) while fighting in Iraq, individual field rations that provided vegetarian options, including vegetable tortellini and veggie burgers. As the American military presence in Iraq became more permanent and intractable, stationed troops began receiving hot meals rather than MREs. For most troops, this was a welcome development. As a result, Reiter's complaints over ground beef in pasta sauce or greens cooked with bacon went largely ignored. Tellingly, Reiter expressed no angst over the potential contradictions between her non-violent diet and role as a member of the military, even thinking wistfully back to her more active combat days because of the better vegetarian options.[46]

Reiter's story unwittingly highlights inherent tensions within modern vegetarianism. As the movement has grown and gained increased public support in the USA, the diet has also become somewhat dislocated from more communal political and ethical concerns. A 2008 survey conducted by *Vegetarian Times*—the largest and oldest print publication of vegetarian news, articles, and recipes—revealed that of the 7.3 million Americans who follow a "vegetarian-based diet," 53% listed personal health as a main motivator. While a similar number listed animal welfare as a motivator as well, the overall results pointed toward an association between the importance of vegetarianism for the individual, rather than vegetarianism and its relationship to society at large (other popular responses included "personal wellness," "food safety concerns," and "weight loss").[47] Reiter personally distanced herself from ethical vegetarianism, explaining that her vegetarianism started at a young age because of "propaganda" spread by animal rights organizations, admitting that her dietary choices eventually became focused more on reasons of personal health rather than ethics.[48] In a society where vegetarianism seems to be loosening from its primary focus on political and ethical concerns, perhaps the seemingly disjointed nature of non-violent dietetics with military violence is not as surprising as one might believe at first glance.

But how does this very modern debate tie together with the questions surrounding vegetarianism during World War I? The question of the relationship between violence and warfare faced vegetarians with the rise of American participation in the war and the eventual proliferation and lauding of Meatless Tuesdays simultaneously by the federal government and private citizens—vegetarians included. Rather than protest the connection between vegetarianism and the war effort, vegetarians embraced the idea, proposing their diet as a patriotic act that could help win the war. Vegetarians at the time understood that the war years provided an opportunity to prove that the diet was no radical proposition, but rather in service of supporting the war cause.[49] Ironically, in the process, vegetarians supported a program that aimed to provide beef to fighting males serving in Europe, contributing to a historical continuity that connected meat consumption with strength, virility, and power.[50]

As a result, the World War I years represented a real breaking point for American movement vegetarianism, a true split from its pacifistic past toward a more politically accommodating future. On the one hand, this new movement gained in its number of followers and popular reputation. However, the new vegetarianism also opened up new opportunities for the diet to be embraced in many different social circles. Nearly a century following the embrace of patriotic militarism by American vegetarians, a far less political vegetarianism has found itself practiced even by members of the American military, with little consideration to the seemingly inherent ethical contradictions at play between dietary and ethical choices. Thus a historical irony was unwittingly produced in 1918 with the launching of Meatless Tuesdays and the corresponding response crafted by American movement vegetarians. A program that aimed to increase meat consumption amongst soldiers serving abroad helped give birth to a new vegetarianism that embraced patriotism and its ties to military service. In the process, Meatless Tuesdays helped separate vegetarianism in America from its pacifistic past. The long-term results also produced the seemingly contradictory marriage between non-violent diets and violent military actions.

Notes

1. Bernarr Macfadden, "What We Need to Win This War," *Physical Culture* 38, no. 3 (September 1917): 1.
2. Melville Durant, "Are Vegetarians Good Fighters," *Physical Culture* 38, no. 3 (August 1917): 90. There is no other record of Durant contributing to *Physical Culture* and no other trace of him within the historical record. It is possible that it was Macfadden writing under a pen name, or a member of his regular writing staff, a practice that occurred frequently throughout the history of *Physical Culture* to make the publication seem more diverse in its contributors.
3. Adam D. Shprintzen, *The Vegetarian Crusade: The Rise of an American Reform Movement, 1817–1921* (Chapel Hill: University of North Carolina Press, 2013), especially Chaps. 6 and 7.
4. Julian E. Zelizer, *Arsenal of Democracy: The Politics of National Security—From World War I to the War on Terror* (New York: Basic Books, 2009), 28; George H. Nash, *The Life of Herbert Hoover: Master of Emergencies, 1917–1918* (New York: W.W. Norton & Company, 1996), 229.
5. For a full copy of Wilson's presidential proclamation, see "President's Proclamation Calling on Citizens for Meatless, Porkless and Wheatless Periods," *New York Times*, Jan. 27, 1918, pg. 2.
6. William Clinton Mullendore, *History of the United States Food Administration* (Palo Alto, CA: Stanford University Press, 1941), 115.
7. United States Food Administration, *War Economy in Food* (Washington, D.C.: Government Printing Office, 1918), 10.
8. Harold L. Platt, *The Electric City: Energy and the Growth of the Chicago Area, 1880–1930* (Chicago: University of Chicago Press, 1991), 225.
9. Ibid., 11.
10. Shprintzen, 130–9.
11. Christopher Joseph Nicodemus Capozzola, *Uncle Sam Wants You: World War I and the Making of the Modern American Citizen* (New York: Oxford University Press, 2008), 94.
12. On the debate over dairy amongst early American vegetarians, see Shprintzen, 64. The USA actually began shipping condensed milk during the World War I years because of a surplus in America. See Debroah Valenze, *Milk: A Global History* (New Haven: Yale University Press, 2011), 188.

13. *Win the War Cookbook* (St. Louis: St. Louis County Unit, Woman's Committee, Council of National Defense, 1918), 76, 106, 139.
14. The recipe can be found as a vegetarian dish in cookbooks such as Francis Emugene Owens, *Mrs. Owens' New Cook Book and Complete Household Manual* (Chicago: Owens' Publishing Company, 1899), 668; Edward E. Howe, *Vegetarian Cook Book* (New York: Squire Publishing, 1887), 74; Sarah Tyson Rorer, *Mrs. Rorer's Vegetable Cookery and Meat Substitutes*, (Philadelphia: Arnold & Company, 1909), 214.
15. "The Festival," *American Vegetarian and Health Journal* 1, no. 10 (October 1851): 176.
16. *Patriotic Cook Book* (Cleveland: Mayor's Advisory War Board, 1918), 1, 27–8.
17. Twentieth Century Club of Pittsburgh, *Twentieth Century Club War Time Cook Book* (Pittsburgh: Pierpont, Siviter & Co., 1918), 47.
18. Eugene Christian, *Meatless and Wheatless Menus* (New York: Alfred A. Knopf, 1917), 5–7, 40–3, 125–6.
19. On Farmer's school, see Laura Shapiro, *Perfection Salad: Women and Cooking at the Turn of the Century* (Berkeley, CA: University of California Press, 2008), 112–5; Alice Bradley, *Wheatless and Meatless Menus and Recipes* (Boston: Miss Farmer's School of Cookery, 1918), i.
20. On the domestic sciences, see Laura Shapiro, *Perfection Salad: Women and Cooking at the Turn of the Century* (New York: Farrar, Straus & Giroux, 2001); Charlotte Biltekoff, *Eating Right in America: The Cultural Politics of Food and Health* (Durham: Duke University Press, 2013), 13–44; and the introduction and first chapter of Megan Elias, *Stir it Up: Home Economics in American Culture* (Philadelphia: University of Pennsylvania Press, 2008).
21. *Putting Meat on the American Table: Taste, Technology and Transformation* (Baltimore: Johns Hopkins University Press, 2006), 32.
22. Hester Martha Conklin and Pauline Dunwell Partridge, *Wheatless and Meatless Days* (New York: D. Appleton and Company, 1918), vii–viii.
23. Ibid., 169–88. Other cookbooks of the time that emphasized meatless living through a patriotic lens include Mary Elizabeth Evans, *Mary Elizabeth's War Time Recipes* (New York: Frederick A. Stokes Co., 1918), and Amelia Doddridge, *Liberty Recipes* (Cincinnati: Stewart & Kidd Company, 1918).
24. *Twentieth Century Club War Time Cook Book*, 44.
25. John Higham, *Strangers in the Land: Patterns of American Nativism, 1860–1925* (Piscataway, NJ: Rutgers University Press, 2002), 208. For

an example of the use of the term "liberty cabbage" in a recipe, see Marion Harris Neil, *The Thrift Cook Book* (Philadelphia: David Mckay Publishers, 1919), 109.

26. David M. Kennedy, *Over Here: The First World War and American Society* (New York: Oxford University Press, 2004), 68; Michael McGerr, *A Fierce Discontent: The Rise and Fall of the Progressive Movement in America, 1870–1920* (New York: Oxford University Press, 2005), 292.
27. Jean Roberts Albert, "A Vegetarian Diet," *Chicago Daily Tribune*, Oct. 13, 1917, pg. 6.
28. "Patriot? Shun Meat Tomorrow," *Chicago Daily Tribune*, Oct. 29, 1917, pg. 1.
29. For examples, see Jane Eddington, "Tribune Cook Book: Many Meatless Menus," *Chicago Daily Tribune*, Nov. 6, 1917, pg. 1917; Jane Eddington, "Tribune Cook Book," *Chicago Daily Tribune*, Nov. 13, 1917, pg. 14; Jane Eddington, "Cream of Carrot Soup," *Chicago Daily Tribune*, Nov. 27, 1917, pg. 14.
30. Jane Eddington, "Tribune Cook Book: This is Meatless Day!" *Chicago Daily Tribune* Mar. 19, 1918, pg. 12.
31. "Have You Ever Dined Upon the Mock Turkey," *Chicago Daily Tribune*, Feb. 20, 1918, pg. 3.
32. "Food, Food, Food, At Great Lakes 'Meatless Day,'" *Chicago Daily Tribune*, Nov. 7, 1917, pg. 7.
33. "Raids Uncover Food Violations on South Side," *Chicago Daily Tribune*, Nov. 29, 1917, pg. 17.
34. "Observance of Meatless Day Causes Beating," *Chicago Daily Tribune*, Dec. 12, 1917, pg. 11.
35. "Quick! A Medal for Mrs. Sturgis," *Chicago Daily Tribune* Dec. 23, 1917, pg. D4.
36. "Meatless Day for Everybody," *Chicago Daily Tribune*, Nov. 21, 1917, pg. 8.
37. "Beefless Day Effective: Hotels in New York and Throughout Country Keep Patriotic Rule," *New York Times*, Oct. 10, 1917, pg. 4.
38. "Extending Mr. Hoover's Powers," *New York Times*, Jan. 17, 1918, pg. 12.
39. "Servants in Big Houses Join For Saving," *New York Times*, February 10, 1918, pg. 44.
40. "No Meat Christmas in the Big Hotels," *New York Times*, Dec. 16, 1917, pg. 28. Vermont turkey as a mock turkey is referenced in "Have You Ever Dined Upon the Mock Turkey," *Chicago Daily Tribune*, Feb. 20, 1918, pg. 3. On the Park Avenue hotel, see Sarah Bradford Landau, Carl

W. Condit, *Rise of the New York Skyscraper: 1865–1913* (New Haven: Yale University Press, 1999), 60.

41. The phrase "muscular vegetarianism" was first coined in studying these vegetarians by James C. Whorton in his essay "Muscular Vegetarianism; The Debate over Diet and Athletic Performance in the Progressive Era," *Sport and Exercise Science: Essays in the History of Sports Medicine*, ed. Jack W. Berryman, Roberta J. Park (Urbana: University of Illinois Press, 1992), 303–15.
42. Shprintzen, 94–114.
43. The United States Food Administration reported that Meatless Tuesdays helped increase the amount of meat shipped to soldiers abroad by over 844 million pounds from fiscal year 1916–1917 to 1917–1918. See Mullendore, 118.
44. Quoted in Max Davis, *The Case for the Vegetarian Conscientious Objector* (Brooklyn, NY: Tostoy Peace Group, 1944), 13. The group is also mentioned in Douglas Goldring, *The Nineteen Twenties: A General Survey and Some Personal Memories* (Longon: Nicholson and Watson, 1945), 140.
45. See "Jacob W. Rose," *World War I Conscientious Objectors Database.* Accessed October 15, 2015. http://fm12.swarthmore.edu/detail.php?dbName=WWI&rid=3393.
46. Kelly Kennedy, "Vegetarians find few choices in mess hall," *Army Times.* March 6, 2013. Accessed December 22, 2015. http://archive.armytimes.com/article/20130306/OFFDUTY03/303060313/Vegetarians-find-few-choices-mess-hall.
47. "Vegetarianism in America," *Vegetarian Times.* Accessed December 22, 2015. http://www.vegetariantimes.com/article/vegetarianism-in-america/.
48. Kennedy, "Vegetarians find few choices in mess hall."
49. Vegetarians were operating during war years that were particularly dangerous and fraught with difficulty for dissenters and political radicals in America, including socialists and anarchists. See Philip S. Foner, *Labor and World War I: 1914–1918*, vol. 7, History of the Labor Movement in the United States (New York: International Publ., 1987), 296–7.
50. On vegetarian physical culture, which stood at the heart of this new movement vegetarianism that emphasized strength, see Shprintzen, 183–203 and Whorton, 303–15.

Veganism and the Politics of Nostalgia

Jessica Carey

To write about food-related nostalgia is to wade into a cultural ocean of common sense, cliché, and iconic narrative: Proust's madeleines, of course; Seders; soul food; mom's cooking; any and all gustatory lifelines to what was once home. Food and memory are, after all, inextricable, even across species lines; most sentient beings actively and continuously need to inhabit the intersection of food and memory in order to survive at all. In this chapter I will compare two different modes of political survival and their reliance on collective memories of food: on the one hand, the corporate food system makes heavy use of nostalgic advertising to keep business profitable, and, on the other, burgeoning vegan foodways are turning to food stories in order to create a sense of community and shared identity. As usual, the political survival of systems and stories has biopolitical consequences: the lives of billions of animals

J. Carey (✉)
Literary and Cultural Studies, Sheridan College, Oakville, ON, Canada
e-mail: jlw.carey@gmail.com

J. Castricano, R.R. Simonsen (eds.), *Critical Perspectives on Veganism*,
The Palgrave Macmillan Animal Ethics Series,
DOI 10.1007/978-3-319-33419-6_11

are at stake in the narratives we choose to live by and the infrastructure we build on those narratives. Given this fact, we need a more robust account of the food memories we are currently being invited to identify with. It is becoming increasingly clear that veganism not only needs to persuade rationally or make ethical appeals—it also needs to tell stories and make memories.

Collective memory and cultural memory are terms theorists often use to denote shared narratives about the past, which not only live in individual minds, but are also reproduced continually in the whole field of discourse.[1] These shared stories help give shape to a group's sense of collective identity, and are thus often a highly politically charged form of memory practice. Many theorists note that cultural memory so often says more about the political needs of the present than about the objective facts of the past. In his seminal work on memory studies, *How Societies Remember*, Paul Connerton notes that "images of the past commonly legitimate a present social order."[2] Literary theorist Edward Said develops this idea further in "Invention, Memory, and Place," arguing that "the processes of memory are frequently, if not always, manipulated and intervened in for sometimes urgent purposes in the present."[3] In other words, what is emphasized, de-emphasized, included, or left out in the necessarily selective account of a shared past so often reflects what is going on right now—the stories we need to hear right now, or the stories that people vying for political dominance want us to hear right now, for reasons that suit their interests.

Yet what is it about nostalgia in particular, as a common form of collective memory that stimulates such desire in people, such yearning? The most common definition of nostalgia involves a kind of temporal homesickness—a form of remembrance that is imbued with pain, pain stemming from the irretrievability of the past. Nostalgia, in other words, is memory structured by desire, desire for something none of us can physically grasp: the past, in either its real or illusory forms. The mapping of desire onto memory makes nostalgia one of the most potent and politically useful forms of collective memory. Because of nostalgia's profound shaping power, it makes sense that it is often deployed in contexts of instability, whether social, economic, political, or ecological. The field of marketing research is well aware of this function of nostalgia. As Vignolles and Pichon summarize such research, "Nostalgia strengthens and augments selfhood; it reduces sources of uncertainty, increases one's ability to deal with the present and restores self-worth by resorting, at least momentarily, to an idealized past."[4] Yet nostalgia can empower both the individual and the collective, separately and in concert: if nostalgia can be marshaled in a particular direction, toward a shared—if currently unrequited—desire for a specific version of the past, then people can be more easily unified under a common political sign, a common political identity and purpose. Of course, there are many fascistic examples of nostalgia's deployment throughout history; the promise of a return to a golden past has served as the fuel for some of history's most deadly political movements. At other moments, this instrument of collective memory has served a decolonizing function—a crucial tool of imagination for oppressed communities whose narratives of the past may have been partially erased or overwritten by a colonizing force. The particular deployment of nostalgia that I am focusing on in this chapter concerns the popular discourse on food in North America, which is nothing if not politically contentious. As anyone who pays attention to food culture can see, contemporary food discourse involves a multitude of passionate perspectives, identities, and principles, rendering it just the kind of unstable social field in which nostalgic cultural memory might be expected to play a key role.

Post-Pink-Slime Food Advertising

Almost as self-evident as the connection between memory and food is the synergy between nostalgia and advertising. Nostalgia's infamous combination of memory and pain is ambrosia for advertising, as it naturalizes a sense of pleasurable yearning rooted in pain. In this mode, advertising functions as a kind of *pharmakon*, the ambiguous ancient Greek conflation of poison and cure that Derrida deconstructs in *Plato's Pharmacy*. Indeed, this kind of advertising introduces the pain and at once promises its alleviation through an act of consumption. Corporate food advertisers are eager to marshal the desires of their audience in a particular direction—namely, toward the idea that the goodness of the past is rooted in traditional, conventional relationships between people and food. Grandmas cook with ground beef; kids drink milk; dads farm our food. This co-articulation of nostalgia, people, and food accomplishes two objectives at once for major food industries and corporations: first, such advertising enables them to overwrite the common narrative that they have ruined traditional foodways through intense industrialization, instead positioning themselves as the trusted guardians of those foodways. Second, through such ads they are able to semiotically link particular foodstuffs—such as meat, eggs, and milk—with goodness, innocence, and positive childhood memories, which implicitly positions food movements advocating for more plant-based diets outside or apart from these virtues and experiences.

The advertisements upon which I base this analysis have all been shown in heavy rotation on major Canadian network television channels, over various periods during the past three years. They are each also currently available for viewing online, either on YouTube or on the websites of the corporations and promotional boards that produced them.[5] The first advertisement is produced by Dairy Farmers of Canada, which, according to their website, is "the national policy, lobbying and promotional organization representing Canada's farmers living on approximately 12,000 dairy farms."[6] This ad follows Brian, a Canadian dairy farmer, showing scenes of his life on the farm spanning from childhood to later life, when he has passed on the family business to his children. Accompanying these scenes is a gentle piano score and a voiceover: "Canadian dairy farmers dedicate their lives to producing milk of the highest quality."[7] The second advertisement is produced by Egg Farmers of Canada, the trade name of the Canadian Egg Marketing Agency, which "manage[s] the national egg supply and promote[s] egg consumption while representing the interests of over 1,000 regulated egg farmers."[8] In this ad, egg farmer Laurent Souligny reminisces over scenes of his family preparing and eating eggs: "Eggs have always been a natural source of protein and nutrients. Ever since our family's been egg farming, I've seen a radio change to a television, a steam engine to a jet plane, a frozen pond to an arena. But in all that time, eggs haven't changed at all. Yes, they've always been naturally good, just the way they are."[9] The third advertisement is part of a series of ads produced by McDonald's Canada, each dedicated to crafting a narrative about a particular ingredient used at McDonald's: potatoes, eggs, and beef. In the ad I examine here, scenes of people preparing and eating beef in various settings is accompanied by this voiceover:

> What type of beef is a McDonald's burger made with? It's the beef you get at your local butcher. It's the type of beef your grandma built a reputation on. The kind reserved for long weekends. It makes men act like men, and gets everyone to the table early. That's right: every McDonald's burger has no additives or fillers, just one hundred percent pure Canadian beef. That's what we're made of.[10]

The final advertisements I am interested in here are part of a series of nostalgic ads produced by Milk Every Moment, a promotional partnership between various milk promotion associations in Canada, including Dairy Farmers of Canada.[11] The ads in this campaign include a series of scenes depicting children playing indoors, often doing playfully strange things such as tying helium balloons to strands of hair, or walking backwards while playing dress-up. One of the most heavily played ads in the campaign has no voiceover, just the scenes of childhood play ending with a caption: "Not everything we did when we were kids made sense. But drinking milk did."[12] All of the ads include some version of this slogan, with some of the ads shaping the nostalgic bent of the campaign more explicitly through use of a voiceover: "logic didn't always rule our lives; our decisions weren't fueled by need, they were driven by want"[13]—"there was a time when we did whatever we wanted, it didn't matter how we looked, all that mattered was how we felt"[14]—"We couldn't be embarrassed any more than we could grow a mustache or tell time".[15] Clearly, all of these ads are baldly nostalgia-driven, attempting to tap into a shared, positive remembrance of consuming animal products, anchored in a romanticized vision of the past.

It makes sense that food advertisers would seek to endow their products with positive associations: of course, it is literally their job to stoke consumer desire for their products. What I find fascinating about this relentless corporate and industry appeal to nostalgia, however, is its unique role in the food politics of the present moment. The meat, dairy, and egg industries of North America, as well as the fast food corporations that make heavy use of these industries, have faced unprecedented public scrutiny in the past 20 years or so: intensive industrial methods of raising and killing animals are becoming increasingly ethically dubious for many members of the public, the animal rights movement has slowly become more mainstream, and scandals and crises from *E. coli* to pink slime have undoubtedly contributed to the growing popularity of various grassroots healthy food movements such as locavorism, whole foods diets, and veganism. Given these political realities, these advertisements I have described seem to offer a depoliticized, soothing alternative imaginary, grounded in nostalgic remembrance of the positive associations that we ostensibly all share regarding fast food, meat, eggs, and dairy. Forget all that upsetting stuff you've heard, they seem to be saying, and rest easy in the fact that eggs haven't changed, milk hasn't changed, beef hasn't changed, and McDonald's hasn't changed. The only thing that has changed is you: you've gotten older, so grab onto the object that has remained constant and let it transport you back in time, if only while you eat it.

These are also representations in which the connotations between food and memory seem to be somewhat overdetermined. In the Dairy Farmers of Canada ad, for instance, Brian the dairy farmer is positioned as a kind of twentieth-century everyman whose whole life has been dedicated to producing milk; thus, the milestones of a late-twentieth-century life—helping one's dad; experiencing teenage moments both of intense yearning for freedom and of a grown-up transition into sacrificing fleeting pleasures for one's longer-term work commitments (both experiences represented by a co-ed group of friends tearing by the farm in a Cadillac, tantalizingly yelling "Hey Brian!"); helping one's own children; coming in for dinner at the end of a long life, job well done—all become connotatively inextricable from the production of dairy products. In the Egg Farmers of Canada ad, the elderly egg farmer frames our shared cultural memory in very broad terms—the shift from radio to television, steam engine to jet plane. It is difficult not to feel included in this field of collective memory because it encompasses society-wide technological shifts that all viewers have experienced or are only one or two generations away from. Against this backdrop, the egg is presented as the one constant to cling to. Perhaps the most explicit and specific links between product and nostalgic recollection are found in the McDonald's ads and the Milk Every Moment campaign. I would argue that the McDonald's ads answering what kind of beef and what kind of eggs the company uses are clearly intended to respond to widespread public controversies regarding the quality of McDonald's ingredients, including, but not limited to, the pink slime PR disaster of 2012, in which an unappetizing photo of chemically treated, highly processed meat-based binder went viral and was popularly linked to industrialized fast food products.[16] The rhetorical framing of these McDonald's ads is one in which connotations are dictated rather than left open to debate. By stating "It's the type of beef your grandma built a reputation on," a collective memory is powerfully constructed in which we are all presumed to know and identify with this kind of grandma, and claim her as our own. The statement "It makes men act like men" is, likewise, a rhetorical call to identify with what Jacques Derrida would call a "carnophallogocentric" vision of masculinity, one in which human subjectivity founds itself in a "virile" effort to distinguish oneself from other animals through a rational and visceral commitment to dominating them.[17] Issuing such a call in this commercial is almost brazen in its erasure of ongoing, broad cultural efforts to deconstruct normative markers of masculinity. Likewise, in the Milk Every Moment campaign, the rhetorical framing strictly defines for us, in advance, a narrative of middle-class childhood experience that "we" are all presumed to share. There is an absolute, almost dictatorial presumption in the voiceovers: this was the way you experienced childhood. Questions about whether every child really did make all their decisions based on wants, not needs, or whether every child really did whatever they wanted or was incapable of feeling embarrassed are foreclosed in advance by the grammar of the script, culminating in the declaration that drinking milk made sense then, and still does now.

Throughout all of these ads, we are prescriptively called upon to share a specific memory of the past, to which certain food items can serve as a portal. Nostalgia becomes a protective salve against the nasty political squabbles of food's present, and a tool of re-enchantment for our memories and our eating experiences. Nostalgia is deployed to these ends in a way that is naturalized, universalized, and more or less rendered rhetorically immune to critical intervention. Here, industries and corporations use nostalgia as an instrument for closing ranks around a specific version of the past, one that may be pleasurably indulged by maintaining their market share in the present.

Playing with Nostalgia in Veganism

How does the nostalgia found in vegan cookbooks compare with the nostalgia in mass-market food advertising? Nostalgia is in fact a considerable force in vegan culture, and just like corporate food advertisers, vegan cookbooks sometimes seem to use nostalgia as a tool to shift food discourse away from negative political connotations in the present. The vegan community, after all, has its own public relations problems, having to do with persistent popular perceptions of veganism as a willfully divisive, politically alienating lifestyle, whose militancy tends to run roughshod over family bonds and friendly civility. Against such popular narratives of the society-shunning vegan wedded to deprivation and isolated austerity, the highest-selling vegan cookbooks are clearly invested in opening up positive associations between vegan food and collective memory. Sometimes, this effort takes the form of embracing a mid-twentieth-century kitschy aesthetic, often including a fetishization of greasy spoon diners and diner culture, though I would argue that this aesthetic approach may soon wane, having completed its arc into the hipster-inflected mainstream. In other instances, vegan cookbook authors strive to situate veganism firmly within narratives of decolonization and resistance, linking vegan practice with collective memories of other struggles against injustice as well as with vegetable-based foodways that are threatened by industrial food production. For instance, in the introduction to his cookbook, *Vegan Soul Kitchen*, vegan chef Bryant Terry discusses African-American collective memories of collard greens as a jumping off point for his work on the book, which he describes as "a succulent gumbo filled with accounts of my life, recipes, and historical notes on what I broadly define as Afro-Diasporic cuisine."[18] However, the cookbooks I will examine in the remainder of this chapter are two of the nine authored or coauthored by Isa Chandra Moskowitz, one of the most popular vegan cookbook authors in the USA since her debut cookbook *Vegan with a Vengeance*, published in 2005. I will focus on the appeals to nostalgia in that cookbook and in her most recent cookbook, *Isa Does It*, published in 2013.

Moskowitz is obviously attuned to the image problem faced by veganism, and she has consistently worked to frame vegan cooking in a positive light. She is the proponent of an approach to vegan advocacy she calls "baketivism," which involves winning people over to vegan food through positive eating experiences; she believes that "tastebuds are the perfect vehicle for change."[19] The anecdotes and recipe headnotes I will discuss here, however, suggest that Moskowitz is interested not only in convincing readers that vegan food tastes good, but also in the fact that it can serve all the other affective functions that we so commonly associate with food, including positive collective memories. Yet, unlike the television ads I have analyzed above, Moskowitz tells nostalgic food stories in a less prescriptive, more open-ended way, that not only multiplies the potential objects and avenues of food nostalgia, but also plays with the notion of nostalgic collective memory itself, accentuating in multiple instances the inherent creativity and constructedness of all collective memory.

First of all, Moskowitz works to open up the field of possible occasions for food nostalgia. Sometimes she accomplishes this by alluding to non-mainstream, non-normative food memories that are, nonetheless, rich with recollections of friends, family bonding, youthful versions of self, and cultural experiences worth treasuring. For instance, in *Isa Does It*, Moskowitz fondly recalls learning to cook:

> So this is where my love of cooking began—in 1989, in a small kitchen in Sheepshead Bay, Brooklyn, with linoleum floors and fluorescent lighting… My best friend would come over and we'd get out all the pots and pans, blast the music, crank up the burners, and have at it. For almost any occasion. I can recall a gigantic Thanksgiving spread with ten different kinds of tofu for every course and strawberries way out of season. A Chinese-inspired Christmas buffet with spring rolls that needed ten layers of brown paper shopping bags to absorb the oil. Everything wasn't always a success, but most of it was, and even if we failed we had fun.[20]

Here, the nostalgia does not rely upon a normatively mainstream sense of sanitized or idyllic innocence, but rather upon an exuberant collective effort pursued in the company of Moskowitz's friends, siblings, and single mom. Significantly, unlike the television ads, she does not presume that her readers' pasts necessarily share all of the particulars, but through sharing her personal historical details she invites, rather than assumes, nostalgic identification on a more general plane: relationships, trial-and-error cooking, celebration. In the process, she demonstrates that making the shift to vegan food does not at all need to mean forgoing the creation or nurturing of nostalgic food memories.

Along these lines, Moskowitz also opens up the field of possible occasions for food nostalgia by loosening the connotative links between particular foods and collective memory. Food objects often still serve as a mnemonic link between the present and the past in her books, but they need not be made of the same exact ingredient or enjoyed in the same exact place to be nostalgically meaningful. By loosening these ties somewhat, Moskowitz rewrites the narrative of deprivation so often attached to veganism, for example, noting, "I do get a little misty eyed when I pass a deli and can't pay a visit to my lost love Reuben but sometimes he is waiting for me at home."[21] Here, Moskowitz emphasizes the affective and pleasurable dimensions of eating as a continuum, between the remembered meals and the present vegan recipe. In a longer example, Moskowitz traces a long, unbroken line of cherished memories related to meatballs:

> If ever there was a dish that screams "family," this is it. Whether it be the family that you were born into, or your six roommates in a Brooklyn loft with no heat, spaghetti and meatballs is what brings everyone together. So many of my best memories revolve around a big juicy meatball and lots of slurpy, garlicky marinara. I picture my grandma's dinner table with a big platter of her meatballs right in the centre of it. There were always a few burned ones that everyone tried to grab first…And later, when I went vegetarian, I remember cooking tofu balls and spaghetti with my mom and sis. Even decades later, when I lived in the aforementioned heatless loft, every Sunday night we'd watch The Sopranos and eat spaghetti and meatballs made from some store-bought soy sausage stuff.
>
> Well, this recipe is none of those exactly, but it draws on all of my spaghetti-and-meatballs memories…A few condiments and pantry spices give me the childhood flavors that I crave. Definitely double this recipe for family occasions, and don't forget to burn a few—those are always the best loved.[22]

Food-related nostalgia becomes redefined in this passage as a form of collective memory that can transcend the vegan/non-vegan political divide. What remains important are the familial and family bonds; the ritual of making and eating a dish made of noodles and ball-shaped protein; the replication of childhood flavors using condiments and spices; and the shared, pleasurable embrace of imperfection. By locating other objects for nostalgic pleasure beyond the presence of beef in the meatballs, and thus implicitly rewriting McDonald's assumption that one ingredient, beef, is the primary referent for all the nostalgia she carries about spaghetti and meatballs, Moskowitz further discursively breaks down the presumed biopolitical barrier between pleasure and veganism.

Yet I want to point also to the ways that Moskowitz's refiguring of food nostalgia is not limited to a claim like "you don't need meat to make good memories," or "you don't need to give up collective memories of non-vegan food when you go vegan"—claims that may only go so far with a skeptical audience. Crucially, she also deploys food nostalgia in ways that encourage critical reflection on the uses and qualities of nostalgia itself, while still indulging in its pleasures. One way Moskowitz invokes such reflection is by calling attention to the constructed and mediated nature of memory and nostalgia that Said and other academics have long noted. For instance, in *Isa Does It*, Moskowitz writes: "I love a classic borscht. It automatically brings to mind my ancestors in Mother Russia, hovered over a wood-burning stove making beet soup. Or maybe that was a scene from *Fiddler on the Roof*? In any case, I have no desire whatsoever to update that mental image. Borscht speaks to my soul, and I like mine chunky with beets (obviously), cabbage, and potatoes."[23] Here, Moskowitz both acknowledges the culturally constructed quality of her narrative of the past and asserts its importance to her sense of self, or her soul, as she puts it. Narrow notions of "authenticity" are thus conceptually decoupled from nostalgia, and not for the last time in Moskowitz's work. Moskowitz even sheds more light on the constructed, yet very real pull of food nostalgia in her recipe headnote for molasses cookies in *Vegan with a Vengeance*: "Maybe these chewy, spicy sweet cookies are good any time of the year but I like to save them for the holidays. They make me nostalgic for the snowy winter days at grandma's cabin, sitting around the fireplace in our flannels singing…OK, fine, my grandma didn't have a cabin. There was no fireplace, no flannels, but these cookies sure make me wish there had been."[24] Here, the cookies serve as the agent of a remembered time that didn't even exist for Moskowitz, but that she, nonetheless, recognizes and yearns for. Again, rhetorically we are invited to indulge in the bittersweet pleasure of nostalgia for its own sake, while consciously recognizing this form of collective memory as a cultural construct, regardless of our personal identification with a Rockwellian scene of the sort that is simply presumed rather than questioned in the ads for meat, milk, and eggs.

In other words, it is not just that what we are really nostalgic for is family, relationships, and happy experiences of our pasts. In the vegan cookbooks I'm examining here, nostalgia is a potentially pleasurable yet hardly infallible exercise in imagination, a playful yet meaningful means of momentarily experiencing connections with other times, places, and people. This form of imagination can, in many other kinds of food media, lead to cultural appropriation: a confidence that one has achieved a full connection and identification with another cultural group by cooking or eating meals one deems authentic to that group. Plenty of insightful and crucial work in food studies continues to examine the link between food and cultural appropriation, and for good reason: it is a pervasive problem in popular food cultures. For Moskowitz, however, while the food serves as the connotative invitation to this and other kinds of connection, she makes no claim to cultural authenticity, and, in fact, several times in her cookbooks, explicitly eschews any notion that she is providing a kind of authentic connection to other cultural groups through food. As she notes in her headnote for Brooklyn Pad Thai, "Is it authentic? Most assuredly not."[25] Instead, Moskowitz keeps the focus on food nostalgia as an imaginative exercise that carries both pleasures and limits—and she suggests that the limits themselves, the things we can't know or access through food, carry their own kind of resonance and meaning. Moskowitz writes in a sidebar on the knish that "I think of my dad whenever I pass [the Yonah Schimmel deli] and I long for that NYC I knew as a child. I can only imagine the NYC my parents remember. I like to stop in once in a while and get nostalgic for a time that I knew and those old days that I didn't know."[26] It is, she seems to suggest, ethically important to engage in a nostalgia that explores various kinds of connection, but that doesn't fool itself into thinking that the past can be fully apprehended by any of us.

Transformative Nostalgia?

Evidently, dairy, meat, and egg industries and vegan cookbooks both use nostalgia for political purposes, as a means of shaping collective memories in ways that might be favorable to the foods they are trying to promote. Consequently, both sites of memory demonstrate the need to turn a critical eye on the function of nostalgia in cultural and collective memory. However, what emerges in Moskowitz's work is a needed demonstration of the potential for creativity, inclusion, and critical reflection in the negotiation of nostalgic pleasure. As a basis for vegan collective memory, this kind of nostalgia is a promising tool of resistance against the discursive efforts of the meat, egg, and dairy industries to attenuate remembrance in a predictably narrow direction. These industries—and their advertising agencies—thoroughly apprehend the powerful affective dimension of cultural politics and the centrality of pleasure in that affective matrix. In this context, it behooves the vegan movement to continue exploring varied ways of fighting pleasure with pleasure, so to speak. The cultivation of cross-species empathy is obviously crucial in vegan politics, but a simultaneous emphasis upon framing memory work on food as a creative form of pleasure is also increasingly necessary in today's intensely nostalgia-oriented and pleasure-driven food cultures.

Notes

1. Many theorists distinguish between cultural and collective memory; for instance, Chris Weedon and Glenn Jordan, in "Special Section on Collective Memory: Introduction." *Cultural Studies* 25, no. 6 (2011): 844, note that cultural memory is primarily concerned with institutionally reproduced forms of memory geared toward reinforcing current relations of power, while collective memory denotes any shared narrative of the past that helps confer a cohesive identity upon a group. For my purposes here, since both concepts have to do with the collective reinforcement of particular stories about the past, I will be using the concepts more or less interchangeably.
2. Paul Connerton, *How Societies Remember* (Cambridge: Cambridge University Press, 1989), 3.
3. Edward Said, "Invention, Memory, and Place," *Critical Inquiry* 26, no. 2 (2000), 179.

4. Vignolles, Alexandra, and Paul-Emmanuel Pichon, "A Taste of Nostalgia: Links Between Nostalgia and Food Consumption," *Qualitative Market Research: An International Journal* 17, no. 3 (2014), 229.
5. See Bibliography for all links. These links are current as of December 2015.
6. Dairy Farmers of Canada, "About Us," *Dairy Farmers of Canada*, accessed November 19, 2015, http://www.dairyfarmers.ca/who-we-are/about-us.
7. Dairy Farmers of Canada, "100% Canadian Quality Milk- 2014 TV Spot," *Youtube*, last modified March 7, 2014, https://www.youtube.com/watch?v=0mjV-3Wzr0A.
8. Egg Farmers of Canada, "About Us," *Get Cracking*, accessed November 19, 2015, http://www.eggs.ca/about-us/.
9. Egg Farmers of Canada, "Promotions: Natural Goodness," *Get Cracking*, accessed November 19, 2015, http://www.eggs.ca/promotions/view/10/natural-goodness.
10. McDonald's Canada, "McDonald's Ads," *McDonald's*, accessed November 19, 2015, http://www.mcdonalds.ca/ca/en/our_story/mcdonalds_ads.html.
11. Milk Every Moment, "About Us." *Milk Every Moment*, accessed November 19, 2015, http://milkeverymoment.ca/about.
12. Milk Every Moment, "Milk Every Moment," *Youtube*, last modified July 24, 2014, https://www.youtube.com/watch?v=XK9MTWBgsLs.
13. Milk Every Moment, "#233 Milk Every Moment anthem best." *Youtube*, last modified January 27, 2014, https://www.youtube.com/watch?v=r5FtWqlvw48.
14. Milk Every Moment, "Milk Every Moment Heart," *Youtube*, last modified June 22, 2013, https://www.youtube.com/watch?v=sfqL8SIVj1c.
15. Ibid.
16. See McDonald's Canada, "Not Without Canadian Farmers: Our Food, Your Questions," *McDonald's*, last modified August 27, 2012, http://yourquestions.mcdonalds.ca/questions/2165; and McDonald's, "FAQs: Our Food, Your Questions," *McDonald's*, accessed November 19, 2015, http://www.mcdonalds.com/us/en/your_questions/our_food/do-you-use-so-called-pink-goop-in-your-chicken-mcnuggets.html. Both McDonald's Canada and its parent company in the USA explicitly deny that they use "pink slime" in any of their products, as part of their ongoing "Our Food, Your Questions" campaign.
17. A fuller description of Derrida's concept of carnophallogocentrism can be found in Jacques Derrida's interview with Jean-Luc Nancy, "'Eating

Well' or the Calculation of the Subject" in *Points…: Interviews, 1974–1994*, eds. Elisabeth Weber and Peggy Kamuf (Stanford: Stanford University Press, 1995), 280.

18. Bryant Terry, *Vegan Soul Kitchen* (Cambridge: Da Capo Press, 2009), xx.
19. Isa Moskowitz, "About." *Post Punk Kitchen*, accessed November 19, 2015, http://www.theppk.com/about/.
20. Isa Moskowitz, *Isa Does It: Amazingly Easy, Wildly Delicious Vegan Recipes for Every Day of the Week* (New York: Little, Brown and Co., 2013), 3.
21. Isa Moskowitz, *Vegan with a Vengeance: Over 150 Delicious, Cheap, Animal-Free Recipes That Rock* (New York: Marlowe and Co., 2005), 102.
22. Moskowitz, *Isa Does It*, 119.
23. Ibid., 41.
24. Moskowitz, *Vegan with a Vengeance*, 192.
25. Ibid., 180. Moskowitz offers a longer explanation of this ethos in the introduction to the "Stirfries and Sautés" chapter of *Isa Does It,* 171: "I think it's important to know what 'authentic' food is, but I definitely don't let the idea box me in. Especially as a vegan chef, I'm constantly playing with traditions and concepts, creating vegan translations of classics and breathing new life into old favorites. But the bottom line is that no matter what culinary roads I travel, my food will always be a reflection of the world as seen by a Jewish-American woman from Brooklyn. As obsessed as I am with how things are really done, I will never be able to do them that way. Will I ever be able to make a stir-fry like a Cantonese chef? Of course not!"
26. Moskowitz, *Vegan with a Vengeance*, 75.

References

Connerton, Paul. 1989. *How societies remember*. Cambridge: Cambridge University Press.

Dairy Farmers of Canada. 2014. 100% Canadian quality milk- 2014 TV spot. *Youtube*. https://www.youtube.com/watch?v=0mjV-3Wzr0A. Last modified 7 Mar 2014.

Dairy Farmers of Canada. 2015. About us. *Dairy farmers of Canada*. http://www.dairyfarmers.ca/who-we-are/about-us. Accessed 19 Nov 2015.

Egg Farmers of Canada. 2015a. About us. *Get cracking*. http://www.eggs.ca/about-us/. Accessed 19 Nov 2015.

Egg Farmers of Canada. 2015b. Promotions: Natural goodness. *Get cracking*. http://www.eggs.ca/promotions/view/10/natural-goodness. Accessed 19 Nov 2015.

McDonald's. 2015. FAQs: Our food, your questions. *McDonald's*. http://www.mcdonalds.com/us/en/your_questions/our_food/do-you-use-so-called-pink-goop-in-your-chicken-mcnuggets.html. Accessed 19 Nov 2015.

McDonald's Canada. 2012. Not without Canadian farmers: Our food, your questions. *McDonald's*. http://yourquestions.mcdonalds.ca/questions/2165. Last modified 27 Aug 2012.

McDonald's Canada. 2015. McDonald's ads. *McDonald's*. http://www.mcdonalds.ca/ca/en/our_story/mcdonalds_ads.html. Accessed 19 Nov 2015.

Milk Every Moment. 2013. Milk every moment heart. *Youtube*. https://www.youtube.com/watch?v=sfqL8SIVj1c. Last modified 22 June 2013.

Milk Every Moment. 2014a. #233 milk every moment anthem best. *Youtube*. https://www.youtube.com/watch?v=r5FtWqlvw48. Last modified 27 Jan 2014.

Milk Every Moment. 2014b. Milk every moment. *Youtube*. https://www.youtube.com/watch?v=XK9MTWBgsLs. Last modified 24 July 2014.

Milk Every Moment. 2015. About us. *Milk every moment*. http://milkeverymoment.ca/about. Accessed 19 Nov 2015.

Moskowitz, Isa C. 2005. *Vegan with a vengeance: Over 150 delicious, cheap, animal-free recipes that rock*. New York: Marlowe and Co.

Moskowitz, Isa C. 2013. *Isa does it: Amazingly easy, wildly delicious vegan recipes for every day of the week*. New York: Little, Brown and Co.

Moskowitz, Isa C. 2015. About. *Post Punk Kitchen*. http://www.theppk.com/about/. Accessed 19 Nov 2015.

Said, Edward W. 2000. Invention, memory, and place. *Critical Inquiry* 26(2): 175–192.

Terry, Bryant. 2009. *Vegan soul kitchen*. Cambridge: Da Capo Press.

Vignolles, Alexandra, and Paul-Emmanuel Pichon. 2014. A taste of Nostalgia: Links between Nostalgia and food consumption. *Qualitative Market Research: An International Journal* 17(3): 225–238.

Weedon, Chris, and Glenn Jordan. 2011. Special section on collective memory: Introduction. *Cultural Studies* 25(6): 843–847.

Is the Moose Still My Brother if We Don't Eat Him?

Margaret Robinson

In Vitro Meat and Mi'kmaq Philosophy

This speculative philosophical essay explores how the advent of in vitro meat might impact Mi'kmaq culture, particularly our understanding of animals as our relatives and our duties toward them in regard to cultural protocols of respect and gratitude. I argue that a Mi'kmaq perspective on cultured meat can be inferred from our philosophy and traditional stories, and from cultural protocols already in place for plants and medicines. I also explore how the Mi'kmaq relationship to the moose may be explored through our growing knowledge of epigenetics.

In the future, eating moose meat may not require the death of a moose. Scientists are putting concerted effort into growing meat in a laboratory setting, and some have called in vitro meat "the future of food."[1] In 2012,

M. Robinson (✉)
Assistant Professor, Department of Sociology & Social Anthropology, Dalhousie University, Halifax, NS, B3H 4R2, Canada
e-mail: margaret.robinson@utoronto.ca

J. Castricano, R.R. Simonsen (eds.), *Critical Perspectives on Veganism*, The Palgrave Macmillan Animal Ethics Series,
DOI 10.1007/978-3-319-33419-6_12

for example, vascular physiologist Dr. Mark Post and colleagues from Maastricht University grew thin strips of cow muscle in a nutrient-rich dish. Once the cultured protein was ready they added lab-grown fat cells and tinted the muscle,[2] eventually producing a product some referred to as "the $325,000 burger."[3] In the on-air taste-test nutritional scientist Hanni Ruetzler stated, "It's close to meat. It's not that juicy, but the consistency is perfect."[4] Dr. Post speculates that such technology could become accessible to the average person, stating, "Potentially, you can do this in your kitchen. You can grow your own meat."[4]

While the methods used to produce in vitro meat today use stem cells from animals killed in a slaughterhouse, a time may come when stem cells are not required to produce in vitro meat.[3] In the meantime, successful commercial meat growth could mean a billion fewer animals raised for slaughter,[2] and may also result in reduced environmental impact.[5] Post's success, combined with 3-D printing technology, invites us to imagine a future where meat comes not from hunting or factory farming, but from a machine similar to the replicator used to synthesize meals on demand in the Star Trek film and television series. In such a future, a moose steak might be as easily available as that of a cow, all at the touch of a button.

As a Mi'kmaq woman, I am vitally concerned with the preservation and transmission of our cultural values, tradition, and practices, and with how that culture changes over time. Our culture binds us together as a people, and binds us to our traditional territories. There is also growing evidence that it keeps us physically and psychologically healthy despite our ongoing exposure to colonial violence.[6-8] I am concerned about the impact that the consumption of in vitro meat might have on my culture, on our relationship with our traditional territories and the animals who live in them, and on the transmission of our culture, philosophy, and values to future generations. While we don't yet know how receptive the Mi'kmaq would be to cultured meat, we do know that the population is becoming increasingly urbanized, with 45% of Aboriginal people in Atlantic Canada now living in an urban center,[9] impacting our food choices and practices. Today most Mi'kmaq purchase our food through the same grocery venues as our non-Indigenous peers, and are exposed to the same dietary products and choices.

As a vegan, I am also interested in how my culture can reduce or eliminate animal suffering, and in how Mi'kmaq philosophy supports our doing so. The emergence of in vitro meat leads me to ask how the relationship of the Mi'kmaq people with the moose would change if we were to consume cultured meat instead of once-living animals. This same question can be asked in relation to many animals traditionally used for food, but I will focus on the moose, whose significance in Mi'kmaq culture places it in a special role. Using a speculative Indigenous philosophy approach, I will explore how changes in food production might impact our relationships with the animals around us, especially if our bond with them is no longer one of dependence.

The Mi'kmaq Cultural Context

The Mi'kmaq are the Indigenous people of the northeastern coast of what is now North America. Our traditional territories include the Canadian provinces of Nova Scotia, Prince Edward Island, New Brunswick, and the Gaspé peninsula of Quebec, although we also have significant populations in the province of Newfoundland and the state of Maine. The Mi'kmaq view of the world is rooted in our relationship with the other animals that share our territories. Half our names for the months of the year refer to the behavior of animals during that time, with April, for example, being egg-laying month.[10, 11] Animals, and my relationship to them, remain foundational to my understanding of myself as a Mi'kmaq woman, despite my having lived in cities for the past 25 years. The Mi'kmaq view human existence as occurring within a web of relations with other beings, and as engaged in relationships of reciprocity and dependence. Although this perspective is rooted in our ancestors' subsistence hunting and gathering practices, the philosophical outlook persists and is relatively common, particular among those of us who are connected with our traditional culture.

Although my ancestors gathered an array of wild fruits and vegetables and also did some farming, traditional Mi'kmaq food culture was heavily focused on meat consumption. My ancestors spent their springs and summers fishing and gathering shellfish on the coast, moved to the rivers in autumn to catch eel, and spent much of the winter inland, hunting animals, including moose.[12] This food-related migration pattern shaped our culture and technology. It is the reason, for example, why our dwelling, the wigwam, is a portable structure, and why we developed the toboggan, the snowshoe, and the birch bark canoe.

Anthropologist Anne-Christine Hornborg describes the Mi'kmaq approach to hunting as a conversation between the animal and the hunter, in which the animals willingly sacrifice themselves to feed the Mi'kmaq.[13] A traditional story tells of a family close to starvation during an early and harsh winter, who prayed for food. In response to their prayers, a moose appeared at their wigwam with a bargain:

> He told them that if they treated the moose with respect by taking a moose only when in need, by making offerings over the body of the moose, by using all parts of the animal and by treating as sacred even the bones of the moose, he would always return to feed the people. If they disrespected the moose, however, then the moose would leave and never return.[14, 15]

This story is a clear example of the belief in animal sacrifice that Hornborg describes, and this dynamic forms the basis of Mi'kmaq hunting protocols.

Due to his immensely large size (an adult moose is 1.5–2 meters high at the shoulder), the moose is considered the chief of all land animals, and is a counterpart to the whale, who is king of the ocean.[11] Traditionally, the killing of a moose was a significant event, and served as a young man's entry into adulthood by signaling that he possessed the skills to support a family and the patience and maturity to participate in political councils.[11] As part of our bargain with the moose the Mi'kmaq used as much of the animal's body as possible. The hide was used to make clothing, moccasins, and to wrap the exterior of our wigwams; tendons were used to create thread; bone and antlers were used to make needles, hunting tools, fasteners, and dice; and moose hair was used for embroidery.[14, 16] Once the bone marrow had been eaten, the Mi'kmaq would pound the moose bones to a powder and boil it to reap fat and produce a medicinal soup.[11] Parts that cannot be used must be treated respectfully and returned to the creator, usually via burial, although some Mi'kmaq hunters now see it as their responsibility to leave edible portions as offering for other animals.[14, 17]

Mi'kmaq hunters are expected to show respect for the hunted moose through traditional ceremony:

> Once [the Moose has been] killed, the traditional harvester lays a circle of tobacco around the moose and says a prayer in gratitude for the earth's offering. The bell or dewlap (the loose skin that hangs on the lower jaw) is removed and hung in a tree as a sign to other animals that the moose was harvested in a sacred way. A pipe ceremony is initiated by the harvester to help release the animal's spirit, to ask forgiveness for taking its life, and to let it know that the gift of its life is appreciated.[14]

The ceremonial interaction with animal spirits is seen as a continuation of our interaction with embodied animals. To fail to show respect to an animal spirit risks future food security, since animal spirits are believed to reincarnate (increasing the animal population) and to communicate with others of their kind, reporting on how we have treated them in life and death.

The importance of showing respect for animal bones is also emphasized in our traditional stories. In the Mi'kmaq story "The Invisible Boy," for example, a celebrated hunter has the moose as his *teomul*, or guardian spirit, and can transform into a moose as a result.[18, 19] In the story, his wife gives birth to a little boy. The hunter's sister warns his wife to take special care of a moose's leg bone, which she is saving for her brother, who will eat the marrow. However, the little boy smashes the leg bone while playing. The hunter's sister realizes that her brother is now injured, and when the women go to him they find him dying. He sends his wife and child home to her father and instructs his sister to kill him with an axe. She obeys and he transforms into a moose, which she consumes.[18, 19] In this story, the moose bone must be treated carefully because of the spiritual relationship between the hunter and the moose, expressed through having the moose as his *teomul*, and in the transformation of human into moose. Metaphorically, the Mi'kmaq reader is in the role of the sister in this story; the moose is our brother who sacrifices himself for us.

In the story "Two Weasels," a pair of sisters encounter a deserted Mi'kmaq village and spend the night in a wigwam there. The older sister warns her sibling not to touch anything, but the younger sister finds the neck bone of an animal lying on the floor and kicks it contemptuously. Whitehead specifies that it is the neck bone of a moose.[19] Later, as they prepare to sleep, the bone shouts indignantly about its treatment and mocks the terrified girls.[18] In the version told by Whitehead, the neck bone transforms into a type of vampire and comes for revenge.[19] These stories embody the importance of respecting animal bones and provide extreme examples of what may happen if that agreement is breached.

The Confederacy of Mainland Mi'kmaq, a council of seven Mi'kmaq communities on mainland Nova Scotia, notes the continuing importance of following protocols of respect:

> Some animals, like moose, give their lives so the Mi'kmaq may have food. They show respect to the moose by treating the remains with respect. The bones of the moose should never be burned or given to household pets, they should be used to make something or buried.[16]

Another protocol of respect is the value of non-interference, which instructs us not to intervene in the lives and choices of others. In the case of animals this means leaving them alone to live their lives, not interfering with their habits or habitats (which may entail active environmental stewardship), and never killing them unless necessary.[11] This is the cultural context in which subsistence hunting occurs among the Mi'kmaq, and it is the context I have in mind when thinking of the development of in vitro meat.

The Decline of Subsistence Hunting

Although cultured meat development has focused on beef, once the technology is perfected there is nothing to preclude so-called "game meats" from being cultured, such as moose, deer, or elk, which are part of the traditional diet of many Indigenous peoples. Cultured meat is not the first development to impact Mi'kmaq food values or practices. Our hunting, gathering, and eating practices have changed considerably as a result of contact with Europeans over the past 500 years. Systemic attempts to eliminate Mi'kmaq culture and colonize or eliminate Mi'kmaq people have been actively resisted with mixed results. The widespread destruction of natural habitats across the Maritime Provinces has significantly reduced the number of animals available for subsistence hunting. For some, tagging, in which one gets close enough to touch the animal with a hand, has replaced hunting as a test of skill.[20] The expansion of human settlements, and practices such as mining, clear cutting, and overhunting have pushed some plant and animal species to endangerment or extinction. There are approximately 1000 moose left in mainland Nova Scotia, for example, making the species officially endangered since 2003,[21] and the population has continued to decline despite conservation efforts since the 1930s.[22] Yet the moose remain culturally important to the Mi'kmaq, even to those who, like me, have never eaten moose. This significance was highlighted recently when Settler trophy hunters visiting Cape Breton killed a white moose that was sacred to us.[23] The Assembly of Nova Scotia Mi'kmaq Chiefs passed a resolution asking the province to issue a ban on hunting white moose.[24]

While the right of the Mi'kmaq to hunt animals such as the moose is protected in treaties signed with the Crown, Mi'kmaq guidelines forbid moose hunting on Mainland Nova Scotia due to its endangered status. In the 1940s moose from the province of Alberta were transported to Cape Breton to supplement the animal population there. There are now estimated to be around 6000 moose on the island and, as a result, limited hunting has been permitted there since the 1980s.[14, 21] The Mi'kmaq prohibit hunting in Cape Breton between January 1 and August 15 of each year, when the female moose are gestating their calves.[15]

The *Unama'ki* (i.e., Cape Breton) Institute of Natural Resources initiated the creation of a Moose Working Group. This group includes the Mi'kmaq Rights Initiative and representatives from various departments in the Provincial and Federal governments, and strives to increase public awareness of Mi'kmaq cultural values related to the moose and to moose hunting.[21] In August of 2009 the Moose Working Group began a program to incorporate moose hunting into cultural training about the Mi'kmaq value of *Netukulimk*, which refers to "the skills and sense of responsibility required to become a protector [or some say hunter] of other species."[25, 26] Historical ecologist Russell Barsh suggests that *Netukulimk* is related to the Mi'kmaq prefix *nutqw-*, indicating insufficiency, and reflects the idea of "avoiding not having enough."[27] In our Mi'kmaq philosophical framework it is not abundance for which we strive, but sufficiency. Our subsistence hunting traditions, framed by ethnocentric Settlers as lacking in industry or commercial vision, is a conscious choice rooted in Mi'kmaq values.

As a population, the Mi'kmaq no longer rely on subsistence hunting to obtain food. A study of Mi'kmaq people living on reserves in the Province of Nova Scotia found that 47% hadn't eaten any traditionally hunted animals, such as moose, caribou, deer, or bear, in the past year.[28] The consumption of larger sea mammals such as whale, seal, or walrus had almost disappeared, and 80% had not even eaten small game animals such as goose, ducks, partridges, rabbits, or muskrat.[28] The most commonly reported traditional foods consumed were bannock, berries or other wild vegetation, and saltwater fish.[28] I find it interesting that these commonly consumed foods are compatible with Settler diets—even bannock is related to the Scottish scone. This leads me to wonder if these choices reflect a conscious effort to incorporate Mi'kmaq foods, or whether these traditional foods appear because they are also part of the Settler diet. The degree to which the Mi'kmaq diet has been colonized is evident from types of food consumed daily by the population living on reserve in Nova Scotia: 71% report having soft drinks, a third report eating fried food, and 17% report having fast food,[28] making their diet remarkably like that of Settler Canadians, 24% of whom report eating fast food on a daily basis.[29] Professor of human ecology Kim Travers reports that for most Mi'kmaq living on reserve the primary sources of protein are peanut butter or processed meats, which is a reflection of both colonization and economic oppression.[30] If this is the case for those living on reserve, which are centers of Mi'kmaq culture and therefore likely to support traditional food practices, then those of us who grew up off reserve, often without status or cultural connection, may face even greater pressure to adopt Settler eating practices.

Animals in Mi'kmaq Philosophy

Despite changes to our diet in the wake of colonization, our recognition that all lives are interconnected has remained constant. This belief is expressed in the Mi'kmaq phrase *M'sit No'kmaq*, which means "All my relations." This phrase includes all living things—birds and fish and frogs and mosquitoes and moose among them. In Mi'kmaq spirituality a great spirit called *Kisu'lk* made the universe and imbued it with life, and as a result almost everything has a spirit, and is alive in some way.[31] The Confederacy of Mainland Mi'kmaq write that "all things—plants, animals, people, and Mother Earth herself—all have the Creator's spirit in them and must be respected."[16] This includes animals, plants, rocks, water, and geographic locations; all of these can have a spirit, an identity, and a personality. The existence of spirit within all things is foundational to the Mi'kmaq view of the world, and is also found in other Indigenous cultures.[32, 33, 34]

As a result of this shared spirit, humans and other animals share the experience of personhood—that is, we are individuals with self-awareness, with the capacity to act, with intrinsic rights, and with responsibilities toward others.[35] Anne-Christine Hornborg describes personhood as "a standard for all life" and as "the common essence of both animals and humans."[13] This shared personhood is reflected in our traditional stories, in which human and animal life is portrayed as interchangeable, with humans becoming animals and animals becoming human.[36, 37] For example, in "The Beaver Magicians and the Big Fish," a hunter meets an elderly man and his family, who invite him in, feed him, and send him home with a gift of moose meat.[38] Once home, the hunter discovers that the moose meat has transformed into the bark of a poplar tree—the favorite food of beavers. He realizes that the family he met were actually beavers, made to seem human by the magic of the eldest beaver.

Because all other living things share in the spirit of *Kisu'lk*, the creator, all beings are deserving of our respect. The Confederacy of Mainland Mi'kmaq explain it succinctly:

> And because everything on Earth is connected, no part should be exploited or abused. Each part must work in harmony with the rest. This does not mean that people cannot cut down trees, or hunt for food, but it does mean that the proper respect must be shown to the Creator for making these resources available to them in the first place.[16]

The Mi'kmaq have a number of protocols for expressing respect, but common ones include prayer, and the ceremonial laying down of tobacco.

The Mi'kmaq are related to other animals because we all share spirit, but we are also related through adoption. The Mi'kmaq creation story recounts the birth of *Kluskap*, the Mi'kmaq cultural hero and prototype of virtuous human life. A number of versions describe the birth of *Kluskap's* grandmother, *Nukumi*, who was formed by the sun shining on a dew-covered rock. *Nukumi* explains to *Kluskap* that due to her advanced age she cannot survive on plants and berries, and will need to eat meat.[39] *Kluskap* calls to Marten (a weasel relative), who was swimming in a river nearby, and asks him to give up his life so that *Kluskap's* grandmother can eat. *Kluskap* is friends with all the animals, and because of this friendship Marten agrees. In exchange for his sacrifice *Kluskap* makes Marten his brother. This story exemplifies the relational basis for meat consumption in Mi'kmaq philosophy. We must only kill an animal for food when it is necessary to do so. The sacrifice of the animal cements a sibling relationship between *Kluskap* and Marten, building on a preexisting friendship between *Kluskap* and all the animals.

In some versions of the story, once Marten is dead, and his body lies on the ground, *Kluskap* is overcome with regret. *Nukumi* intervenes, and calls on *Kisúlkw* to return Marten to life. Marten returns to his home in the river, leaving a second body behind for them to eat. At this juncture, *Nukumi* explains that from this point on the animals would be "willing to provide food and clothing, shelter and tools but always they must be treated with the respect given a brother and friend because they would only be there to provide what is necessary for life."[39] This story embodies the Mi'kmaq regret at animal death, the belief in animal sacrifice, and the belief that dead animals somehow regenerate.[36, 37] All of these elements can be brought forward into our interactions with in vitro meat.

In Vitro Meat and Mi'kmaq Culture

While I am excited by the prospect that in vitro meat will reduce or eliminate animal slaughter and suffering, the creation of in vitro meat also raises concerns for me around our respect for animals. Colonization has already impacted how the respect directive is lived. Today, most Mi'kmaq get our food from the same mass-market industry channels as our Settler neighbors, and, increasingly, our respect protocols are reserved for practices that resemble those of our ancestors. So, for example, while a Mi'kmaq might smudge, pray, or put down tobacco when hunting, I know none who make such an offering when they go to the grocery store. The advent of in vitro meat might mean that moose meat comes to be seen as a thing—a despirited secular object—rather than the sacrificial gift of a brother and a friend. If the mechanisms by which sacrifice is construed—the moose permitting themselves to be caught—is removed from the equation, and moose meat is grown and harvested, will our relationship with our food, and with the moose from which it originates, still be grounded in respect and gratitude?

The commodification of animals is not a traditional Mi'kmaq value,[14] but was absorbed through our relationship with Settlers. Hornborg writes that Mi'kmaq culture "shows a rupture in their traditional ethic when it is challenged by the European interest in furs."[13] For many Mi'kmaq, our relationship with Settlers has usurped the importance that animals once held in our lives.[37] Removing the living animal from meat production may cement our assimilation to Settler values and further distance us from our traditional intimacy with animals.

A second concern I have about the development of in vitro moose meat is the transition it would support from a communal value system to an individualistic value system. The traditional moose harvest is about more than just meat. It's about community. When harvesting a moose, there is value in the moose itself, there is value in our relationship with moose as a species, and with the local ecosystems we both inhabit. Even now, when a moose is killed the meat is shared with members of the human and animal communities who need food support. Lefort and colleagues note that some Mi'kmaq communities in Cape Breton have a community meat freezer, which enables families to take meat according to their need.[14] Returning to the traditional moose harvest has revived an appreciation for traditional ways, and an interest in knowledge rooted in our traditional territories.[14] In a future where lab-cultured moose meat is bought at a store or generated in the kitchen, what we risk losing is the experience of being a community, and of engaging in communal activities that express and reinforce our identity as a people. In vitro meat promises a future where I need not interact with anyone to have moose meat—especially not the moose itself.

In thinking of how to positively shape the impacts of in vitro meat on Mi'kmaq culture, I thought about how the values embedded in our relationship with the moose—such as *Netukulimk*, brotherhood, and respect—could be transferred to a context in which moose is cultured in a lab or kitchen rather than hunted. One possibility is that our relationship with the moose is incorporated into the epigenetic code of the moose (and into ours as well).

While our DNA contains our genetic blueprint, not all aspects of our DNA are active at any given time. Epigenetics examines how interactions with our environment activate some genes and leave other dormant. I'll start with the example of mice. Researchers Brian G. Dias and Kerry J. Ressler conducted experiments to change the nervous system of mice.[40] Using electroshock, they conditioned male mice to respond anxiously to acetophenone (a scent reminiscent of cherry blossoms). Two weeks later those mice were bred with female mice. Their offspring, never exposed to the shocks or to the cherry scent, also responded anxiously when smelling the acetophenone for the first time. Dias and Ressler then used in vitro fertilization to breed rat pups from the conditioned mice and placed the pups with non-conditioned mice parents to determine that information about the dangerous smell was inherited genetically and not transmitted socially. The experiment had even changed the neurons in the nose of the offspring mice and the structure of their brain to help them better identify the dangerous smell. Dias and Ressler conclude that "the experiences of a parent, before even conceiving offspring, markedly influence both structure and function in the nervous system of subsequent generations."[40]

Inheriting the experiences of previous generations is familiar to many Indigenous people through the concept of blood memory. Indigenous author Dr. N. Scott Momaday is quoted as saying that "each of us bears in his genes or in his blood or wherever a recollection of the past."[41] Indigenous Studies professor Dr. Bonita Lawrence, explains that "blood memory promises a direct link to the lives of our ancestors, made manifest in flesh, which she suggests can be appealing to "a people who have had much of their knowledge of the past severed," as we have.[42] Yet the experiment with the mice suggests that we may indeed carry genetic sequences that encode the experiences of our ancestors, ancient and recent. A number of researchers have examined the impact of historical trauma on subsequent generations of Indigenous people, terming this effect "intergenerational trauma."[43, 44, 45, 46] Although generally framed as socially transmitted, some have speculated that there may be an epigenetic element involved in the transmission of intergenerational trauma as well.[47, 48]

If the moose of *Mi'kma'ki* (the Mi'kmaq territories) experience their history with us as traumatic, then this may be encoded into their DNA, and might be present if not activated, in cultured meat created from the cells of a moose. Even if that moose itself had never been hunted, the "memory" of being hunted (so to speak) could be inherited epigenetically from its ancestors. Some evidence suggests that moose find pursuit stressful. A study in Sweden measured cortisol levels in 78 moose shot by rifle.[49] Cortisol is a chemical released by the adrenal gland during times of stress. Cortisol levels were significantly higher for moose who had been pursued for more than ten minutes compared with those pursued for under ten minutes. There were no significant differences in cortisol levels between moose who were pursued for less than ten minutes and those killed unawares. That the stress relates to pursuit specifically (and not to, say, death) is supported by the high levels of cortisol also found in moose who were pursued but tranquilized instead of killed.[49]

There is limited evidence that interactions with human beings may already have been encoded into the moose. Zoologist Dr. Samuel K Wasser did a chemical analysis of animal scat in the oil sands area of Alberta and notes that moose appear to have a lower physiological stress response to human activity than do animals such as caribou, suggesting that acquired familiarity with humans may be passed on to future generations.[50] So it seems likely that the genetic code of the moose in our traditional territories contains elements that have been shaped by their interactions with us. In short, we may be in their blood (and other cells). Such a relationship would presumably not be embodied by moose from outside *Mi'kma'ki*, and may not be apply to moose whose genetic inheritance was shaped in another territory.

Does this epigenetic mechanism also apply to positive things such as relationships of significance? There's some evidence to support this idea as well. Weaver and colleagues found that baby rats who were groomed frequently by their mothers were more relaxed as adults.[51] Biologists Cornelius Gross and Rene Hen established that the offspring of high-grooming mothers remained relaxed, even if fostered from birth by a low-grooming rat mother, suggesting that the impact of grooming habits had entered their epigenetic code.[52] Within humans, neuroscientist Rachel Yehuda and colleagues showed that receiving psychotherapy for PTSD may result in similar epigenetic changes in how we process stress.[53] If the moose of *Mi'kma'ki* value their relationship with us, as our traditional stories claim they do, then this too could be epigenetically transmitted. In short, meat grown from the cells of such animals might contain remnants of our relationship with the moose; such memories may be written into their genetics, and into our own as well.

Culturally Embedding In Vitro Technology

Without question, Mi'kmaq philosophy, ceremony, or practice may change with the advent of cultured meat, and it is our responsibility to actively shape such change in ways that protect our cultural continuity. If our relationship with animals is no longer one of dependence then we have at least two philosophical models upon which to build.

One model is the gathering of traditional plants and medicines. When harvesting, for example, we must approach plants with respect, ask their permission to gather them, and avoid overharvesting.[54] James Youngblood Henderson notes that "a person gathering roots, leaves or bark for medicinal purposes pleases the life-force of each plant by placing a small offering of tobacco at its base."[55] Such protocols could be transferred, effectively treating cultured meat as a type of plant.[56] Such a protocol would have to compete with habits that instrumentalize food, but in philosophical terms the gathering of plants and medicines offers a starting point for a protocol around cultured meat that is rooted in our traditional values.

A second model emerges from our oral traditions. In the story of *Nukumi* and Fire, for example, *Kluskap* is friends with the animals, and in some versions of the story it seems as if he didn't eat meat prior to the birth of his grandmother (although in other stories he is a skilled hunter).[39] So *Kluskap* could serve as a type of model for our relationship with animals, even in a context where they are "friends, not food." In that same story, *Kluskap* regrets the death of his friend Marten, and *Nukumi* communicates with *Kisúlkw*, who brings Marten back to life. Marten is alive as both the friend and brother of *Kluskap*, but he is also present as the dead body that can be eaten.[37] In a future where in vitro meat technology requires only cells from a living animal then the moose enters a situation akin to Marten—available to eat, yet also alive as a friend and brother. The choice to eat in vitro meat could embody our own regret at animal death, and at our failure to live out the value of *Netukulimk* (avoiding not having enough). Thinking back to the story in which the moose makes a bargain with the starving Mi'kmaq family, we might reimagine a situation in which we renegotiate that agreement to one in which the moose provide stem cells rather than laying down their lives. The consumption of in vitro meat could be framed an expression of *Netukulimk*, and also as an expression of non-interference, since it reduces our impact on animal life.

In a future where I can order a moose steak from a 3-D printing machine, or purchase one grown in a lab, my relationship with actual living moose becomes freed from a relationship of sacrifice, as well as dependence, and can begin to approach something akin to that of relatives who, after a long period of tension, have finally become friends.

Notes

1. Brandon Griggs, "How Test-Tube Meat Could Be The Future Of Food," *CNN.com*, May 5, 2014. Accessed October 15, 2015. http://www.cnn.com/2014/04/30/tech/innovation/cultured-meat/.
2. Helen Shen, "Mark Post, Tissue Engineer." *University of Santa Cruz, Science Communication Program*. March 31, 2012, accessed October 15, 2015. http://scicom.ucsc.edu/publications/QandA/2012/post.html.
3. Henry Fountain, "Building a $325,000 Burger," *The New York Times*, May 12, 2013, accessed October 15, 2015. http://www.nytimes.com/2013/05/14/science/engineering-the-325000-in-vitro-burger.html?pagewanted=all.
4. BBC, "What Does A Stem Cell Burger Taste Like?" *BBC.com*, video, 3:03, August 5, 2013, accessed October 15, 2015. http://www.bbc.com/news/science-environment-23529841.
5. Hanna L Tuomisto and M. Joost Teixeira de Mattos, "Environmental impacts of cultured meat production," *Environmental Science & Technology* 45, no. 14 (2011): 6120. DOI: 10.1021/es200130u.

6. Richard T Oster, Angela Grier, Rick Lightening, Maria J. Mayan, and Ellen L. Toth. "Cultural Continuity, Traditional Indigenous Language, And Diabetes In Alberta First Nations: A Mixed Methods Study," *International Journal for Equity in Health* 13, no. 1 (2014): 92, DOI: 10.1186/s12939-014-0092-4.
7. Michael J. Chandler and Christopher E. Lalonde, "Cultural Continuity as a Moderator of Suicide Risk Among Canada's First Nations," in *Healing Traditions: The Mental Health of Canadian Aboriginal Peoples*, ed. Lawrence J. Kirmayer and Gail Guthrie Valaskakis (Vancouver, BC: University of British Columbia Press, 2009), 221–248.
8. Michael J. Chandler and Travis Proulx, "Personal persistence and persistent peoples: Continuities in the lives of individual and whole cultural communities," in *Self-Continuity: Individual And Collective Perspectives*, ed. Fabio Sani (East Sussex, UK: Psychology Press, 2008), 213–226.
9. Service Canada, "Aboriginal Peoples: Atlantic Canada," Employment and Social Development Canada, 2011, accessed December 29, 2015. http://www.esdc.gc.ca/eng/jobs/lmi/publications/csp/abor/atlantic/abor_march2014.pdf.
10. Patsy Paul-Martin, "Mi'kmaq Months of the Year," Truro, NS, Canada: Eastern Woodland Publishing, n.d., accessed October 15, 2015, http://firstnationhelp.com/ali/posters/pdf/months.pdf.
11. Wilson D Wallis and Ruth Sawtell Wallis, *The Micmac Indians of Eastern Canada* (Minneapolis, MN, USA: University of Minnesota Press, 1955).
12. Ralph Pastore, "Traditional Mi'kmaq (Micmac) Culture," *Heritage Newfoundland and Labrador*, 1998, accessed October 15, 2015. http://www.heritage.nf.ca/articles/aboriginal/mikmaq-culture.php.
13. Anne-Christine Hornborg, *Mi'kmaq Landscapes: From Animism to Sacred Ecology* (Hampshire, UK: Ashgate, 2013).
14. Nadine Lefort, Clifford Paul, Ernest Johnson, and Charlie Dennis, "*Tiam*. Mi'kmaq Ecological Knowledge: Moose in *Unama'ki*." *Unama'ki Institute of Natural Resources*, 2014, accessed October 15, 2015, http://www.eecapacity.net/sites/default/files/fellows/docs/Moose%20in%20Unama'ki.pdf
15. Assembly of Nova Scotia Mi'kmaq Chiefs, "*Tia'muwe'l Netuklimkewe'l. Unama'ki* Moose Harvesting According to *Netukuliml*." *Mi'kmaq Rights Initiative*, 2009, accessed October 15, 2015, http://mikmaqrights.com/uploads/MooseGuidelines.pdf.
16. Confederacy of Mainland Mi'kmaq, *Kekina'muek (Learning): Learning About the Mi'kmaq of Nova Scotia* (Truro, NS, Canada: Eastern Woodland Print

Communication, 2007), accessed October 15, 2015. http://hrsbstaff.ednet.ns.ca/clambert/KekinamuekEntiredocumentprintable.pdf.

17. Lacia Kinnear, "Contemporary Mi'kmaq Relationships Between Humans And Animals: A Case Study of the Bear River First Nation Reserve in Nova Scotia" (Master's thesis, Dalhousie University, 2007).
18. Silas T. Rand, *Legends of the Micmacs*. Volume I, 1893, Reprint (West Orange, NJ, USA: Invisible Books, 2005).
19. Ruth H. Whitehead, *Stories From The Six Worlds: Micmac Legends* (Halifax, NS, Canada: Nimbus, 1988).
20. Evan T. Pritchard, *No Word For Time: The Way of the Algonquin People* (San Francisco, CA, USA: Council Oak Books, 2001).
21. Province of Nova Scotia, "Current Status," *NovaScotia.ca*, October 13, 2013, accessed October 15, 2015, http://novascotia.ca/abor/office/what-we-do/negotiations/current-status/.
22. Province of Nova Scotia, "Species At Risk Overview," *Nova Scotia.ca*, May 28, 2014, accessed October 15, 2015. http://novascotia.ca/natr/wildlife/biodiversity/species-list.asp.
23. Ross Brooks, "Nova Scotia's Sacred Albino Moose Killed by Visiting Hunters," *Inhabitat.com*, October 16, 2013, accessed October 15, 2015, http://inhabitat.com/nova-scotias-sacred-albino-moose-killed-by-visiting-hunters/.
24. Jayson Baxter, "Mi'kmaq Chiefs Want Killing White Moose to be Banned," *Atlantic CTV News,* October 18 2013, accessed October 15, 2015, http://atlantic.ctvnews.ca/mi-kmaq-chiefs-want-killing-white-moose-to-be-banned-1.1503590.
25. Annamarie Hatcher, Cheryl Bartlett, Albert Marshall, and Murdena Marshall, "Two-Eyed Seeing In The Classroom Environment: Concepts, Approaches, And Challenges," *Canadian Journal of Science, Mathematics, and Technology Education* 9 (2009): 141–153.
26. Kwilmu'kw Maw-klisiaqn Negotiation Office, Mi'kmaq Rights Initiative, "Have We Had Any Success?" *KMKNO News* 1, no. 2 (2013): 3, accessed October 15, 2015, http://mikmaqrights.com/uploads/KMKNONewsletterMarch2013.pdf.
27. Russel L. Barsh, "Netukulimk Past And Present: Miqmaw Ethics And The Atlantic Fishery," *Journal of Canadian Studies* 37, no. 1 (2002): 15–38.
28. Charlotte Loppie and Fred Wein, "The Health of the Nova Scotia Mi'kmaq Population." *Mi'kmaq Health Research Group*, January 7, 2007, accessed October 15, 2015, http://www.unsi.ns.ca/wp-content/uploads/2015/07/ns-rhs-report-07.pdf.

29. Didier Garriguet, "Canadians' eating habits," *Health Reports* 18, no. 2 (2007): 17, accessed December 29, 2015, http://www.statcan.gc.ca/pub/82-003-x/2006004/article/habit/4148989-eng.htm.
30. Kim D. Travers, "Using Qualitative Research to Understand the Sociocultural Origins of Diabetes among Cape Breton Mi'kmaq," *Chronic Diseases in Canada* 16 (1995): 140–143.
31. Jean Muin'iskw and Dan Crowfeather, "Mi'kmaw Spirituality 101." *Mi'kmaq Spirit*, accessed October 15, 2015, http://www.muiniskw.org/pgCulture2.htm.
32. Joseph E. Couture, "*A Metaphoric Mind: Selected Writings of Joseph Couture*" (Athabasca, AB, Canada: Athabasca University Press, 2013).
33. Jean-Paul Restoule, "Everything Is Alive And Everyone is Related: Indigenous Knowing and Inclusive Education," *Federation for the Humanities and Social Sciences*, January 25, 2011. accessed October 15, 2015, http://www.ideas-idees.ca/blog/everything-alive-and-everyone-related-indigenous-knowing-and-inclusive-education.
34. Marie Battiste and James (Sákéj) Youngblood Henderson, *Protecting Indigenous Knowledge and Heritage: A Global Challenge* (Saskatoon, SK, Canada: Purich, 2000).
35. Charles Taylor, "The Concept of a Person," in *Philosophical Papers. Volume 1, Human Agency and Language*, ed. Charles Taylor (Cambridge, UK: Cambridge University Press, 1985) 97–114.
36. Margaret Robinson, "Veganism and Mi'kmaq Legends," *Canadian Journal of Native Studies* 33, no. 1 (2013): 189–196.
37. Margaret Robinson, "Animal Personhood in Mi'kmaq Perspective," *Societies* 4, no. 4 (2014): 672–688.
38. Silas T. Rand, *Legends of the Micmacs: Volume II*, 1893, Reprint (West Orange, NJ, USA: Invisible Books, 2004).
39. Laura Redish and Orrin Lewis, "Nukumi and Fire," *Native Languages of the Americas*, 2015, accessed October 15, 2015, http://www.native-languages.org/mikmaqstory2.htm.
40. Brian G. Dias, and Kerry J. Ressler, "Parental olfactory experience influences behavior and neural structure in subsequent generations," *Nature Neuroscience* 17, no. 1 (2014): 89–96.
41. Pauline Turner Strong and Barrik Van Winkle, "Indian Blood: Reflections On The Reckoning And Reconfiguring Of Native North American Identity," *Cultural Anthropology* 11.4 (1996): 547–576.
42. Bonita Lawrence, *'Real' Indians and Others: Mixed-Blood Urban Native Peoples and Indigenous Nationhood* (Vancouver, BC, Canada: University of British Columbia Press, 2004).

43. Amy Bombay, Kimberly Matheson, and Hymie Anisman, "The Intergenerational Effects Of Indian Residential Schools: Implications For The Concept Of Historical Trauma," *Transcultural Psychiatry* 51, no. 3 (2014): 320–338.
44. Maria Yellow Horse Brave Heart, "*Oyate Ptayela*: Rebuilding The Lakota Nation Through Addressing Historical Trauma Among Lakota Parents," *Journal of Human Behavior in the Social Environment* 2, no. 1–2 (1999): 109–126.
45. Maria Yellow Horse Brave Heart and Lemyra M. DeBruyn. "The American Indian Holocaust: Healing Historical Unresolved Grief," *American Indian and Alaska Native Mental Health Research* 8 (1998): 56–78.
46. Eduardo Duran and Bonnie Duran, *Native American Postcolonial Psychology* (Albany, NY, USA: State University of New York Press, 1995).
47. Karina L Walters, Ramona E. Beltran, David Huh, and Teresa Evans-Campbell, "A Dis-placement and Dis-ease: Land, Place and Health among American Indians and Alaska Natives," in *Communities, Neighborhood, and Health: Expanding the Boundaries of Place*, ed. Linda M. Burton, Susan P. Kemp, ManChui Leung, Stephen A. Matthews, and David T. Takeuchi (Philadephia, PA, USA: Springer ScienceBusiness Media, LLC, 2011), 163–199.
48. Karina L Walters, Selina A. Mohammed, Teresa Evans-Campbell, Ramona E. Beltrán, David H. Chae, and Bonnie Duran, "Bodies Don't Just Tell Stories, They Tell Histories," *Du Bois Review: Social Science Research on Race* 8, no. 1 (2011): 179–189.
49. Line Gertrud Lundstein, "Can Cortisol Be Used To Assess Acute Stress In Moose?" (Bachelor thesis, Hedmark University College, 2014).
50. Samuel K. Wasser. Jonah L. Keim, Mark L. Taper, and Subhash R. Lele, "The Influences Of Wolf Predation, Habitat Loss, And Human Activity On Caribou And Moose In The Alberta Oil Sands." *Frontiers in Ecology and the Environment* 9, no. 10 (2011): 546–551.
51. Ian C. G. Weaver, Nadia Cervoni, Frances A. Champagne, Ana C. D'Alessio, Shakti Sharma, Jonathan R. Seckl, Sergiy Dymov, Moshe Szyf, and Michael J. Meaney, "Epigenetic Programming By Maternal Behavior," *Nature Neuroscience* 7, no. 8 (2004): 847–854.
52. Cornelius Gross and Rene Hen, "The Developmental Origins Of Anxiety." *Nature Reviews Neuroscience* 5, no. 7 (2004): 545–552.
53. Rachel Yehuda, Nikolaos P. Daskalakis, Frank Desamaud, Louri Makotkine, Amy L. Lehrner, Erin Koch, Janine D. Flory, Joseph

D. Buxbaum, Michael J. Meaney, and Linda M. Bierer, "Epigenetic Biomarkers As Predictors And Correlates Of Symptom Improvement Following Psychotherapy In Combat Veterans With PTDS," *Frontiers In Psychiatry* 4 (2013): 118, DOI:10.3389/fpsyt.2013.00118.
54. Marilyn Walker, *Wild Plants of Eastern Canada* (Halifax, NS, Canada: Nimbus, 2008).
55. James Youngblood Henderson, *The Mi'kmaq Concordat* (Halifax, NS, Canada: Fernwood Publishing, 1997).
56. See, for example, Impossible Foods, "The Impossible Cheeseburger," *Impossible Foods,* 2015, accessed December 29, 2015. http://www.impossiblefoods.com

References

Assembly of Nova Scotia Mi'kmaq Chiefs. 2009. *Tia'muwe'l Netuklimkewe'l. Unama'ki* moose harvesting according to *Netukuliml. Mi'kmaq Rights Initiative.* http://mikmaqrights.com/uploads/MooseGuidelines.pdf. Accessed 15 Oct 2015.

Barsh, Russel L. 2002. Netukulimk past and present: Miqmaw ethics and the Atlantic fishery. *Journal of Canadian Studies* 37(1): 15–38.

Battiste, Marie, and James (Sákéj) Youngblood Henderson. 2000. *Protecting indigenous knowledge and heritage: A global challenge.* Saskatoon: Purich.

Baxter, Jayson. 2013. Mi'kmaq chiefs want killing white moose to be banned. *Atlantic CTV News,* October 18. http://atlantic.ctvnews.ca/mi-kmaq-chiefs-want-killing-white-moose-to-be-banned-1.1503590. Accessed 15 Oct 2015.

BBC. 2013. What does a stem cell burger taste like? *BBC.com,* video, 3:03, August 5. http://www.bbc.com/news/science-environment-23529841. Accessed 15 Oct 2015.

Bombay, Amy, Kimberly Matheson, and Hymie Anisman. 2014. The intergenerational effects of Indian residential schools: Implications for the concept of historical trauma. *Transcultural Psychiatry* 51(3): 320–338. doi:10.1177/1363461513503380.

Brave Heart, Maria Yellow Horse. 1999. *Oyate Ptayela*: Rebuilding the Lakota Nation through addressing historical trauma among Lakota parents. *Journal of Human Behavior in the Social Environment* 2(1–2): 109–126. doi:10.1300/J137v02n01_08.

Brave Heart, Maria Yellow Horse, and Lemyra M. DeBruyn. 1998. The American Indian Holocaust: Healing historical unresolved grief. *American*

Indian and Alaska Native Mental Health Research 8: 56–78. doi:10.5820/aian.0802.1998.60.

Brooks, Ross. 2013. Nova Scotia's Sacred Albino moose killed by visiting hunters. *Inhabitat.com*, October 16. http://inhabitat.com/nova-scotias-sacred-albino-moose-killed-by-visiting-hunters/. Accessed 15 Oct 2015.

Chandler, Michael J., and Christopher E. Lalonde. 2009. Cultural continuity as a moderator of suicide risk among Canada's first nations. In *Healing traditions: The mental health of Canadian aboriginal peoples*, ed. Lawrence J. Kirmayer and Gail Guthrie Valaskakis, 221–248. Vancouver: University of British Columbia Press.

Chandler, Michael J., and Travis Proulx. 2008. Personal persistence and persistent peoples: Continuities in the lives of individual and whole cultural communities. In *Self-continuity: Individual and collective perspectives*, ed. Fabio Sani, 213–226. East Sussex: Psychology Press.

Confederacy of Mainland Mi'kmaq. 2007. *Kekina'muek (Learning): Learning about the Mi'kmaq of Nova Scotia*. Truro: Eastern Woodland Print Communication. http://hrsbstaff.ednet.ns.ca/clambert/KekinamuekEntiredocumentprintable.pdf. Accessed 15 Oct 2015.

Couture, Joseph E. 2013. *A metaphoric mind: Selected writings of Joseph Couture*. Athabasca: Athabasca University Press.

Dias, Brian G., and Kerry J. Ressler. 2014. Parental olfactory experience influences behavior and neural structure in subsequent generations. *Nature Neuroscience* 17(1): 89–96.

Duran, Eduardo, and Bonnie Duran. 1995. *Native American postcolonial psychology*. Albany: State University of New York Press.

Fountain, Henry. 2013. Building a $325,000 burger. *The New York Times*, May 12. http://www.nytimes.com/2013/05/14/science/engineering-the-325000-in-vitro-burger.html?pagewanted=all. Accessed 15 Oct 2015.

Garriguet, Didier. 2007. Canadians' eating habits. *Health Reports* 18(2): 17. http://www.statcan.gc.ca/pub/82-003-x/2006004/article/habit/4148989-eng.htm. Accessed 29 Dec 2015.

Griggs, Brandon. 2014. How test-tube meat could be the future of food. *CNN.com*, May 5. http://www.cnn.com/2014/04/30/tech/innovation/cultured-meat/. Accessed 15 Oct 2015.

Gross, Cornelius, and Rene Hen. 2004. The developmental origins of anxiety. *Nature Reviews Neuroscience* 5(7): 545–552. doi:10.1038/nrn1429.

Hatcher, Annamarie, Cheryl Bartlett, Albert Marshall, and Murdena Marshall. 2009. Two-eyed seeing in the classroom environment: Concepts, approaches, and challenges. *Canadian Journal of Science, Mathematics, and Technology Education* 9: 141–153. doi:10.1080/14926150903118342.

Henderson, James Youngblood. 1997. *The Mi'kmaq Concordat*. Halifax: Fernwood Publishing.

Hornborg, Anne-Christine. 2013. *Mi'kmaq landscapes: From animism to sacred ecology*. Hampshire: Ashgate.

Impossible Foods. 2015. The impossible cheeseburger. *Impossible Foods*. http://www.impossiblefoods.com. Accessed 29 Dec 2015.

Kinnear, Lacia. 2007. *Contemporary Mi'kmaq relationships between humans and animals: A case study of the bear river first nation reserve in Nova Scotia*. Master's thesis, Dalhousie University.

Kwilmu'kw Maw-klisiaqn Negotiation Office, Mi'kmaq Rights Initiative. 2013. Have we had any success? *KMKNO News* 1(2): 3. http://mikmaqrights.com/uploads/KMKNONewsletterMarch2013.pdf. Accessed 15 Oct 2015.

Lawrence, Bonita. 2004. *'Real' Indians and others: Mixed-blood urban native peoples and indigenous nationhood*. Vancouver: University of British Columbia Press.

Lefort, Nadine, Clifford Paul, Ernest Johnson, and Charlie Dennis. 2014. *Tiam*. Mi'kmaq ecological knowledge: Moose in *Unama'ki*. *Unama'ki Institute of Natural Resources*. http://www.eecapacity.net/sites/default/files/fellows/docs/Moose%20in%20Unama'ki.pdf. Accessed 15 Oct 2015.

Loppie, Charlotte, and Fred Wein. 2007. The health of the Nova Scotia Mi'kmaq population. *Mi'kmaq Health Research Group*, January 7. http://www.unsi.ns.ca/wp-content/uploads/2015/07/ns-rhs-report-07.pdf. Accessed 15 Oct 2015.

Lundstein, Line Gertrud. 2014. *Can cortisol be used to assess acute stress in moose?* Bachelor thesis, Hedmark University College. http://brage.bibsys.no/xmlui/bitstream/handle/11250/196259/Lundstein.pdf?sequence=3&isAllowed=y. Accessed 15 Oct 2015.

Muin'iskw, Jean, and Dan Crowfeather. Mi'kmaw spirituality 101. *Mi'kmaq Spirit*. http://www.muiniskw.org/pgCulture2.htm. Accessed 15 Oct 2015.

Oster, Richard T., Angela Grier, Rick Lightening, Maria J. Mayan, and Ellen L. Toth. 2014. Cultural continuity, traditional indigenous language, and diabetes in alberta first nations: A mixed methods study. *International Journal for Equity in Health* 13(1): 92. doi:10.1186/s12939-014-0092-4.

Pastore, Ralph. 1998. Traditional Mi'kmaq (Micmac) culture. *Heritage Newfoundland and Labrador*. http://www.heritage.nf.ca/articles/aboriginal/mikmaq-culture.php. Accessed 15 Oct 2015.

Paul-Martin, Patsy. n.d. *Mi'kmaq months of the year*. Truro: Eastern Woodland Publishing. http://firstnationhelp.com/ali/posters/pdf/months.pdf. Accessed 15 Oct 2015.

Pritchard, Evan T. 2001. *No word for time: The way of the Algonquin people*. San Francisco: Council Oak Books.

Province of Nova Scotia. 2013. Current status. *NovaScotia.ca*, October 13. http://novascotia.ca/abor/office/what-we-do/negotiations/current-status/. Accessed 15 Oct 2015.

Province of Nova Scotia. 2014. Species at risk overview. *Nova Scotia.ca*, May 28. http://novascotia.ca/natr/wildlife/biodiversity/species-list.asp. Accessed 15 Oct 2015.

Rachel, Yehuda, Nikolaos P. Daskalakis, Frank Desamaud, Louri Makotkine, Amy L. Lehrner, Erin Koch, Janine D. Flory, Joseph D. Buxbaum, Michael J. Meaney, and Linda M. Bierer. 2013. Epigenetic biomarkers as predictors and correlates of symptom improvement following psychotherapy in combat veterans with PTDS. *Frontiers In Psychiatry* 4: 118. doi:10.3389/fpsyt.2013.00118.

Rand, Silas T. 2004. *Legends of the Micmacs*, vol. II. 1893. Reprint, West Orange: Invisible Books.

Rand, Silas T. 2005. *Legends of the Micmacs*, vol. I. 1893. Reprint, West Orange: Invisible Books.

Redish, Laura, and Orrin Lewis. 2015. Nukumi and fire. *Native Languages of the Americas*. http://www.native-languages.org/mikmaqstory2.htm. Accessed 15 Oct 2015.

Restoule, Jean-Paul. 2011. Everything is alive and everyone is related: Indigenous knowing and inclusive education. *Federation for the Humanities and Social Sciences*, January 25. http://www.ideas-idees.ca/blog/everything-alive-and-everyone-related-indigenous-knowing-and-inclusive-education. Accessed 15 Oct 2015.

Robinson, Margaret. 2013. Veganism and Mi'kmaq legends. *Canadian Journal of Native Studies* 33(1): 189–196.

Robinson, Margaret. 2014. Animal personhood in Mi'kmaq perspective. *Societies* 4(4): 672–688. doi:10.3390/soc4040672.

Service Canada. 2011. Aboriginal peoples: Atlantic Canada. *Employment and Social Development Canada*. http://www.esdc.gc.ca/eng/jobs/lmi/publications/csp/abor/atlantic/abor_march2014.pdf. Accessed 29 Dec 2015.

Shen, Helen. 2012. Mark post, tissue engineer. *University of Santa Cruz, Science Communication Program*, March 31. http://scicom.ucsc.edu/publications/QandA/2012/post.html. Accessed 15 Oct 2015.

Strong, Pauline T., and Barrik Van Winkle. 1996. Indian blood: Reflections on the reckoning and reconfiguring of native North American identity. *Cultural Anthropology* 11(4): 547–576.

Taylor, Charles. 1985. The concept of a person. In *Philosophical papers. Human agency and language*, vol. 1, ed. Charles Taylor, 97–114. Cambridge: Cambridge University Press.

Travers, Kim D. 1995. Using qualitative research to understand the sociocultural origins of diabetes among Cape Breton Mi'kmaq. *Chronic Diseases in Canada* 16: 140–143.

Tuomisto, Hanna L., and M. Joost Teixeira de Mattos. 2011. Environmental impacts of cultured meat production. *Environmental Science & Technology* 45(14): 6117–6123. doi:10.1021/es200130u.

Walker, Marilyn. 2008. *Wild plants of Eastern Canada*. Halifax: Nimbus.

Wallis, Wilson D., and Ruth S. Wallis. 1955. *The Micmac Indians of Eastern Canada*. Minneapolis: University of Minnesota Press.

Walters, Karina L., Ramona E. Beltran, David Huh, and Teresa Evans-Campbell. 2011a. A dis-placement and dis-ease: Land, place and health among American Indians and Alaska natives. In *Communities, neighborhood, and health: Expanding the boundaries of place*, ed. Linda M. Burton, Susan P. Kemp, Man Chui Leung, Stephen A. Matthews, and David T. Takeuchi, 163–199. Philadephia: Springer Science Business Media, LLC.

Walters, Karina L., Selina A. Mohammed, Teresa Evans-Campbell, Ramona E. Beltran, David H. Chae, and Bonnie Duran. 2011b. Bodies don't just tell stories, they tell histories. *Du Bois Review: Social Science Research on Race* 8(1): 179–189.

Wasser, Samuel K., Jonah L. Keim, Mark L. Taper, and Subhash R. Lele. 2011. The influences of wolf predation, habitat loss, and human activity on caribou and moose in the Alberta oil sands. *Frontiers in Ecology and the Environment* 9(10): 546–551. doi:10.1890/100071.

Weaver, Ian C.G., Nadia Cervoni, Frances A. Champagne, Ana C. D'Alessio, Shakti Sharma, Jonathan R. Seckl, Sergiy Dymov, Moshe Szyf, and Michael J. Meaney. 2004. Epigenetic programming by maternal behavior. *Nature Neuroscience* 7(8): 847–854. doi:10.1038/nn127.

Whitehead, Ruth H. 1988. *Stories from the six worlds: Micmac legends*. Halifax: Nimbus.

Part IV

New Media Is the Message

From Seitan Bourguignon to Tofu Blanquette: Popularizing Veganism in France with Food Blogs

Ophélie Véron

Once disparaged, ridiculed, or just plain ignored by the general public, vegans are a fast-growing group in French society. Until quite recently, meatless products could be found only at organic shops and were viewed by the general public as something for macrobiotic hippies who adore flavorless tofu. But today, these products represent a market that is expanding at an impressive pace. Although most restaurant owners are still dragging their feet when it comes to adding vegan or even vegetarian options to their menus and school cafeterias are still required by law to include animal protein in every dish they serve, it cannot be denied that veganism has arrived in France. The most-watched evening news shows regularly feature segments on the so-called "veggie trend," vegan restaurants seem to be popping up like mushrooms all over Paris, food manufacturers are experimenting with vegetarian meat alternatives, and no fewer than 15 French-language cookbooks with the word "vegan" in

O. Véron (✉)
University College London, London, UK
e-mail: ophelie.ei.veron@gmail.com

J. Castricano, R.R. Simonsen (eds.), *Critical Perspectives on Veganism*,
The Palgrave Macmillan Animal Ethics Series,
DOI 10.1007/978-3-319-33419-6_13

their title appeared in the year following the first one in April 2014. In a country whose national cuisine has been listed as a UNESCO world heritage, and where a traditional dish is defined first and foremost by the type of meat used, this shift has been far from easy.

In this chapter, I will be analyzing the rise of veganism in a society that is still marked by the prevalence of what Melanie Joy has termed "carnism"[1] in its culture, public institutions, and daily practices. I will focus in particular on the role vegan food blogs play in these changes in societal perception and behavior. The first French blogs of this kind appeared in 2006–2007, and today there are more than 50. With their recipes and articles, these bloggers work to educate their readers, promote vegan cuisine, and facilitate a transition to an animal-free diet for as many people as possible. I will argue that by revisiting traditional dishes, highlighting the culinary delights offered by vegan cuisine, and presenting it as a healthy and delicious alternative to meat-based food, these blogs have increased awareness of veganism among people outside their usual readership, and have thus helped expand acceptance of veganism in French society. Although some fear that this popularity could weaken the radical impetus of veganism as a politics, I will highlight the effects it has had on the growing awareness of issues related to the welfare and rights of animals.

I will first examine how blogs, and particularly food blogs, are involved in community-building and the formation of subcultures. Next, I will look at the growth of vegan food blogs in France. I will then demonstrate that although these blogs' initial main audience was the existing vegan community, they later expanded beyond this small circle, reaching new audiences, and thus helping popularize veganism in French society and, arguably, beyond. Finally, I will measure the role of these blogs against the increase in general awareness of animal ethics issues. While evaluating the possible risks of awareness-raising efforts centered more on culinary enjoyment than issues of justice and animal rights, I will conclude by presenting the relatively positive impact these blogs have had on the general public's growing interest in the animal cause.

This paper is based on research related to vegan food blogging, in which I participated first as a blogger and activist, and later as a researcher, combining my two identities as I became increasingly involved in the movement. Drawing upon cyber-ethnography methods, interviews with vegan food bloggers, and a survey of their readers, whom I approached via social networks and my own blog, I provide examples of how animal-free diets and lifestyles are becoming normalized in French society. Finally, I suggest that vegan blogs can provide insight into new ways of viewing and practicing veganism in the twenty-first century.

Food Blogging Communities and Subcultures

Although a certain number of researchers have examined issues of food and culture,[2,3,4,5] (Belasco, 2008; Parasecoli 2008; Brun and Jacobs, 2006; Pence 2002; Meigs, 1998) and the role of blogs in society,[6] few sociological studies have focused on food blogs, much less vegan food blogs. Recipe-sharing is, however, a long-standing practice. Sharing recipes and food-related information is a way of expressing one's "experiences, preferences, observations, and desires."[7] On a blog, this sharing is public and reaches a wider audience, as is the case with books and magazines, and thus becomes a way of reflecting a culture and defining a community, thus "inscribing the self with a sense of place, belonging and achievement."[8] (Gallegos 2005) Because blogs allow for interaction going beyond the individual or family level, they often lead to the creation of a community founded on a sense of shared identity.[9] (Ferguson 2012) Offering a more dynamic relationship than cookbooks, blogs allow people to come into contact regardless of the physical distance that may separate them.[10] (Lofgren 2013) These opportunities for sharing and discussion, which break with traditional notions of passive media spectatorship,[11] (Jenkins 2008:3) make these blogs a vector for participatory culture. Many bloggers offer advice to their readers who, in return, comment, ask questions, and share their personal experiences, thus allowing for the emergence of a type of dynamic and collective expertise.

A virtual community created in this way can be considered a subculture. According to Nancy Baym (2010), these groups are composed of "like-minded individuals" and can be compared to "semi-organized grassroots social movements." Blogs may thus share many features with non-virtual groups, in which individuals "develop identities through performances that build distinctive styles."[12] Jeff Bishop and Paul Hoggett note that subcultures are a key vector through which "dominant values are transmitted, resisted, or negotiated and new sets of values, which may take as their point of origin a different mode of production and social organization, emerge".[13] Food blogs can therefore be a way for bloggers and their readers to question and challenge certain food-related norms, habits, or dominant representations. The very nature of the social networks on which they depend sometimes lends them an influence extending beyond just the food blogosphere.[14]

In the next section, I will explore the aspects of French vegan food blogs that relate to subculture, identity, and community, and will examine their potential impact on French society and its dominant representations—in particular, as they are related to speciesist and carnist ideologies.

Building a Vegan Subculture Community Online

French vegan food blogs appeared around the same time than the general French food blogosphere emerged. Many in the general public associated animal-rights activists with the lightning-raid tactics of the Animal Liberation Front (ALF) and thought of people following a vegan diet as ascetics who belong to cults and live almost exclusively on soy burgers and sprouts.[15] The only vegan recipes available were found on black-and-white brochures distributed at stands run by a handful of animal-protection associations, and many people followed a vegan diet on their own without ever meeting any other vegans or activists. Vegan food blogs thus played a special role in bringing people together and strengthening the community. Drawing on content analysis, participant observation, interviews with 16 bloggers, and a survey of 276 vegan food blog readers, I show how this vegan blogosphere has helped build and develop the identity of the French vegan community.

According to Nancy Baym (2010), a community, virtual or not, is based on a certain number of characteristics including a sense of belonging and space; shared/social identities, resources and practice; sociability and interpersonal relationships; information and support.[16] The notions of belonging and shared identity can be seen in most of the names chosen for French vegan blogs—a large number of them include a direct reference to their vegan identity: *VG-Zone, Végébon, Vegansfield, Enfant Végé, 100 % Végétal, Ma Cuisine Végétalienne*, etc. The idea of providing support and information is also clear in the mission statements of many of these blogs. One of the first French vegan blogs to appear was *VG-Zone*, which is "first and foremost designed to make daily life easier for Parisian vegans, but [is] also for anyone passing through our lovely capital city."[17] This is not a blog that aims to "explain to newbies what strict vegan and ovo-lacto-vegetarian diets are" but rather "an urban survival guide for grocery shopping," suggesting places to dine out "without the risk of finding a bone in your food" and "making a quick meal in your tiny kitchen at home." The blog's goal of offering its readers assistance is clearly stated: "Don't panic! We're here to help!"[18] The bloggers behind *VG-Zone*, Laura and Sébastien, pointed out that at the time they created their blog, this type of information "didn't exist" and that "the few [vegan blogs] that were around didn't do a great job of promoting veganism."[19] The idea of a community founded on a common practice, shared resources, and a network of mutual support is clearly echoed in the comments left by their very first readers. One of them wrote, "Your site is really great! Both for the recipes and the wealth of information about eating well as a vegan … It's so nice to finally find a good source like this one in the chaos of the blogosphere. Many thanks to both of you for everything you're doing for the veg community!"[20] The blog has thus become a space for sociability and interpersonal relationships, since readers share their own practices and ask for advice. The authors often reply to these comments, but like on a forum, other readers also add their thoughts.

The results of my survey of readers of vegan blogs reveal similar motivations. Several respondents stated that they began reading this kind of blog after going vegetarian or vegan, with a view to "finding recipes and tips to make things easier," "diversifying [their] diet and getting guidance," "having support for the transition" or seeing "that this lifestyle was really possible." In places where veganism is less common, blogs are often the only—or at least the first—direct contact people have with others who share their values. One survey respondent said that for her, these blogs allowed her to "see that [she] wasn't the only one who had these beliefs." Another wrote that she was "at a loss as far as what to eat" and contacted some bloggers who "very kindly mentored [her] through the first steps." A desire to escape from a form of social isolation by meeting other vegans can be seen in the experience of another reader, for whom the blogs were like "a breath of fresh air" and provided "support and education in terms of cooking and nutrition when everyone else you know swears by a meat-based diet and knows nothing about other ways of eating." Some respondents stated that the blogs made them feel "reassured about [their] choices" and "less alone." Vegan food blogs therefore seem to have played a major role in terms of identity- and community-building at a time when veganism was not well known or widely accepted in French society.

These blogs are also a core part of a subculture that defies dominant ideological norms. By presenting alternative lifestyles and consumer models that raise questions about the traditional French culinary landscape, bloggers regularly challenge the speciesism and carnism that are prevalent in French society. In this sense, their actions can be compared to what Michel de Certeau (1984) called "cultural poaching."[21] Using digital tools, bloggers are like "poachers" who, slipping through breaches in the dominant culinary landscape, redesign their daily eating habits and inspire their readers to make changes of their own. Revisiting traditional recipes to create egg-free crêpes, seitan bourguignon, and tofu blanquette recall the *détournement* tactics of the Situationists, offering individuals the opportunity to question a dominant system's rules and to reappropriate its existing codes. This can be an enjoyable challenge, offering the satisfaction of creating something new as well as a treat for the taste buds, and furthermore shows that anyone can reappropriate and revisit the great classics of French cuisine. One of the bloggers interviewed described vegan cuisine as "fun and creative" and said that it provides a chance to "question the merits of traditional meat-based cuisine."[22] The name of one of these blogs, *Pigut*,[23] an acronym standing for *Petites Idées pour Grandes Utopies* (Small Ideas for Big Utopias), testifies to this tactical aspect, since its author, who goes by the name Melle Pigut, shows that with relatively few resources—an "old computer," a "trusty camera" and "a couple of reliable cooking utensils"—it is possible to create "big utopias … day after day, together with you and our small ideas."[24] The recipes posted by these bloggers represent transgressions against the traditional paradigm, ways of rejecting the inevitability of consuming animal products, creating different cultural references and devising a new system built on alternatives to the dominant practices. Vegan food blogs help strengthen and develop the vegan subculture and have also popularized it outside its community of origin.

Vegan Blogs: Inspiring Behavioral and Societal Change

Some vegan blogs make no secret of their efforts to reach out to those outside their circle of supporters and to help veganism become better known in French society. Here, I will examine the two main strategies blogs use to help popularize the vegan diet: updating the image of veganism in France and putting this way of eating within everyone's reach.

To change the way veganism is perceived, it is essential to do away with certain misconceptions—namely, that vegans are marginal members of society at best, dangerous extremists at worst; that they are usually pale and nutritionally deficient; that they follow a bland, restrictive, and monotonous diet. The idea, in one blogger's words, is to show that "vegans [are] not daisy-smoking hippies."[25] To this end, many bloggers focus on how diverse, refined, and delicious vegan food can be. The blog *100 % Végétal* makes this clear right away in its mission statement:

> Here, raw food, flaxseed and coconut oil rub shoulders with burgers, nuggets and even a vegan version of fish and chips. In the holiday season, there are recipes such as chestnut-stuffed seitan roast, hazelnut roulades and even a pâté inspired by foie gras. When summer comes around, there are colorful homemade ice creams packed with fruit and cakes that may skip the eggs but certainly don't compromise on flavor. Basically, you won't find any deprivation in this kind of cuisine. It's more ethical, eco-friendly and healthy, more in tune with the seasons, with even more different flavors and new combinations that you'll want to experience ASAP.[26]

As an example of this diversity, the blog *Végébon*[27] regularly features photos of meals made at home, ordered at restaurants, or taken to work in a lunch box, while the blog *Au Vert avec Lili*[28] offers no fewer than 26 types of recipes ranging from salads, crudités, and wraps to quiches, brioches, and muffins. According to Gaëlle, the blogger behind *Better than Butter*,[29] the idea is to "show people it's possible to skip meat and still eat a variety of delicious and nutritionally balanced dishes."[30]

To debunk the misconceptions, it is essential to show that vegan cuisine can be delicious and elegant. As Laura and Sébastien see it, vegans "have unfortunately inherited the image of the 1970s macrobiotic movement: ascetic, restrictive and unglamorous."[31] Their blog offers elegant dishes and desserts on par with the creations of top chefs, underscoring how important culinary excellence and visual appeal are when it comes to vegan cuisine. For this reason, vegan bloggers tend to take a great deal of care with food styling and photography. Many of them have galleries of their culinary creations, and some even work as professional photographers. Special attention is also paid to blog design, since, as Gaëlle put it, "attractive food blogs are appealing and show the dishes in their best light, which can help get omnivores interested."[32]

Putting veganism within everyone's reach is the second goal of vegan food blogs. This strategy is founded upon a pragmatic approach aimed at making life easier for people by offering simple, everyday recipes. Since vegan cuisine is often burdened by a reputation for being complex, requiring exotic ingredients found only at organic food stores, the idea here is to make veganism accessible to everyone. Sophie, the author of *Enfant Végé*, said that she feels one of the best ways to promote veganism is to "show that being vegan today is EASY!"[33] The blogger of *L'Aventure Lavable*[34] said that "low-cost recipes are an initial entry point" that can win people over. She therefore tries to offer budget-friendly recipes so that readers will see that animal issues do not have to "take a back seat" if they ever experience "financial difficulties."[35] Sandrine of *Végébon* summarizes the general idea as follows: "veganism can be for anyone."[36]

These strategies have clearly affected the readers of these blogs. According to the results of my survey, although 38 % of the readers were omnivores before discovering these blogs, only 6 % of them still are now. And while 9 % of them were already vegan, 37.5 % of the readership now identifies this way. Among the blog followers, 72 % read only French-language vegan food blogs. This diminishes the influence other types of blogs, particularly English-language ones, may have on their changes in eating habits. Several respondents mentioned the role French blogs played in their discovery of veganism: "My children have food allergies. I was looking for new recipes so I could make a greater variety of dishes. That's how I stumbled upon a few vegetarian and vegan blogs, which helped me learn about another way of living and also inspired me to explore a new path myself and become vegan." Others spoke of the blogs making something "click" inside them. One person said that the blogs opened the way to "discovering another world" and "taking a new look at [his or her] lifestyle, adopting more critical thinking and realizing that cooking vegetarian or vegan isn't as hard as all that." Another indicated that the blogs allowed her to "take the plunge" while yet another said, "Happening upon a vegan food blog by chance is what led to my going vegan. I was wondering, 'why are there vegans?' and I looked it up. And I went vegan too." Several respondents named one or more blogs that directly inspired their decision: "I went vegan because of the blog *Au Vert avec Lili*." Others stressed that if they had not learned about vegan cooking through these blogs, they might never have brought their actions into line with their beliefs: "If I hadn't discovered that you can eat great food without milk, eggs or honey, I would never have taken the leap," and "At the beginning, veganism seemed extreme. Vegan blogs … introduced me to vegan cuisine, and soon enough my transition to veganism had been made, almost without my thinking about it!"

This last aspect is particularly important, since it suggests that information alone (about the conditions of farm animals or the impact of animal agriculture on the environment) is not enough to trigger changes in beliefs. Here, it seems that preferences (habits, emotions, desires, etc.) can play a central role in accepting and processing information. This is confirmed by one blog reader's comment: "When I wasn't yet ready to consider veganism, I avoided information on animal ethics … Once I realized that vegan food was great, I became more receptive to the ethical part and began learning as much as possible so that I could make my choices with full knowledge of the facts!" This phenomenon demonstrates the need to take the emotional rationalities of individuals into account and to "prepare" them for the information by working on preferences with a view to breaking down the psychological barriers and the cognitive dissonance (Festinger 1957; Gibert 2015) that lead to attitudes of resistance to information and change. It seems here that vegan bloggers pay particular attention to these cognitive biases not only by offering content centered on information, but by taking into account the emotions, beliefs, desires, or habits that may lead individuals to develop resistance mechanisms.

In just a few years, it seems that the French food culture landscape has changed quite a bit and that vegan cooking blogs have played a major role in this evolution. Many are the bloggers who, after their blogs proved successful, have been asked by publishers to write vegan cookbooks (as was the case for Marie Laforêt of the blog *100 % Végétal*, who in April 2014 published France's first openly vegan cookbook)[37] or have gone on to professionalize their involvement in the culinary world by offering cooking classes or online coaching. Laura and Sébastien said that between the time when their blog was just getting started and today, the image of veganism in French society has improved so much that "it's like night and day."[38] For Sophie, the change "in the media [is] obvious," while Gaëlle considers that veganism "is becoming more and more accessible … and 'scaring' people less."[39, 40] Almost all of them agree that this change in the French cultural landscape is a positive one. Around 42 % of respondents felt that blogs have an "enormous" impact in terms of popularizing veganism, and 38.5 % said that they play a "large" role. One reader said that "the many vegan blogs that exist today have made veganism synonymous with modernity, youth, energy and gourmet cuisine, in my opinion. They have crushed the old stereotypes of a meatless, animal-product-free meal being 'not a real meal,' 'flavorless,' 'not filling,' 'boring' or 'outdated.'" While the concept of veganism seems foreign to the majority of the French population, the term is, according to a reader, "beginning to enter the vocabulary of the average person, along with certain information."

The impact vegan food blogs have had on the French culinary landscape is major. Yet the growing popularity of veganism gives rise to certain questions: by focusing more on taste than on animal ethics, do we not run the risk of detracting attention from the movement's political side? It is this question that I will address and attempt to answer in the following section.

French Vegan Blogs and the "Veganist" Strategy

Several voices have recently been raised in the French anti-speciesist and equalitarian community against what they call the "vegetarianist" or "veganist" strategy of the animal-rights movements. According to Pierre Sigler (2014), this type of strategy is based on the following ideas: observing a vegan diet is the biggest thing we can do to help the animals; the best way to weaken the meat industry is to increase the number of vegetarians and vegans; trying to convince others to become vegetarians or vegans is the most effective way to increase the number of vegetarians or vegans.[41] In the view of Anushavan Sarukhanyan (2013), this conversion strategy is ineffective because it makes the animal issue a question of personal choice and not of justice—two aspects he seems to consider as mutually exclusive and which make him demand, in a provocative manner, "the abolition of veganism."[42] Furthermore, as he sees it, the use of arguments other than ethical, focusing either on health or on the environment, is immoral and tends to convey an implicitly speciesist message. For Bonnardel,[43] we must strive to change society, not individuals—the promotion of vegetarianism and veganism "is so important that it overshadows political demands."

Perhaps more than any other form of activism, vegan food blogs embody this veganist strategy. While animal ethics is a major factor in their food choices, a certain number of bloggers never address this topic on their blogs. One of them admitted that she prefers to "inspire people to eat plant-based food … rather than openly engaging in activism."[44] Laura and Sébastien said that they aim "to never engage in any proselytism" and are "careful not to talk about the various reasons for which someone may want to adopt a strictly plant-based diet."[45] Marie Laforêt acknowledged that "promoting a vegan diet for health or environmental reasons while sweeping the ethics argument under the rug seems … problematic from a strategic point of view."[46] According to another blogger, "in a perfect world, we would center everything on ethics." She compared it to the abolition of slavery: "it would be ridiculous to say, 'Stop buying black slaves—it's bad for the environment' or 'Buying too many black slaves is bad for your budget!'"[47] Melle Pigut felt that the use of arguments other than ethics can be "dangerous" if presented on their own.[48] For Sophie, however, these other arguments must be used, "because they are also positive for the animals, and if people are already eating less meat for other reasons, they'll be more likely to listen to the ethical arguments without taking offense. But ethics must be kept at the forefront!"[49]

Yet although 78 % of the blog readers surveyed felt that vegan blogs have played a "fairly important," "important," or "major" role in their food choices, this figure fell to 59 % when it came to the role blogs have played in their awareness of animal ethics, and the percentage of those who said this role was not very significant rose from 16 % to 28.5 %. Similarly, while 42 % of them considered that the blogs made an "enormous" contribution to popularizing veganism in society, and 38.5 % thought they had a "large" impact, these figures declined to 16 % and 31 %, respectively, when it came to the role they play in the general public's awareness of animal ethics. The percentage of those saying that the blogs help "a little" rose from 14 % to 38 %, while the "not at all" answers increased from 3.5 % to 11.5 %. One reader lamented that "the blogs … too often shy away from justice issues," adding: "I find it a shame that many of them focus on nutrition and the environment, which conveys the erroneous message that we're vegan for our health or the climate. I'm not against the idea of addressing these issues—they're important—but I would like to see it said more often that we are vegan first and foremost for reasons of justice." Another respondent worried about a trend that reflects the perception of veganism in society: "Most people that I meet ask me first if I'm vegan for my health, and not if it's for the animals. Few people truly realize that there is a dead animal on their plate when they eat meat, which for me is problem number one."

However, more respondents seem to condemn the "green & healthy" trend, which is to say a certain tendency throughout the blogosphere to focus on health and nutrition and to make a lot of fuss over green smoothies and "detoxing" salads. Gaëlle is not happy about "this 'healthy eating' fad in which veganism is thought of as a 'weight-loss diet' and no mention is even made of the ethical aspects."[50] Melle Pigut said that this can "introduce more people to veganism" but that "it's a double-edged sword because the goals are not the same and the message gets obscured."[51] Marie Laforêt commented that "it isn't possible to effectively fight animal exploitation by using arguments or campaigns that reproduce forms of oppression or discrimination," such as fat-shaming.[52] Many bloggers and blog readers alike condemn the idea of veganism as a passing trend focused on health and nutrition. One reader spoke of "confusion" and "discredit" that risks making veganism look like "something for fashionable grannies." Another pointed out that "this association with a trend could lead to people going vegan only for a short time (what will happen when this green & healthy trend, like all trends, goes out of style?) and never making any connection with animal ethics." A third commented that "this trend may popularize eating less meat and feeling better about yourself, but does not directly promote veganism, which is a political struggle."

These testimonies pose the question of a solely veganist approach, such as apparently exemplified by blogs that offer "only" vegan recipes. Does this strategy weaken the radical impetus of veganism as a politics?

A Pragmatic Complementary Approach

While it is useful to assess the pitfalls of such an approach, it seems, however, that vegan food blogs, on the one hand, are part of a context of complementarity and, on the other hand, opt for a pragmatic approach that does not conceal the ethical aspect of veganism but takes the psychological reality of their audience into account. Finally, we must take note of the diversity of the content offered by some vegan bloggers—rather than posting only recipes, they also offer articles featuring information or thoughts about animal welfare and rights. In this sense, their blogs support the political demands of the animal-rights movement.

Vegan food blogs are not the only form of animal-rights activism in France. A number of organizations, groups, and networks of activists have formed over the past few years working, like bloggers, to end speciesism and carnism. The French abolitionist organization L214 has made animal ethics its key focus, circulating "disturbing" content such as videos filmed undercover at animal production facilities and events aimed at making people aware of animal suffering. It orients sympathizers to vegan food blogs to help them put their philosophy into practice.

Indeed, some blog readers mentioned that although their awareness of animal ethics is due to content presented by these organizations, they regularly visit vegan food blogs to get recipe ideas and inspiration. In this sense, blogs play a fundamentally pragmatic role, providing their readers with concrete support and helping them put anti-speciesism into practice in their daily lives.

Furthermore, because "disturbing" content and events can encounter resistance in French society, it may seem easier to bring the general public closer to ethical issues by normalizing veganism. One blogger said that "showing shocking images of animals being mistreated often puts people off—they stop listening to what we have to say!" In her opinion, it is essential to "identify the type of person you're dealing with"—a cultural, psychological, and interpretative approach that takes into account the resistance mechanisms that act as obstacles blocking acceptance of the information.[53] This approach also recognizes the possibility of gradual individual evolution. A certain number of the bloggers interviewed admitted that they did not become interested in veganism for ethical reasons, but rather due to environmental or nutritional motivations. At the same time, they all felt that they had been made aware of animal ethics issues and that these were now core to their efforts. The possibility of this kind of evolution is echoed in the comments of some survey respondents. One said, "I became vegetarian mainly because of environmental and health issues. I was aware of the animal cause, but not enough for me to take action. The vegan blogs I read made me think more about it, in particular the idea of speciesism."

Finally, not all vegan food blogs focus only on food, but even the ones that do often include the blogger's thoughts about ethical and political issues alongside the recipes. Sophie said that she addresses the issue of animal ethics "as often as possible in [her] articles, since that's the most important part for [her]."[54] Melle Pigut published a two-part article on "the myth of happy meat" in which she deconstructs the different strategies aimed at suppressing the feeling of cognitive dissonance.[55] Lili wrote an article exposing the force-feeding of ducks and geese to make foie gras. Finally, Sandrine covered animal ethics in a series of articles with titles such as "Why go vegetarian?," "Why eat eggs?," and "Why eat honey?"[56,57,58] This diversity was mentioned by several survey respondents, according to whom "vegan food blogs don't necessarily offer recipes only; they also provide information and figures … that can help change mindsets."

It thus seems that vegan food blogs are not founded upon a purely veganist strategy, which could weaken the political message of the anti-speciesist movement. More generally, it appears that efforts to get more recognition for animal rights is inseparable from a pragmatic approach taking into account the cultural and psychological aspects surrounding acceptance of social and ethical demands and the implementation of political action.

Conclusion

Vegan food blogs have played a key role in helping veganism grow in France over the past few years. Far from addressing only activists who are already dedicated to the cause, bloggers have begun targeting a wider audience, posting recipes, cooking techniques, tricks, and tips to put veganism within everyone's reach and, by introducing sophisticated, modern vegan cuisine, updating the image of veganism in French society. If their direct role in raising the general public's awareness of animal ethics seems limited, it is, nevertheless, important, since blogs have often used their popularity to circulate messages about justice and animal rights to a population that had previously been almost completely unaware of these issues. I will thus conclude this chapter with the positive impact these blogs have had on promoting veganism as a political struggle, while underscoring the need to avoid limiting ourselves to one type of strategy, whether it is based on conversion or public debate. Just as a purely veganist strategy is not enough to generate political and institutional evolution, anti-speciesism as a social movement cannot do without practical daily support taking into account the cultural and psychological considerations of the target audience. For this reason, we should ensure that veganism should not be "abolished" but help support all demands for the abolition of animal exploitation. An approach solely centered on culinary enjoyments or health runs the risk of conveying on erroneous message on veganism, which would therefore be deprived of its philosophical component and reduced to a mere plant-based lifestyle and diet.

Because veganism is first and foremost an enacted way of opposing speciesism and other ideologies of oppression, ethics should always be kept at the forefront of the vegan movement.

Notes

1. Melanie Joy, *Why we Love Dogs, Eat Pigs and Wear Cows—An Introduction to Carnism*. San Francisco: Conari Press, 2010.
2. Warren Belasco, *Food: The Key Concepts*, Oxford: Berg, 2008.
3. Anna Meigs, Food as a Cultural Construction. In *Food and Culture: A Reader*, eds. C. Counihan and P. Van Esterik, New York: Routledge, 1997.
4. Fabio Parasecoli, *Bite Me: Food in Popular Culture*, Oxford: Berg, 2008.
5. Gregory E. Pence, Introduction: The Meaning and Ethics of Food. In T*he Ethics of Food: A Reader for the Twenty-First Century*, ed. G. E. Pence, Lanham: Rowman & Littlefield, 2002.
6. Alex Bruns and Joanne Jacobs, Uses of Blogs, ed. S. Jones, *Digital Formations*, New York: Peter Lang Publishing, 2006.
7. Belasco, *Food: The Key Concepts*, 44.
8. Danielle Gallegos, Cookbooks as manuals of taste. In *Ordinary Lifestyles: Popular Media, Consumption and Taste*, eds. D. Bell and J. Hollows, Maidenhead: Open University Press, 2005:99.
9. Kennan Ferguson, Intensifying Taste, Intensifying Identity: Collectivity through Community Cookbooks, *Signs*, 37 (3) 2012:695–717.
10. Jennifer Lofgren, Food Blogging and Food-related Media Convergence, M/C Journal, 16:3, 2013.
11. Henry Jenkins, *Convergence Culture: Where Old and New Media Collide*, New York: New York University Press, 2008:3.
12. Nancy Baym, *Personal Connections in the Digital Age. Digital Media and Society Series*, 2010:16.
13. Jeff Bishop and Paul Hoggett. *Organizing Around Enthusiasms: Patterns of Mutual Aid in Leisure*. London: Comedia Publishing Group, 1986:44.
14. Jennifer Lofgren, *ibid.*
15. Ample evidence of these common prejudices are the TV shows and press articles which depicted veganism as an extreme and marginal movement. The official nutritional guidelines in France still refer to veganism as a dangerous diet, which causes deficiencies and may lead to death.

16. Nancy Baym, *ibid.*
17. VG-Zone, "A propos de VG-Zone", *Vg-Zone.net,* http://vg-zone.net/a-propos-du-site/ [accessed: 01/10/2015].
18. VG-Zone, *ibid.*
19. Interview, October 2015.
20. VG-Zone, *ibid.* http://vg-zone.net/a-propos-du-site/#comment-1152/ [accessed: 01/10/2015].
21. Michel de Certeau, *The Practice of Everyday Life. Vol. 1 and 2.* Minneapolis: University of Minnesota Press, 1984.
22. Interview, October 2015.
23. http://pigut.com
24. Melle Pigut, "Qui Comment Quoi Pourquoi?", *PIGUT,* http://pigut.com/a-propos/ [accessed: 01/10/2015].
25. Interview, October 2015. The original text mentions "des hippies fumeurs de pâquerettes" and refers to the stigma often attached to the vegan movement in France. Vegans are often stereotyped as marginal non-conformists and pacifists, wearing long hair, keeping an unkempt appearance and using recreational drugs.
26. Marie Laforêt, "A propos", *100% Végétal,* http://www.100-vegetal.com/p/blog-page_19.html [accessed: 01/10/2015].
27. https://vegebon.wordpress.com
28. http://auvertaveclili.fr/
29. http://betterthan-butter.blogspot.fr/
30. Interview, October 2015.
31. Interview, October 2015.
32. Interview, October 2015.
33. Interview, October 2015.
34. http://l-aventure-lavable.over-blog.com/
35. Interview, October 2015.
36. Interview, October 2015.
37. Marie Laforêt, Vegan. Paris: La Plage, 2014.
38. Interview, October 2015.
39. Interview, October 2015.
40. Interview, October 2015.
41. Pierre Sigler, De l'appel à la vertu à l'exigence de justice pour les animaux. In *L'exploitation animale est une question de société,* eds. P. Sigler and Y. Bonnardel 2014:3–26.

42. Anushavan Sarukhanyan, "Pour l'abolition du véganisme, pour l'abolition de l'esclavage", *Pour l'abolition du véganisme, pour l'abolition de l'esclavage*, http://abolitionduveganisme.blogspot.fr/2013_03_01_archive.html, 2013 [accessed: 10/07/2015].
43. Yves Bonnardel, La question de la viande est un problème de société. In *L'exploitation animale est une question de société*, eds. P. Sigler and Y. Bonnardel 2014:39.
44. Interview, October 2015.
45. Interview, October 2015.
46. Interview, October 2015.
47. Interview, October 2015.
48. Interview, October 2015.
49. Interview, October 2015.
50. Interview, October 2015.
51. Interview, October 2015.
52. Interview, October 2015.
53. Interview, October 2015.
54. Interview, October 2015.
55. Melle Pigut, "Le mythe de la viande heureuse et autres contes merveilleux", *PIGUT*, http://pigut.com/2013/01/14/le-mythe-de-la-viande-heureuse-autres-contes-merveilleux-12/ [accessed: 01/10/2015].
56. Végébon, "Pourquoi devenir végétarien ? Raison 3 : l'éthique animale", *Végébon*, https://vegebon.wordpress.com/2010/05/09/pourquoi-devenir-vegetarien-ethique-animale/ [accessed: 01/10/2015].
57. Végébon, "Pourquoi manger des œufs", *Végébon*, https://vegebon.wordpress.com/2010/09/13/pourquoi-manger-des-oeufs/ [accessed: 01/10/2015].
58. Végébon, "Pourquoi consommer du miel ?", *Végébon*, https://vegebon.wordpress.com/2010/11/07/pourquoi-consommer-du-miel/ [accessed: 01/10/2015].

References

Baym, N.K. 2010. *Personal connections in the digital age. Digital media and society series*. Cambridge: Polity Press.

Belasco, W. 2008. *Food: The key concepts*. Oxford: Berg.

Bishop, J., and P. Hoggett. 1986. *Organizing around enthusiasms: Patterns of mutual aid in leisure*. London: Comedia Publishing Group.

Bonnardel, Y. 2014. La question de la viande est un problème de société. In *L'exploitation animale est une question de société*, ed. P. Sigler and Y. Bonnardel, 27–60. http://www.reseau-antispeciste.org/wp-content/uploads/pourpolitisation_cahier.pdf. Accessed 1 July 2015.

Bruns, A., and J. Jacobs. 2006. Uses of blogs. In *Digital formations*, ed. S. Jones. New York: Peter Lang Publishing.

de Certeau, M. 1980/1984. *The practice of everyday life*, vol. 1 and 2. Minneapolis: University of Minnesota Press.

Ferguson, K. 2012. Intensifying taste, intensifying identity: Collectivity through community cookbooks. *Signs* 37(3): 695–717.

Festinger, L. 1957/1976. *A theory of cognitive dissonance*. Stanford: Stanford University Press.

Gallegos, D. 2005. Cookbooks as manuals of taste. In *Ordinary lifestyles: Popular media, consumption and taste*, ed. D. Bell and J. Hollows, 99–110. Maidenhead: Open University Press.

Gibert, M. 2015. *Voir son steak comme un animal mort*. Montréal: Lux.

Jenkins, H. 2008. *Convergence culture: Where old and new media collide*. New York: New York University Press.

Joy, M. 2010. *Why we love dogs, eat pigs and wear cows: An introduction to carnism*. San Francisco: Conari Press.

Laforêt, M. 2014. *Vegan*. Paris: La Plage.

Lofgren, J. 2013. Food blogging and food-related media convergence. *M/C Journal* 16(3).

Meigs, A. 1997. Food as a cultural construction. In *Food and culture: A reader*, ed. C. Counihan and P. Van Esterik, 95–106. New York: Routledge.

Parasecoli, F. 2008. *Bite me: Food in popular culture*. Oxford: Berg.

Pence, G.E. 2002. Introduction: The meaning and ethics of food. In *The ethics of food: A reader for the twenty-first century*, ed. G.E. Pence. Lanham: Rowman & Littlefield.

Sarukhanyan, A. 2013. *Pour l'abolition du véganisme, pour l'abolition de l'esclavage*. http://abolitionduveganisme.blogspot.fr/2013_03_01_archive.html. Accessed 10 July 2015.

Sigler, P. 2014. De l'appel à la vertu à l'exigence de justice pour les animaux. In *L'exploitation animale est une question de société*, ed. P. Sigler and Y. Bonnardel, 3–26. http://www.reseau-antispeciste.org/wp-content/uploads/pourpolitisation_cahier.pdf. Accessed 01 July 2015.

Lisa Simpson and Darlene Conner: Television's Favorite Killjoys

Juawana Grant and Brittni MacKenzie-Dale

Mainstream television writers are beginning to reflect the historical link between feminism and animal liberation. Portraying actively vegan and feminist characters on television is a recent phenomenon, but the connection between feminism and veganism is long-standing. Feminists like Margaret Sibthorp and Edith Ward in the nineteenth century saw a thread woven between the subjugated status of both nonhumans and women, and many advocated vegetarian diets alongside their feminist activism.[1] Following Sibthorp, Ward, and other feminists in the nineteenth century, the 1990s saw vegetarian ecofeminist scholars theorizing about the similarities between the oppression of women and animals. At the same time,

---

J. Grant
Dept. of Communications, Concordia University
e-mail: juawana@gmail.com

B. MacKenzie-Dale (✉)
Dept. of Creative and Critical Studies,
University of British Columbia University
e-mail: brittnimdale@gmail.com

J. Castricano, R.R. Simonsen (eds.), *Critical Perspectives on Veganism*,
The Palgrave Macmillan Animal Ethics Series,
DOI 10.1007/978-3-319-33419-6_14

the world was introduced to two willful daughters on primetime television. Widely watched and award-winning programs, *The Simpsons* and *Roseanne* introduced us to Lisa Simpson and Darlene Conner, two pop icons who brought vegetarian ecofeminism into America's living rooms. Lisa and Darlene barreled into primetime television, unapologetic about their views, asking their families and two large, loyal fan bases tough questions about gender relations and animal ethics. This chapter looks closely at the intermingling of their feminist and vegetarian identities, particularly during moments of conflict in their fictional communities.

Lisa Simpson is an intellectual, progressive, environmentalist, feminist vegetarian and a foil to her meat-eating family. Her portrayal on the show is paradoxical. In some episodes, she is celebrated for her ethical stance on meat-eating, and in others she is vilified. Darlene Conner is also a progressive role model, but more countercultural—a kind of dark, sardonic Lisa Simpson. They are both killjoys—relentlessly spoiling the pleasures of others by pointing out unethical behaviors others take for granted, particularly meat-eating and problematic representations of women. As sleeper ecofeminists, Lisa and Darlene serve to disrupt both patriarchy and what Dr. Melanie Joy calls "carnism"—the invisible ideology of meat-eating—clueing in viewers to the link between the two. We argue that it is inevitable that feminism and veganism appear together, because patriarchy and carnist culture share an inherent connection as dominant ideologies that rely on differential power relations between groups. Lisa and Darlene, as written, can be seen to resist patriarchy and carnist culture, as the two ideologies share an inherent and intuitive connection. Carnism and patriarchy function as a kind of ideological wallpaper—that is, their connection is not obvious on first inspection. They are what Maria Popova calls the "invisibilia of life."[2] When pop culture is countercultural, it seeks to render the invisible visible, often through comedy or satire. Our essay aims to showcase the ways in which these prominent television shows raise the uncomfortable question of vegan-feminism. In what ways do these characters—as both feminist and vegetarian—ask the question: can you be one without the other?

We have chosen to focus our analysis upon the characters of Lisa and Darlene despite them identifying as vegetarian instead of vegan; this is because they still function as vegan-feminist killjoys, in the sense that Richard Twine and Sara Ahmed articulate in their theoretical works. Ahmed, in her feminist killjoy theory, describes the experience of alienation feminists encounter by recognizing and calling out problematic situations that are typically deemed acceptable by the status quo. Coining the term "affect aliens," Ahmed's insight reveals how, in relational experiences with others, the feminist is made to be the problem, rather than sexism (or another injustice), by not "experienc[ing] happiness from the right things."[3] Twine relates Ahmed's theory to the vegan experience adding that "the overt presence of animals especially in the form of 'meat' invites disruption of the everyday 'normality' of animal consumption."[4] Twine incorporates Ahmed into the vegan experience by pointing out that, like feminists, vegans are also made to be the problem, rather than the commodification of animal bodies, if they refuse to take pleasure consuming animal products with those close to them. We argue that although Lisa and Darlene are not explicitly vegan, their vegetarian diets serve to alienate them similarly and, in combination with their gender politics, this makes them "vegan-feminist killjoys" as Twine and Ahmed define it.

In addition to considering theoretical terminology, in the 1980s and 1990s, when *The Simpsons* and *Roseanne* rose to prominence, the word "vegan" was not prevalent in the cultural lexicon. Many animal rights scholars and philosophers during this time used the word vegetarian synonymously with vegan, or dubbed vegetarians who skipped dairy "strict vegetarians."[5] It is necessary, therefore, to keep this in mind when considering the context in which these characters were first created. By applying the vegan-feminist killjoy theory, we focus on the ways in which Lisa Simpson and Darlene Conner function rather than the material realities of their diet. We understand Lisa and Darlene's refusal to eat meat as performing the same role that one might expect of a vegan today—the role of the killjoy. Both of these television shows are still on the air: *The Simpsons* is presently enjoying its 27th season and *Roseanne* is in syndication on major network channels. In order to properly examine the vegan-feminist connection as portrayed in popular culture media, we thought it necessary to choose characters who were at once well-known, well-loved, and consistent in their politics. Many vegetarian or vegan characters today are either missing a feminist component or used as merely episodic tools of conflict, and often their abstinence from meat is temporary or convenient. Rachel Berry from hit television show *Glee* serves as an apt example of this phenomenon—she self-identifies as vegan, but later in the series is shown to be eating animal products when her veganism is no longer needed for plot progression. For Darlene and Lisa, their vegetarianism and their feminism are long-standing and integral to who they are as characters and what they represent in carnist society.

Carnism, Killjoys, and Pop Culture Icons

Ecofeminists, critical animal studies scholars, and activists continue to draw connections and explore the similarities between the logics of domination that uphold sexism, racism, colonialism, speciesism, and other axes of oppression.[6] Our understanding of the vegan-feminist is strongly informed by scholarship presented by ecofeminists around the same time Lisa and Darlene had their anti-carnist awakenings. Carol J. Adams' concept of the "absent referent" in her groundbreaking text, *The Sexual Politics of Meat*, is central to the understanding of carnist ideology. Adams theorizes that we psychologically distance the meat on our plate from the living animal it once was, effectively rendering nonhumans invisible or "absent" when we exploit them, consume them, and wear them. When Lisa and Darlene remind their community that they are eating individuals, they force them to imagine the living creatures their food came from—bringing the nonhuman back into focus, having been lost to objectification, fragmentation, and consumption.[7] In one episode, Lisa quite literally imagines a living lamb at the dinner table, calling back the absent referent for the viewers. For our analyses, we rely on three main frameworks: psychologist Melanie Joy's theories on carnism; killjoy theories articulated by feminist scholar Sara Ahmed and reframed by critical animal studies scholar Richard Twine; and media scholar Brian L. Ott's understanding of how pop culture affects identity politics.

After Melanie Joy spent two decades critically examining the complex and incongruous relationships between humans and nonhumans, she coined the term "carnism." She defines it as "the invisible belief system that conditions us to eat (certain) animals."[8] Joy describes carnism as both a dominant and violent ideology, which enables otherwise moral and compassionate humans to engage in behaviors that directly and indirectly causes unnecessary harm and death to other sentient animals. Psychic numbing, the mental event that allows people to uphold this moral dissonance without feeling any overt stress, is the harmful consequence of carnism. Effectively, carnism serves as a gap in consciousness that renders the nonhuman invisible and morally insignificant. Socially, carnism is accepted and rewarded, as meat consumption is deemed natural, necessary, and normal.[9] As in Adams' absent referent, psychological defenses are necessary to uphold carnist ideology. Cognitive processes such as objectification, dichotomization, and overgeneralization help individuals maintain an inconsistent view toward nonhumans. Socialization allows us to compartmentalize our views of different species, encouraging us to eat some while nurturing others.[10] As with all dominant ideologies, carnistic views are upheld by the media. Media scholar Carrie Packwood Freeman uses the term "carnonormativity" to describe the dominant anthropocentric view of animals as objects, primarily food. She points out that, across media, most people are assumed to be carnists unless identified otherwise. Further, she cites a UK study that found news media to consistently represent vegans as "ascetic, sentimental, or extreme."[11] Lisa Simpson and Darlene Conner counter that norm, disrupting carnism and revealing the moral inconsistency the status quo allows, insisting on making visible again the absent referent. Further, they do so in conjunction with their feminism, embodying a long-standing feminist-animal liberation connection. We explore further the implicit complexities that inform Lisa and Darlene's multiple political identities.

At the base of our analysis lies the vegan-feminist killjoy, as depicted by Twine and Ahmed. In her original piece "Feminist Killjoys (And Other Willful Subjects)," Ahmed describes the feminist killjoy as being instrumental in a critique of "the happiness order": the social pressure to avoid discomfort that is used to justify social norms.[12] Ahmed says, "To be willing to go against a social order, which is protected as a moral order, a happiness order is to be willing to cause unhappiness, even if unhappiness is not your cause." This willingness, or willfulness, turns the killjoy into an "affect alien," causing them to be "unseated at the table of happiness."[13] Ahmed claims that embodying anti-patriarchal politics will lead to feelings of alienation and exclusion. This, she says, is the inevitable consequence of becoming politicized and challenging the status quo. In his article "Vegan Killjoys at the Table," Twine places Ahmed's feminist killjoy at the dinner table, arguing that ethical veganism, too, is an act of resistance against norms of the happiness order. Eating meat is a cherished social order; when choosing to abandon animal products, one becomes an affect alien at Ahmed's "table of happiness." According to Twine, the vegan is a potentially more obvious killjoy, even more so than the feminist, as it is easy for the vegan to be identified by their refusal of non-vegan food. Twine recognizes what is at stake in such culinary moments where the vegan is often interrogated: "Why are you vegan?" "What do you eat?" and, finally, "How do you do that?"[14] By resisting the dominant order, Darlene and Lisa's politics often cause familial or community conflict. As explored later in this chapter, Lisa ruins the mood of her father Homer's block party by offering a vegetarian dish at a barbeque; Darlene, too, insists on not participating in her family's "loose meat sandwich" business, creating conflict with her attempts to serve meatless sandwiches instead. Twine concludes that the vegan-feminist killjoy is a figure that encourages intersectional alliances, having the potential to remake the "happiness order" as more inclusive and less oppressive.

The importance of examining popular culture icons like Lisa and Darlene cannot be underestimated. Brian L. Ott expounds on pop culture icons, concluding that they do far more for viewers than simply entertain them:

> Television furnishes consumers with explicit *identity models*, models not of who to be but *how to be* … [Television] furnishes consumers with the symbolic resources—the actual cultural bricks—with which to (re)construct identity.[15] (emphasis in original)

Ott's remarks help us to understand that the televised figures of Lisa and Darlene are, indeed, "cultural bricks," which thus allow for (re)constructed identities. Similarly, bell hooks argues that "pop culture is where the pedagogy is … it's where the learning is."[16] In other words, the observant viewer is not simply consuming but also reinventing. By studying these pivotal characters, we point to a connection between resistance to patriarchy and carnism—that uncovering inherent but constructed hierarchies in one oppressive system will inevitably lead to discovering others. Accepting that popular culture is a place of learning, we suggest that Lisa and Darlene are mentors in social justice.

"Normally I'd Say Speak Your Mind ... but You've Been Doing that an Awful Lot Lately": Lisa as a Threat to Carnism

Award-winning comedy show *The Simpsons* has been on the air since 1989 and revolves around the eponymous five-member animated family. The show has tackled sensitive issues such as religion, sexuality, bullying, and meat-eating. At only eight years old, Lisa became one of the Western world's most iconic vegetarians.[17] In 1995, her decision to stop eating meat paralleled the same decision of producer David Mirkin. As a pop culture identity figure, Lisa has been highly influential. "So many people have told me that [Lisa] was responsible for their own switch to vegetarianism," Mirkin says.[18] Lisa is constructed as champion of ethical concerns, making her decision to adopt vegetarianism, in Mirkin's words, "so true to who she is."[19] In fact, she is the only truly politicized member of her family; she is a critical thinker. Arguably, she is consistently both the voice of reason and wisdom and a foil to the ethical blindness in her unenlightened family.[20] In contrast to Lisa, Homer, Marge, and Bart are often hyperbolized into caricatures of a dysfunctional family, serving to demonstrate that Lisa is the moral compass of the sitcom.

America's favorite spiky-haired eight-year old is construed as a progressive thinker early on in the long-standing series. In 1993, viewers watched Lisa tackle Malibu Stacy for the first time and assert herself as a feminist.[21] Lisa is frustrated by the limitations of the doll: how it represents only one type of womanhood, a certain ideal of femininity, and, most worrying to Lisa, an anti-intellectual archetype of a passive housewife. She describes the doll as embodying a perverse version of what young girls should strive to be and in dismissing Malibu Stacy as a "vacuous [ninny] whose only goal is to look pretty," Lisa laments that Malibu Stacy's fans will aspire to follow in her vapid footsteps. To combat this, Lisa proposes a new type of doll to Stacy's creator, Ms. Lovell: Lisa Lionheart, a figure that Lisa designs to encapsulate "the wisdom of Gertrude Stein ... the wit of Cathy Guisewite [and] the tenacity of Nina Totenberg."[22] Although this is a first and powerful episode in forming Lisa's identity as a feminist, other narratives throughout the series perpetuate her identity in this same manner. In a season seventeen episode, "Girls Just Want to Have Sums," Lisa and

Bart's grade school is split into two groups (according to gender) to avoid unnecessary arguing. Because the "girl's side" of the school does not hold a real math class—instead, they talk about how they feel about numbers; in a move that largely undermines the feminist theme of the episode—Lisa dresses like a boy to be taken seriously. By the end of the episode, she is the best math student in the class and also reveals herself, proudly and publicly, to be a girl. Repeatedly, we see Lisa bolstered to the position of the moral agent. She is the ethical spokeswoman that other characters look to in order to solve their problems. This becomes important when, in 1995, Lisa eschews meat-eating.[23] Feminism and vegetarianism—written together in one bold, confident, unapologetic, and progressive character—create a killjoy who is resistant to patriarchal carnist culture.

On the show, Lisa's role is complicated: she not only sits in disavowal of carnist culture but also becomes the butt of jokes in numerous episodes. The narrative simultaneously mocks and celebrates Lisa's vegetarianism. We are at once told to trust Lisa and also to see her as part of a bizarre outgroup. Lisa first shuns meat in the fifth season's "Lisa the Vegetarian." After attending a petting zoo with her family, Lisa makes the connection between the lamb she meets on the farm with the cooked lamb on her plate. By individualizing the lamb, Lisa brings the absent referent to the table. In her mind, she continuously makes whole the animal parts that she is accustomed to eating: rump roast falls off a cow, a chicken's breast is removed, and she reimagines the lamb chop on her plate as the sentient animal from which it came.

Lisa serves as a revolutionary figure in relation to the carnist status quo. Since, according to Joy, carnism relies on meat-eating being natural, normal, and necessary, Lisa acts as a champion for vegetarianism. When Lisa first goes vegetarian, she is immediately ostracized. Mockingly, Homer and Bart dance around the living room, singing "you don't win friends with salad."[24] They imply that by foregoing meat, one becomes a killjoy; no one will want to socialize with her because her ethical choices make people uncomfortable. Homer is appalled that Lisa does not want to bring meat to a barbecue: "Normal people love meat," he balks. He is embarrassed when Lisa brings out gazpacho, a cold vegetarian dish. By undermining the tradition of a barbecue, Lisa's choice dampens the party atmosphere, allowing the other attendees to dismiss her as a killjoy and thus to use her as a scapegoat.

Although the surrounding characters respond with disdain toward Lisa's newfound vegetarianism, the narrative itself portrays her as having done the morally correct thing. Freeman outlines the situation in her analysis of the episode:

> Lisa's classmates laughed at her vegetarianism on two occasions using childish taunts such as "are you gonna marry a carrot?" and also repeating back the insults verbatim that they learned in a meat council marketing film that Principal Skinner showed to combat Lisa's dissent (there, her classmates called her "crazy" and a "grade A moron"). These particular examples make Lisa appear smarter and more of a critical-thinker than the other children as they prove her point that they have been "brainwashed by corporate propaganda."[25]

Lisa, made sympathetic in contrast to her dim-witted classmates, stands alone in her boycott of the meat industry. Near the end of this episode, Lisa runs away to Apu's Kwik-E-Mart in search of meat, desperate to fit in again. Feeling the consequences of her disruption, she longs to restore the happiness order. Apu, sympathetic to her feelings of alienation, brings her to a secret oasis on top of his store. In this lush garden, he confesses that he is a vegan. Celebrities Paul and Linda McCartney make an appearance in Apu's rooftop garden where they, too, confide that they are vegan; all three identify this way, acting as mentors for Lisa's "extreme" positioning, thus providing Lisa with moral backup. Feeling guilty, Lisa says to Apu that he must think she is a monster for still eating dairy products. Apu admits that he does, but that he has learned to be tolerant. Lisa confesses: "I guess I have been pretty hard on a lot of people. Especially my dad." Lisa is remorseful for disrupting the happiness order; however, she does not give up her ethics. Instead, with the moral guidance of Apu, she learns at once to be tolerant of carnism (and its proponents) while still remaining critical. Essentially, she is allowed her convictions as long as she keeps them to herself. She compromises the strength of her convictions in the face of the happiness order, limiting the scope and perhaps influence of her ethics on other Springfield community members.

Some of Lisa's strongest killjoy moments occur in "Lisa's Wedding," wherein she and her vegetarian fiancé are portrayed as uptight wet blankets.[26] Lisa becomes enamored with Hugh, whom she meets at a prestigious university. She is impressed by his intelligence and discipline, and the two quickly become romantically involved. The episode uses montage to show what they bond over: a desire to achieve intellectual greatness and a concern with social justice issues. Again, a mediator character is used to render Lisa's vegetarianism as not *too* extreme. Lisa's fiancé is portrayed as even more of a crank than Lisa because, while Lisa is generally tolerant of her meat-eating companions, Hugh outwardly expresses his contempt for carnism. "Eating animals is wrong," Hugh contends at one point during this montage. "When will the world learn?" Feeling similarly isolated and alienated due to her political positioning in such a largely carnist family and community, Lisa responds positively to this. "We're both studying the environment and are both utterly *humorless* about our vegetarianism," Hugh says at one point, reflecting on the ways in which he and Lisa are compatible for marriage.[27] The important word here is "humorless," as even a character who clearly believes passionately in nonhuman ethics admits that he is rigid and perhaps unhappy. The episode pokes fun at both the carnists and the vegetarians, in order to elucidate a tough cultural conversation about carnism.

To be humorless is to make others feel uncomfortable or at least to be a source of awkwardness, merely by being in proximity to others. Ahmed says that a killjoy disrupts the order of the status quo: "A killjoy [is] the one who gets in the way of other people's happiness. Or just the one who is in the way—you can be in the way of *whatever*, if you are already perceived as being in the way."[28] The problem here is not that Lisa and her fiancé Hugh may or may not be morally correct in their decision not to eat animals, but that their dietary decision makes others in *The Simpsons* universe uncomfortable as they become aware of their own carnistic choices. At the end of this episode, the internal contradiction rears its head again. Hugh, who is deemed too rigid, too unforgiving, too cold of a person (too much of a killjoy, even for Lisa), is sent back to his homeland (a move which, by and large, reinstates the hierarchy, with Lisa being simultaneously brilliant but ignored). Homer reaffirms Lisa's identity as the moral center of the show. "You're my greatest accomplishment," he says to his daughter, "From the day you could pin your own diapers, you've always been smarter than me."[29] What does this mean for Lisa's role as a vegan-feminist killjoy, if even the pro-status quo Homer can admit that Lisa's convictions are worthy of upholding, all the while not adhering to them himself?

It comes back around to Marge's biting statement in "Lisa vs. Malibu Stacy", where Lisa is first shown to be questioning the oppressive status quo: "Normally I'd say speak your mind … but you've been doing that an awful lot lately."[30] Lisa is allowed to speak with an anti-patriarchal, anti-carnist voice only when it does not disturb the happiness order. Twine writes that "sometimes even the known presence of a vegan will be enough to trouble the prevailing happiness order"[31]; Lisa's body is a site of confrontation in which she is simultaneously lauded for her convictions and told to keep them at bay. Her presence at Ahmed's figurative table (a place where members of society perform their roles[32]) reveals the ego defenses employed by the rest of the Simpson family. In response, the supporting characters strive to invalidate Lisa's position if it threatens their way of life too directly. The social negotiation outlined between this vegetarian character and her carnist counterparts is as follows: Lisa is allowed to kill her own joy (she can, like Hugh, acknowledge that she is "humourless"[33]), but when it begins to infringe on others, she is ostracized. Here, we see Lisa, functioning as the vegan-feminist killjoy, successfully othered in order to be rendered morally insignificant.

Lisa's vegetarianism is used not solely as a source of conflict, but also as a way to reveal buffoonery and lazy thinking. Freeman points this out by explaining that "the social dynamic represents such a strong contrast between Lisa as the smart, committed, caring vegetarian and almost everyone else as somewhat shallow and unreflective meat-eaters, it makes Lisa's animal-friendly stance seem *right* and ethically preferable"[34] (emphasis original). At one point, Homer tells Lisa that he "used to believe in things when [he] was young"[35]; the joke here is that as we grow older, we become jaded and cynical, particularly about changing the status quo. This one-liner points to socialization (in this case, carnism), and the ways in which we are defeated when our challenges to hegemony are met with dismissal. We are at once told to brush Lisa off as embarrassingly quixotic and forced to examine our own mental apathy.

Different episodes frame the question of the nonhuman in such a way as to foster moral confusion: should we eat meat or should we not? Most citizens in Springfield would contend that we should consume nonhumans, but, through the use of satire, we see that the reason for this is socialization (via the aforementioned meat propaganda video) and not necessarily through careful and thoughtful observation. Lisa, our moral agent, is one of the few characters allowed to question violence toward nonhumans. Most episodes take a nonpartisan approach to the act of meat-eating, framing Lisa as a killjoy when necessary for the episode's conflict, but using other, more extreme killjoys to soften the blow when desiring to make her more sympathetic. Lisa's pretentious fiancé, Hugh, is one example of this, as are Apu and Paul and Linda McCartney in their veganism. Season twelve's "Lisa the Tree Hugger" showcases another instance of this technique. Once more, we see Lisa as empathetically concerned for most ways of life. In this episode, Lisa falls for the leader of a radical environmentalist group named Jesse Grass. Grass chastises Lisa for not taking the leap to veganism. "I'm a level five vegan," he says snidely, "I don't eat anything that casts a shadow." Because this is impossible, the idea of eschewing all animal products, too, becomes ridiculous juxtaposed with this hyperbole. In this episode, it is Jesse who is the killjoy.[36] The use of these mediating characters demonstrate that it is the killjoy's disruption to the happiness order in and of itself that is the source of conflict; when Lisa is the killjoy of the episode, she becomes the problem, not carnism and not patriarchy.

"Okay, Okay, So You Have Convictions. Just Shut Up About Them": Darlene and Roseanne, Feminist Killjoy Competitors

Roseanne's nine seasons aired from 1988 to 1997; the show is lauded for being one of the first situational comedies to feature a female-led working-class family. The woman-centered show stars Roseanne Barr and revolves around the Conners and Jacky Harris, Roseanne Conner's sister. The show is known for taking on issues of class and gender explicitly, with some episodes covering other controversial topics such as the hidden, violent history of Thanksgiving, race, and sexuality. Janet Lee's article "Subversive Sitcoms: Roseanne as Inspiration for Feminist Resistance," analyzes the show as a site for feminist resistance. Lee frames Roseanne as a feminist character who expertly uses humor to offer messages of feminist resistance and also to mock the patriarchal and class systems in which it takes place.[37] In a recent article describing her experiences with sexism during the show's production, Roseanne asserts that she "created, wrote, and starred in television's first feminist and working-class-family sitcom (also its last)."[38] It is no secret that *Roseanne* was intended to be an explicitly feminist program—a gem amongst the postfeminist sentiment, coloring popular culture during the 80s and 90s.

Roseanne's younger daughter, Darlene, is arguably most like her, in terms of both her humor and her feminism. Darlene begins the series as a tomboy, playing sports and bonding with her father, Dan Conner, over her athleticism. She rejects the feminine hobbies of her older sister, Becky, asserting that things like makeup and fashion are superficial pursuits unworthy of celebration. She rejects the performance of femininity for the sake of male attention, even stating in one third season episode: "I mean, I like dating and boys and everything, it's just that I don't think you should have to go through all the crap to get to the good stuff."[39] In his 2011 article exploring the tensions between feminism and postfeminism in *Roseanne*, Taylor Cole Miller reveals that the character of Darlene was intentionally written in opposition to the postfeminist ethics of her sister Becky and, sometimes, Roseanne herself. He cites an interview he conducted with Roseanne Barr in which she explains that the television executives thought Roseanne Conner's politics should be contained because of her abrasive persona:

> So when studio executives decided to mute or "soften" Roseanne's character with a loving husband in John Goodman, Roseanne responded by creating a new kind of feminist hero, a pre-teen daughter for Roseanne Conner named Darlene. With Darlene, Roseanne was able to deliver her political feminist intentions and distinct voice through the witty and precocious repartee of the young character.[40]

Darlene maintains her feminist voice throughout the series even after she leaves sports behind for creative writing and enters into a long-term romantic relationship with artistic collaborator and best friend, David Healy. Roseanne Conner may be the feminist heart of the still lauded sitcom, but Darlene is the feminist backbone. Darlene is tasked with being more radical than Roseanne, who still holds feminist space in the show. To construct Darlene's political identity, the show moves her character from athletic tomboy to a brooding and countercultural young woman—a killjoy, in other words.

It comes as no surprise that Darlene is the one who takes up vegetarian politics in the show's fourth season. In an episode titled "Darlene Fades to Black," Darlene shocks her family by becoming morose and idle.[41] She quits the basketball team, starts wearing all black, and begins her transition into a teenaged killjoy. This season presents a series of moments wherein Darlene disrupts the happiness order. She becomes increasingly politicized in this season, denouncing high school as a place "for learning lies and telling lies."[42] She refuses to participate in Thanksgiving, calling it a holiday that "celebrate[s] the exploitation of Indians by a group of religious fanatics,"[43] and challenges her mother's belief in God with a defiant, "What would you do if I told you I didn't believe in God?"[44] Her outbursts in these episodes are addressed as behavioral problems. Rather than looking closely at the colonial violence she is speaking against, the storylines unseat Darlene from the table, in Ahmed's sense, and she becomes the center of conflict. It is during this transition that she becomes vegetarian; at first, this is casually mentioned in a voiceover during the opening scene to the seventh episode (Becky: "Who stole my bacon?. Darlene: "Don't ask me, I don't eat meat anymore."[45]) and then again in subsequent episodes throughout the season, sometimes framed as acts of resistance ("I'm not gonna eat some animal's face"[46]). Darlene's feminism and political angst packaged with her vegetarianism serve to both lend credence to her identity as a woman with conviction, while simultaneously allowing her choices to be easily dismissed as the inevitable expression of a rebellious teenager.

Her vegetarian politics are not used to create conflict until an episode titled "Lanford Daze," premiering late in the fifth season, when Darlene protests outside The Lunch Box, the family's loose meat sandwich restaurant. Darlene's vegetarianism remains consistent throughout the rest of the series (she is often seen wearing her now iconic "Meat Stinks" T-shirt, including at the grand opening of The Lunch Box). It is only in this one episode that her beliefs function to create conflict rather than acting as a benign character trait.

The episode begins with Darlene asking Roseanne and Dan for $300 to cover the printing costs of the comic book she is writing with David. When she is refused the money, she complains to him while they work on their comics together in the kitchen. David takes a break to microwave a snack, and Darlene notices that he is no longer vegetarian. She asks him to at least respect her beliefs by not eating meat in front of her, but David challenges her to admit what he sees as her hypocrisy:

> David: You're such a hypocrite—you won't eat meat but your family makes its living selling meat. Where do you think your mom's gonna get the money for our comics?
> Darlene: Oh please, that's completely different.
> David: No, it's not, and what about your allowance? Face it, Darlene, you are a part of the giant meat industrial complex.[47]

The next day she stages an artistic protest outside The Lunch Box by drawing chalk outlines of cows with red ink spilled at their heads. By protesting the family business, Darlene directly challenges the morality of her family's first stable economic endeavor. Roseanne assumes Darlene's protest was done out of revenge for being denied the $300 she asked for the day before, but Darlene insists that she was "making a statement" against the "Conner family money-making death machine."[48] Roseanne does not believe Darlene, and, as punishment, Darlene is made to serve loose meat sandwiches at the annual Lanford Days festival. This punishment is particularly offensive as it is a blatant disrespect for Darlene's politics. Roseanne is convinced Darlene's reason for protesting is about the money and not a genuine concern for the lives of animals. This reveals a gap between the representation of Darlene as merely a rebellious teenager and how she sees herself as a woman with "principles despite everything [Roseanne] taught [her]."[49] Unlike Lisa Simpson, who is the obvious and only ethical voice on *The Simpsons*, Darlene is one among many ethical characters in *Roseanne*. Darlene's vegan-feminist killjoy functions differently than Lisa's in the already feminist imaginary created by *Roseanne.* There is less room for Darlene's other politics outside of class and gender.

Later at Lanford Days, we see Darlene's Aunt Jacky serving the customers while Darlene reads Anne Rice's *Interview with the Vampire.* When Roseanne returns, she suspects Darlene has not been working and demands she serve a sandwich to the next customer—country music darling, Loretta Lynn. Darlene refuses and tensions rise as Roseanne tries to balance being star struck and disciplining her willful daughter. This is where Loretta Lynn serves as the celebrity mediator (much like Paul and Linda McCartney in "Lisa the Vegetarian"), stating, "I don't mean to interfere or nothin', but I gotta respect the girl for saying what she feels, and I do, honey. Now gimme a big ol' hunk of that meat 'cause I'm starvin' to death."[50] Humor serves to break the tension and Roseanne folds, but not without directing Darlene to be less of a killjoy. After Darlene says, "What did you think when you saw me as a baby? 'God she's cute, I hope she never has the courage to stand up for her convictions.'" Roseanne replies: "Okay, okay so you have convictions. Just shut up about them."[51] Like Lisa, Darlene is rewarded for believing in something, but only as long as there is no lasting disruption. In this way, both *The Simpsons* and *Roseanne* fail to take a fixed stance on vegetarianism, while still reinforcing the link between feminism and a disavowal of carnism.

Both Lisa and Darlene function as vegan-feminist killjoys in these episodes, but the reactions of Homer and Roseanne, both breadwinners, differ in significant ways. Homer concedes and supports Lisa's beliefs, though he will never follow her lead. Roseanne resists Darlene's disruption outright based on class consciousness. We see this when Darlene asks her mother, "Doesn't it bother you that you make a living exploiting animals?" Roseanne contends that it does not bother her: "You don't get it, do ya? We are too low on the food chain to exploit people—all that's left for us is animals."[52] Roseanne's class and feminist consciousness act as barriers to extending ethical consideration to the plight of animals. Darlene's role as the more radical feminist voice, or feminist killjoy, in the program adds to the weight of her vegetarian convictions, causing Roseanne to fumble at the end of their argument in the booth at Lanford Days. In an attempt to regain power by revealing Darlene to be a hypocrite, Roseanne tells Darlene that the ink in her pens comes from tortured baby squids. Roseanne's grasp to usurp Darlene's killjoy position, though, is disingenuous and serves only to elicit sarcastic apathy in Darlene, who responds with, "Why breathe?" The audience is amused because they realize Roseanne is forced to lie to regain power in the conflict. In this way, and similar to Lisa's position at the end of "Lisa the Vegetarian," the vegan-feminist killjoys disrupt carnism by essentially "winning" the battle.

"Why Breathe?": Creating Space for the Nonhuman

Roseanne's reaction suggests that there is no room for animals within the frameworks of feminism and class justice. Darlene's concern for animals and her position as the backbone of feminism in the show call into question the mainstream feminist movement's carnonormativity. In contrast to Roseanne's active resistance, Homer's quiet complacency suggests that there is something intuitively connected between feminism and veganism in Lisa's politics: when he resists her, he resists all of her politics (be it environmentalism, feminism, or vegetarianism), and when an episode revolves around Lisa, all of her politics tend to come to the forefront. As a critical thinker, Lisa views all oppression as intrinsically linked. Because Lisa is the moral compass of the show she demonstrates the tacit implication of how deconstructing one hierarchy while leaving others intact might point to ethical contradictions. This is different from Darlene's position, when juxtaposed with Roseanne's, as Roseanne is often the moral compass, and it is not a given that Darlene's choice is the ethical one in the way Lisa's is. In either case, the killjoy's disruption is seen as the problem in and of itself. Lisa and Darlene are allowed to have beliefs (they are even lauded for their convictions), but when they cause too much of a disruption it is they who become the source of conflict. Yet, with Darlene, it is only her dietary choices, and the ways she functions in her role of vegan-feminist killjoy, that are silenced. Regardless of Lisa and Darlene's differing moral authority within their familial communities, the fact that their vegetarian politics align with their feminist politics implies a commonality between the two forms of oppression. Being affect aliens, Lisa and Darlene are "contesting an *intersecting* set of entrenched social norms," which, according to Twine, "constitutes an especially poignant and potentially threatening killjoy position."[53] Lisa and Darlene are exposing the anthropocentrism of the happiness order. Twine calls the vegan-feminist killjoy "integral to the reflexivity of (intersecting) social movement and prefigurative politics."[54] Lisa and Darlene resist carnism, but more than that, they serve as a call to feminists to address their own anthropocentrism and "enable a more systemic critique of political uses of 'happiness' and 'joy.'"[55]

Final Thoughts: A Call for More Lisas and Darlenes

Lisa and Darlene are clear examples of vegan-feminist killjoys. Many twenty-first century vegans and vegetarian characters are written away as a source of moral pretension, used as pure comedic fodder, or as a tool for subplot conflict. Popular movie *Scott Pilgrim vs. the World* features a prominent vegan character, but he is seen as both vapid and pompous. Brie Larson's Envy Adams says at one point that "[veganism] just makes you better than most people."[56] Other popular TV shows and movies featuring vegetarian or vegan characters include *Jurassic Park*, *Legally Blonde*, and *Seven Pounds*. However, they are rarely if ever coupled with a feminist consciousness. More often than not, these side characters are labeled vegetarian/vegan, yet not shown to be disrupting carnist culture or challenging other characters' convictions. When coupled with feminism, Lisa and Darlene become the ultimate "willful subjects" and both their feminism and choice to advocate for animals disrupt in tandem. Historically, Lisa and Darlene serve as markers for bringing the vegetarian/feminist connection to primetime and upholding their identities as strong, smart, progressive women for the duration of their respective series.

Although, reasonably, it was a conscious decision to invent politicized women-identified characters in Lisa and Darlene, it was ostensibly an unconscious decision to construe them as opponents to multiple oppressions; however, the very fact that they wind up being both feminist and vegetarian speaks to an intuitive understanding of how different social justice movements are linked. Lisa and Darlene, ethical mentors for vegetarianism and radical thinking, do what Ott hopes the best television characters will do for consumers: "become models of not who to be but *how* to be."[57] Darlene and Lisa tacitly ask the question: Can you be a feminist and still eat meat? Like Lisa and Darlene, and the ecofeminists before them, we recognize the need for greater consideration of nonhuman animals in our feminist politics. As writer and activist Calvin Neufeld writes about his gender identity: "As a transsexual, I claim my dignity, I claim my right to freedom, health, happiness, life. As a vegan, I refuse to deny those rights to anyone else."[58] As feminists, we consider bodily autonomy fundamental to the social justice project. As vegans, we seek an end to hierarchical thinking and include nonhumans in our call to justice.

Notes

1. See, for more on this connection, Leah Leneman, "The Awakened Instinct: Vegetarianism and the Women's Suffrage Movement in Britain." *Women's History Review* 6.2 (1997): 271–87.
2. Maria Popova, introduction to *The Best American Infographics 2015*, Ed. Gareth Cook. (New York: Houghton Mifflin Harcourt, 2015), x.
3. Sara Ahmed, "Feminist Killjoys (And Other Willful Subjects)". *The Scholar & Feminist Online*, no. 8.3 (2010). http://sfonline.barnard.edu/polyphonic/ahmed_01.htm.
4. "Richard Twine. "Vegan Killjoys at the Table—Contesting Happiness and Negotiating Relationships with Food Practices." Societies 4, no. 4: (2010).
5. See, for examples, Peter Singer, *Animal Liberation: A New Ethics for our Treatment of Animals* (New York: Avon Books, 1975); and Carol J. Adams, *The Sexual Politics of Meat (20th Anniversary Edition): A Feminist-vegetarian Critical Theory* (New York: Continuum, 2000).
6. See, for example, Carol J. Adams, and Lori Gruen, *Ecofeminism: Feminist Intersections with Other Animals and the Earth* (New York: Bloomsbury, 2014).
7. Carol J. Adams, "Why feminist-vegan now?" *Feminism & Psychology* 20 no. 3 (2010): 304–305.
8. "Melanie Joy—Carnism: The Psychology of Eating Meat," YouTube video, 1:00:43, posted by "John McDougall," February 27, 2012, https://www.youtube.com/watch?v=7vWbV9FPo_Q.

9. Ibid.
10. Melanie Joy, "Psychic numbing and meat consumption: the psychology of Carnism," (Ph.D. dissertation, Saybrook Graduate School, 2002), 2–3.
11. Carrie Packwood Freeman, "Lisa and Phoebe, Lone Vegetarian Icons: At Odds with Television's Carnonormativity," in *How Television Shapes Our Worldview*, ed. Deborah A Macey, Kathleen M Ryan, and Noah J Springer, (Lanham, Maryland: Lexington Books, 2014), 197.
12. Ahmed, "Feminist Killjoys"
13. Ibid.
14. Richard Twine, "Vegan Killjoys," 626.
15. Brian L. Ott, "'I'm Bart Simpson, Who the Hell Are You?' A Study in Postmodern Identity (Re)Construction," *The Journal of Popular Culture* 37, no. 1 (2003): 58.
16. "bell hooks: Cultural Criticism & Transformation." YouTube video, 6:02 posted by "ChallengingMedia," October 3, 2006, https://www.youtube.com/watch?v=zQUuHFKP-9s.
17. *The Simpsons,* "Lisa the Vegetarian," Fox Television, first aired Oct. 15, 1995, written by David. S. Cohen, directed by Mark Kirkland.
18. Alexa Joy Sherman, "Prime Time for Veggies: Television Is Showing Vegetarian Characters In A New Light," *Vegetarian Times*, March 2014, 53.
19. Ibid., 53.
20. Freeman, "Lisa and Phoebe," 208.
21. *The Simpsons,* "Lisa vs. Malibu Stacy," Fox Television, first aired February 17, 1994, written by Bill Oakley and Josh Weinstein, directed by Jeff Lynch.
22. *The Simpsons*, "Lisa vs. Malibu Stacy."
23. *The Simpsons*, "Lisa the Vegetarian."
24. *The Simpsons*, "Lisa the Vegetarian."
25. Freeman, "Lisa and Phoebe," 200.
26. *The Simpsons,* "Lisa's Wedding," Fox Television, first aired March 19, 1995, written by Greg Daniels, directed by Jim Reardon.
27. *The Simpsons*, "Lisa's Wedding."
28. Ahmed, "Feminist Killjoys."
29. *The Simpsons*, "Lisa's Wedding."
30. *The Simpsons*, "Lisa vs. Malibu Stacy."
31. Twine, "Vegan Killjoys," 626.
32. Ahmed, "Feminist Killjoys."
33. *The Simpsons*, "Lisa's Wedding."

34. Freeman, "Lisa and Phoebe," 201.
35. *The Simpsons,* "Lisa the Vegetarian."
36. *The Simpsons,* "Lisa the Tree Hugger," Fox Television, first aired November 19, 2000, written by Matt Selman, directed by Steven Dean Moore.
37. Janet Lee. "Subversive sitcoms: Roseanne as inspiration for feminist resistance," *Women's Studies: An Interdisciplinary Journal* 2, no. 1 (1992): 19.
38. Roseanne Barr, "And I Should Know," NYMag.com. May 15, 2011. http://nymag.com/arts/tv/upfronts/2011/roseanne-barr-2011-5/.
39. *Roseanne,* "Dances with Darlene," ABC, first aired April 31, 1991, written by Brad Issacs, directed by Gail Mancuso.
40. Taylor Cole Miller. "Too Short To Be Quarterback, Too Plain To Be Queen," *Gnovis Journal,* Georgetown University, (2011).
41. *Roseanne,* "Darlene Fades to Black," ABC, first aired April 31, 1991, written by Jeff Abugov, directed by Andrew Weyman.
42. *Roseanne,* "Stressed to Kill," ABC, first aired November 19, 1991, written by Maxine Lapiduss and Jeff Abugov, directed by Andrew Weyman.
43. *Roseanne,* "Thanksgiving '91," ABC, first aired November 26, 1991, written by Brad Issacs, directed by Andrew Weyman.
44. *Roseanne,* "Santa Claus," ABC, first aired December 17, 1991, written by Chuck Lorre and Maxine Lapiduss, directed by Andrew Weyman.
45. *Roseanne,* "Vegas," ABC, first aired November 5, 1991, written by Sid Youngers and Don Foster, directed by Andrew Weyman.
46. *Roseanne,* "Looking for Loans in all the Wrong Places," ABC, first aired October 20, 1992, written by Eileen Heisler and DeAnn Heline, directed by Andrew Weyman.
47. *Roseanne,* "Lanford Daze," ABC, first aired January 26, 1991, written by Eileen Heisler and David Raether, directed by Andrew Weyman.
48. *Roseanne,* "Lanford Daze."
49. *Roseanne,* "Lanford Daze."
50. *Roseanne,* "Lanford Daze."
51. *Roseanne,* "Lanford Daze."
52. *Roseanne,* "Lanford Daze."
53. Twine, "Vegan Killjoys," 626.
54. Twine, "Vegan Killjoys," 626.
55. Twine, "Vegan Killjoys," 626.
56. *Scott Pilgrim* Vs. *the World,* directed by Bryan Lee O'Malley, (Universal Pictures, 2010), DVD.
57. Ott, ""I'm Bart Simpson."
58. Calvin Neufeld, "Trans Veganism," *Geez,* no. 38 (2015): 51.

Acknowledgement We would like to thank Cole Mash for his extensive and detailed knowledge of far too many episodes of The Simpsons, and also for his encouragement. We would also like to extend gratitude to G.E. McKinnon for giving us feedback on an earlier draft.

References

Adams, Carol J. 2000. *The sexual politics of meat (20th anniversary edition): A feminist-vegetarian critical theory*. New York: Continuum.

Adams, Carol J. 2010. Why feminist-vegan now? *Feminism & Psychology* 20(3): 302–317.

Adams, Carol J., and Lori Gruen. 2014. *Ecofeminism: Feminist intersections with other animals and the earth*. New York: Bloomsbury.

Ahmed, Sara. 2010. Feminist Killjoys (and other willful subjects). *The Scholar & Feminist Online* 8(3). http://sfonline.barnard.edu/polyphonic/ahmed_01.htm.

Barr, Roseanne. 2011. And I should know. *NYMag.com*, May 15. http://nymag.com/arts/tv/upfronts/2011/roseanne-barr-2011-5/.

bell hooks: Cultural Criticism & Transformation. YouTube video, 6:02 posted by "ChallengingMedia," October 3, 2006. https://www.youtube.com/watch?v=zQUuHFKP-9s.

Freeman, Carrie Packwood. 2014. Lisa and Phoebe, lone vegetarian icons: At odds with television's carnonormativity. In *How television shapes our worldview*, ed. Deborah A. Macey, Kathleen M. Ryan, and Noah J. Springer, 193–212. Lanham: Lexington Books.

Joy, Melanie. 2010. *Why we love dogs, eat pigs, and wear cows: An introduction to carnism: The belief system that enables us to eat some animals and not others*. San Francisco: Conari Press.

Lee, Janet. 1993. Subversive sitcoms: Roseanne as inspiration for feminist resistance. *Women's Studies: An Interdisciplinary Journal* 21(1): 87–101.

Leneman, Leah. 1997. The awakened instinct: Vegetarianism and the women's suffrage movement in Britain. *Women's History Review* 6(2): 271–287.

Melanie Joy. 2012. Carnism: The psychology of eating meat. YouTube video, 1:00:43, posted by John McDougall, February 27. https://www.youtube.com/watch?v=7vWbV9FPo_Q.

Miller, Taylor Cole. 2001. Too short to be quarterback, too plain to be queen. *Gnovis Journal*. Georgetown University, April 4. http://www.gnovisjournal.org/2011/04/04/too-short-to-be-quarterback-too-plain-to-be-queen/.

Neufeld, Calvin. 2015. Trans veganism. *Geez* 38: 50–51.

Ott, Brian L. 2003. I'm Bart Simpson, who the hell are you?' A study in post-modern identity (re)construction. *The Journal of Popular Culture* 37(1): 56–82.

Popova, Maria. 2015. Introduction. In *The best American infographics 2015*, ed. Gareth Cook. New York: Houghton Mifflin Harcourt.

Roseanne. Dances with Darlene. First aired April 31, 1991 by ABC. Written by Brad Issacs and directed by Gail Mancuso.

Roseanne. Darlene fades to Black. First aired April 31, 1991 by ABC. Written by Jeff Abugov and Directed by Andrew Weyman.

Roseanne. Lanford Daze. First aired January 26, 1991 by ABC. Written by Eileen Heisler and David Raether and directed by Andrew Weyman.

Roseanne. Looking for loans in all the wrong places. First aired October 20, 1992 by ABC. Written by Eileen Heisler and DeAnn Heline and directed by Andrew Weyman.

Roseanne. Santa Claus. First aired December 17, 1991 by ABC. Written by Chuck Lorre and Maxine Lapiduss and directed by Andrew Weyman.

Roseanne. Stressed to kill. First aired November 19, 1991by ABC. Written by Maxine Lapiduss and Jeff Abugov and directed by Andrew Weyman.

Roseanne. Thanksgiving'91. First aired November 26, 1991 by ABC. Written by Brad Issacs and directed by Andrew Weyman.

Roseanne. Vegas. First aired November 5, 1991 by ABC. Written by Sid Youngers and Don Foster and directed by Andrew Weyman.

Scott Pilgrim Vs. the World. DVD. Directed by Bryan Lee O'Malley. Universal Pictures, 2010.

Sherman, Alexa J. 2014. Prime time for veggies: Television is showing vegetarian characters in a new light. *Vegetarian Times*, March: 52–55.

Singer, Peter. 1975. *Animal liberation: A new ethics for our treatment of animals.* New York: Avon Books.

The Simpsons. Lisa the Tree Hugger. First aired November 19, 2000 by Fox Television. Written by Matt Selman and directed by Steven Dean Moore.

The Simpsons. Lisa the vegetarian. First aired Oct. 15, 1995 by Fox Television. Written by David. S. Cohen and directed by Mark Kirkland.

The Simpsons. Lisa vs. Malibu Stacy. First aired February 17, 1994 by Fox Television. Written by Bill Oakley and Josh Weinstein, directed by Jeff Lynch.

The Simpsons. Lisa's Wedding. First aired March 19, 1995 by Fox Television. Written by Greg Daniels and directed by Jim Reardon.

Twine, Richard. 2014. Vegan killjoys at the table—Contesting happiness and negotiating relationships with food practices. *Societies* 4(4): 623–639.

The Carnivorous Mission of the Celebrity Chef

Francesco Buscemi

This chapter focuses on how celebrity chefs support meat consumption. Every day in Europe and in the USA, on myriads of food shows, celebrity chefs, TV presenters, journalists, and amateurs cook, recommend, and eat meat. On the face of it, this seems benign in a "carnist" culture,[1] as people are free to decide what to cook and eat in front of the camera. However, what television hides in these frequent representations is that for the last 20 years the scientific community has been considering meat as a problem for both the external environment and the human body. I will return to this division between the external environment and the body at the end of this chapter. What is important to note here is that disease, pathogens, medical costs, pollution, and animal suffering are among the issues linked to meat consumption. These issues are hard to question, as many studies adopting different methodologies, perspectives, and theories point to the fact that meat eating contributes significantly to

F. Buscemi (✉)
Media and Communication, Bournemouth University, Bournemouth, UK
e-mail: buscemifrancesco@hotmail.it

J. Castricano, R.R. Simonsen (eds.), *Critical Perspectives on Veganism*,
The Palgrave Macmillan Animal Ethics Series,
DOI 10.1007/978-3-319-33419-6_15

all of them. Among the many examples, Joseph DesJardins, Sjur Kasa, and Philip Lymbery and Isabel Oakeshott[2] offer a good overview and provide useful literature on all of this. The present chapter focuses on why these studies, and many others demonstrating the dangerousness of meat-eating, have not changed the media perception of meat.

Even though scientists have issued warnings about the health risks of meat consumption for at least 20 years, television and other mainstream media have continued to represent it in a positive way. Consequently, meat consumption has not decreased in most Western countries,[3] whereas illness and pollution have continued to spread. This chapter offers a critique of the support by celebrity chefs to encourage carnism in spite of the alarming results provided by scientific research on meat.

To fully understand the reasons why celebrity chefs support meat, comparing this situation with a similar practice is useful. Meat consumption has not been the only dangerous practice positively represented by the media. Smoking, for example, was depicted for years as a means to enhance one's "independence, excitement, and sexuality,"[4] reinforcing both masculine and feminine identities[5]; at times, smoking was even depicted as an act of transgression.[6] A clear example of the latter trend is provided by the many iconographic photos of James Dean sensuously holding a cigarette between his lips.[7] Yet, when in the 1960s the scientific community highlighted the dangers of smoking, the media changed their approach in only a few years. It became taboo to smoke in front of cameras and in TV studios. The film industry no longer portrayed smoking as a seductive, countercultural, or courageous action[8]; further, in 1968, "an anti-smoking film aimed at children" was produced.[9] (Only recently, in some American series smoking is on the rise again, just as a symbol of political incorrectness.)

What is it about meat, then? Drawing on Melanie Joy, this chapter argues that this is a cultural issue, and that eating or not eating meat is not only a food choice, but a philosophical stance relating to personal attitudes and beliefs.[10] Therefore, before analyzing four cases to illuminate the issue, in the next section I give a broad account of the many theoretical perspectives on meat consumption and its cultural meanings within Western society.

Theoretical Perspectives

Meat

To investigate the ways that meat consumption affects daily life beyond its dietary value, food studies offer a helpful perspective. Food studies is an academic discipline that analyzes all the fields connected to food apart from its nutritional aspects. Food is a subject for anthropologists, historians, sociologists, ethnologists, psychologists, philosophers, economists, and other social scientists. Jeff Miller and Jonathan Deutsch define food studies as "the study of the relationships between food and the human experience."[11] While nutritionists, physicians, chemists, and other scientists explore how food impacts health, food studies researchers consider how it influences social lives. There is a split in our knowledge and perception between what happens inside and outside us. What is important to underline here is that food studies researchers have always emphasized the relevance of meat in terms of how we structure society, relate to others, and position ourselves in a community.

In his well-known work, Claude Lévi-Strauss[12] found that how meat is cooked structures social relationships. Nick Fiddes[13] argues that meat has symbolical meanings linked to masculinity, blood, and the supposed human supremacy over animals. Norbert Elias,[14] Jack Goody,[15] Noëlie Vialles,[16] and Stephen Mennell[17] underline the ideological gap between the idea of the living animal and the meat we eat. I have argued elsewhere[18] that, so far, this gap has occurred on the stages of commercialization, preparation, and consumption, and that in the future this separation will also be found on the stage of production, for example, in the case of cultured meat. Carolyn Marvin and David Ingle[19] relate meat to the role of the soldier and to "touching death" and I have analyzed[20] how cooking shows draw on these connections. Carol Adams[21] sees meat as a symbol of male violence toward women, and I have shown[22] how meat shaped Nazi propaganda to fuel anti-Semitism. Finally, more broadly, Pierre Bourdieu[23] finds that certain foods are associated with social prestige and social distinction, and Peter Naccarato and Kathleen LeBesco[24] focus on the accumulation of "culinary capital." To summarize, even though these scholars approach the topic of meat from different disciplines, points of view, and methodologies, they all agree that meat is a constitutive element of our society. What is more, for Melanie Joy it is an ideology,[25] a set of beliefs which our society is based on.

The Celebrity Chef

Since the 1990s, another element has contributed to supporting the foundational role of meat in daily life and, more specifically, in the mediascape: the celebrity chef. The celebrity chef is the embodiment of choice, preparation, and consumption in food culture. In fact, far from considering food as simply something to eat, she/he uses food to figure something else, that is, perhaps an ideology; a stereotypical assumption; a cultural value; a nation; or, more often than we imagine, commercial interests. Frequently, the celebrity chef is the bearer of political and ideological elements,[26] and in a perfectly symbiotic relationship, she/he fuels and draws on those cultural and ideological connections that have been outlined by the theorists cited above. As a result, in only a few years, celebrity chefs have become tastemakers, opinion-leaders able to affect the food choices of millions of people, who consider them experts.[27]

Drawing on the theories expressed so far and on their intersections with the figure of the celebrity chef, this chapter sheds light on why popular chefs support the production, preparation, and consumption of meat. To do so, it analyzes four TV shows according to the methods explained in the next section.

Methodology

In terms of methodology, the first issue was determining how to select the celebrity chefs to analyze. In fact, I did not want to superficially look at the category in general; rather, I was interested in analyzing a number of chefs in detail. The idea that other unhealthy practices, such as smoking, have been banned from the media made me wonder why meat-eating has been treated differently. Moreover, some celebrity chefs have healthy approaches to food. For example, they ban from their programs unhealthy, fatty, or sweet foods, and often support health-related campaigns and scientific guidelines. Why have they banned these foods and continue to support meat? I found these celebrity chefs appropriate to this study, because it is relevant to explore how they behave in relation to meat and how this behavior is different and in contradiction with what they say and do in relation to other unhealthy practices.

I soon found that some of the most popular celebrity chefs had already adopted healthy points of view in their cooking. In the end, I chose Jamie Oliver, who has fought for eliminating fatty school dinners[28]; Heston Blumenthal, who focuses on salt reduction in his cooking[29]; Gordon Ramsay, who has suggested reducing butter and cream[30]; and Rachael Ray, who has supported Michelle Obama's anti-obesity campaign.[31]

Once I chose the chefs, I focused on one show for each of them, centering on examples related to meat. Specifically, in analyzing Jamie Oliver, I found that his show, *Jamie's Great Britain,*[32] frequently represents meat to support the ancestral character of the nation. The show takes viewers on a journey through Britain, and Oliver often focuses on the rough-and-tumble side of his nation. This form of "banal nationalism"[33] is rarely supported by cakes or salad; rather, meat is the main ingredient. In *Heston's Christmas Feast,*[34] the fifth and last episode of a series on various feasts, Blumenthal serves meat to some participants in the show, both celebrities and ordinary people. They believe that they are eating what Melanie Joy considers allowed meat (pork, beef, etc.). They do not know that they are actually eating a dead dormouse, which in Western food culture is rarely considered food. Moreover, they are also eating the legs and head of the animal, breaking another food taboo of our culture, which wants the animal on the plate de-animalized.[35] Blumenthal plays with these participants who attempt to distinguish meat from the whole animal. In this way, the chef visualizes what usually meat-eaters do not want to see, that "absent referent"[36] that for Adams permits many people to eat meat. Regarding Gordon Ramsay, I analyze a short episode of *The F Word*,[37] where he comments on the fact that an old butchery has been replaced by a *Holland & Barrett*, a British chain that provides its customers with very healthy foods. Finally, I analyze a long scene in an episode of the *Rachel Ray Show,*[38] in which the female chef interviews the actor Donnie Wahlberg, representing him as an example of masculine well-being and strength, an authentic champion of barbecuing.

Concerning the chapter's methodology, the analysis is carried out through textual, image, and semiotic analysis of the shows mentioned above, in order to catch their symbolic or connotative meanings hidden behind their images. Finally, it is thanks to a branch of semiotics, biosemiotics, that I reach a further level of detail, as I explain in the conclusion of the chapter.

The Programs

Jamie Oliver: A Meaty Nation

In *Jamie's Great Britain*, Jamie Oliver visits his nation by driving an army truck that contains a small pub with a stone oven. As I have explained elsewhere,[39] in semiotic terms, a kind of military code undergirds the entire series. Certainly the army truck is an element of this military code. Moreover, Oliver often adorns his dishes with military signs, such as the RAF Wings, and pays homage to national institutions, such as the Monarchy, in order to celebrate the nation. In the end, all of this fits with the role of the soldier that Oliver plays in the series. As a makeshift soldier, he holds the nation together through food and, especially, through meat. Throughout the six episodes, in fact, Oliver hunts and shows carcasses of animals without reserve; at one point, he even slaughters a lamb on camera.[40] The crudity of these images is the reason why on Channel 4's website these episodes are preceded by a warning to particularly sensitive people.[41] It is interesting that in other shows, Oliver has rarely shown meat in this way. In fact, this crudity is necessary to Oliver's project, as the chef plays the role of the soldier holding the nation together. It is by showing blood and dead animals that Oliver, the soldier, touches death[42] and celebrates the masculine side of Britain.[43]

In another study,[44] I have underlined the importance of the animal origins of meat. A "civilising process"[45] tends to ban from the plate the parts of the animal that suggest that the meat was once a living being. In a sort of counterprocess, provocatively, many chefs focus on the link between meat and the animal, and display legs, heads, and tails to remind us that humans are stronger than Nature. In fact, by killing and eating animals, humans reaffirm their power over the other creatures, becoming "masters of the world."[46] It is as if Oliver asked us: are you courageous enough to watch and, therefore, to eat this meat? By showing his audience the dead animal, Oliver seeks to position humans as beings stronger than Nature, and to encourage the view that Britain is built by the courageous men who have the nerve to touch death.

Finally, as often happens with celebrity chefs, any idealistic stance has a more or less hidden commercial motive. During the same time as *Jamie's Great Britain* was being broadcast, Oliver opened a restaurant chain called *Union Jack*, serving the same meat-based, masculinized, and idealized British dishes shown on the air.[47] The restaurant and the show were strongly linked to each other, as the restaurant was named after the British national flag, which also was the logo of *Jamie's Great Britain*. This is an evident, cunning combination of the rhetoric of nationalism and the hunger for profit of capitalism.

Heston Blumenthal: The Unspeakable Truth about Meat

The strategy of showing the animal origins of meat in order to remind the audience that humans are stronger than Nature may also be found in *Heston's Christmas Feast*, Blumenthal's show that interestingly turns animal corpses into *haute cuisine*. In the program, the animal origins of meat are explicitly shown to the audience, but hidden to the participants, both celebrities and ordinary people. These people do not know what they are eating. Only after they have swallowed the last bite does the chef tell them what they have eaten: in one instance, they consumed an entire dormouse, with head and legs. In this moment, the camera focuses on the faces of the participants, to catch their disgust that causes anger and, in some cases, vomit.

Semiotically, what Blumenthal does here is to shift the link between signifier and signified. In semiotic terms, the word "dormouse" is the signifier relating to the meaning of the animal dormouse, which is the "signified." In traditional Saussurian semiotics, the two elements are indissolubly bound to each other as the two sides of a sheet of paper. Charles Peirce's semiotics argues that meanings are not fixed but change in relation to the context, and introduces the notion of "unlimited semiosis,"[48] which sees that the relationships between signifier and signified may change and are always in a state of flux. This is exactly what happens on Blumenthal's show, as the same signifier relates to two different meanings. The chef offers his customers meat as the signifier of something to eat, but later he shifts the signified. After they eat, the meat becomes the signifier of another signified, which is "dead animal." Joy focuses on the same shift when she imagines that a woman who is hosting an elegant party suddenly unveils the recipe of the meat dish that her guests are enjoying. She tells her guests that they are eating "five pounds of golden retriever meat, well marinated."[49]

The socially accepted relationship between signifier and signified is that meat is a kind of food, and that the animal origins of meat must be hidden. Blumenthal provocatively challenges all of this, and a new, upsetting signified is suddenly associated to the signifier "meat": meat is a dead animal. Once the unconscious truth is unveiled, the diners react with disgust. By associating the same signifier with a new signified, and thus by creating a new semiosis, Blumenthal has achieved his aim. He has created scandal and has ridiculed the eaters, who wanted to forget the unspeakable truth: meat is made up of the flesh of a dead animal.

Gordon Ramsay: The Good Old Meat

Gordon Ramsay is the embodiment of the celebrity chef more popular for his way of performing the role of chef rather than for preparing food. No viewer will ever forget the scenes where he shouts at wannabe chefs looking up to him as a sort of god.

In *The F Word*, Ramsay confirms his aggressive approach to cooking by stating that one of the aims of the show is to get women back in the kitchen. In the show there is a strong divide between male and female cooking, and meat is frequently represented as a male product. Many episodes of the show received complaints for the disturbing images of animals killed and slaughtered.[50] As with Oliver's show, crudely displaying the animal origins of meat is a male exclusive and a macho attitude. The focus on animal killing and slaughter has the effect of creating hierarchical difference between genders.

In Episode 9 of Series 2, however, something happens that pushes the boundaries of this distinction. Rather than splitting gender roles, meat is used to split the good old days from the present. To position himself on the side of the meat eaters and meat lovers, Ramsay nostalgically recalls that his grandfather was a butcher and lived in Scotland. In order to pay homage to his grandfather, the celebrity chef decides to search out the old butcher shop owned by his ancestor. Ramsay discovers that his grandfather's shop has been replaced by a branch of the chain *Holland & Barrett. Holland & Barrett* is a British brand selling health foods such as cereals, nuts, and vitamins. Ramsay defines this reality as "quite sad," the demonstration that yesterday's food was better than what we eat today, and seizes the opportunity to celebrate the past, when being a butcher was a rewarding activity. To highlight the difference between the meaty past and the veggie present, the chef interviews some young people and asks them if they would like to be butchers. They answer "no," and the only boy who offers an explanation for this rejection says that "it smells." As was the case with Oliver, there is also a commercial reason for all of this. While reminiscing of the good old days, Ramsay reminds his audience that he owns *Steak*, a meat restaurant in Las Vegas.

In his seminal *Tourism and the Semiotics on Nostalgia*, John Frow sheds light on how postmodern society, in order to soothe its anxieties, constructs pre-modernity as a mythological Other that is "rough, differentiated, lacking the homogeneity of the commodity."[51] This is what Ramsay's grandfather embodies, and is also part of that "sentimental nostalgia"[52] constructed around meat which Lori Gruen and Robert Jones refer to. However, being aware of the good old days may improve our present. As with the glorious replica of Britain built by *Jamie's Great Britain*, here meat has the precise role of recalling a mythical past in which life was probably ruder and tougher, but that we must, nevertheless, remember if we want to keep in touch with our roots. Meat is among the archaic elements of our society, and, by celebrating this area of food, Ramsay wants to celebrate the myth of the old Britain.

Rachel Ray: The Joy of Meat

Given the masculine, tough, and crude representations of meat put forward by the three chefs already analyzed, it is relevant to also explore how a female celebrity chef deals with the practice of eating meat. Rachael Ray is one of the most popular celebrity chefs in the USA, and one of the most innovative.[53] She supported Michelle Obama's campaign on eating healthily and often advises her audience on how to eat well.[54]

Even though Ray is clearly concerned with health issues, she does not have a critical approach to meat. Joy[55] underlines that in cases like this meat-eating has been normalized, the issues linked to it have been hidden and thus the practice has been allowed by society. In the *Rachael Ray Show* of 27 September 2012, in fact, Ray celebrates meat and, interestingly, does it from a female point of view by relating meat to male beauty and sexuality. The special guest of the episode is the American actor Donnie Wahlberg. Ray asks him to identify a male dish, and the actor celebrates the hamburger as the quintessential "male food" and barbecuing as a rough and sexy male activity. During the scene, once more, cultural values and commercial interests intertwine as Wahlberg promotes the hamburger restaurant chain that in 2011 he opened together with his brother. It is called *Wahlburgers*, and while the actor promotes it, the show displays the logo and the venue of the restaurant.

Ray adds a different perspective compared to those of the three male celebrity chefs analyzed above. In the case of the three men investigated above, the animality of meat served the purpose of representing toughness to celebrate the nation, the shift of the relationship signifier/signified, and the past, respectively. Here, instead, meat aims to celebrate masculinity in front of a public of women. In his analysis of meat as a powerful source of symbols, Fiddes finds that "meat is widely reputed to inflame the lustful passions, particularly in men."[56] For Fiddes, moreover, there is also a matter of language, as "men are referred to as meaty or beefy ('he's a real beefsteak') … in the sense that meat is full of power."[57] Finally, "the penis is quite often referred to as a man's sausage."[58]

In the end, meat's meaning and value do not change, but perfectly adapt to the various contexts in which this food is represented. Meat may mean nationalism, truth, past, or sex, but in each of these contexts it never challenges sedimented mindsets and ideologies, as I will further elaborate in the next section.

The Two Levels of the Results

The list of celebrity chefs who promote meat consumption could be expanded, but demonstrating that celebrity chefs support meat is not the main goal of this chapter. What this study aims to discover, instead, is why this happens, and why even chefs concerned with health-related problems (from salt reduction to school dinners, as shown above) change their approaches when it comes to dealing with meat. The stark contrast between what they usually do in relation to other unhealthy practices and what they do with meat deserves attention.

There are two levels of results in this study. The first is that on these shows meat is construed as a means of reviving elements and values that we are in danger of losing. This is clearly a social construction because, actually, meat as an item of food does not inherently relate to national identity, unconscious truths, nostalgia, or masculinity. In this sense, meat becomes "normal, natural and necessary"[59] to the preservation of society. However, by adopting meat not as something to eat, but as "something else," we have seen that these shows construct this item of food in relation to these elements. Thus, meat connects us to things the viewer is losing in all the analyzed cases. Firstly, in a global age, Oliver points out that we are losing contact with the nation, our ancestral home[60] that we must be willing to die for. In Oliver's show, meat helps us to rediscover Britain. Secondly, in an age of processed food, Blumenthal suggests that we are losing contact with the origins of what we eat, and his show focuses on the origins of meat to remind us of this. Thirdly, in an accelerated society, old traditions risk being forgotten, and Gordon Ramsay, through his advocacy of meat, attempts to retrieve them. Fourthly, in a world where masculinity seems to be threatened by feminism, Rachael Ray resorts to advocating meat consumption to retrieve the authentic man. In all of these cases, meat is constructed as the medicine to relieve social anxieties.

I argue that this fully explains why these four celebrity chefs (and the many others that this chapter has not examined) change their healthy approaches when it comes to talking about meat. Through their shows, the chefs are telling the viewers that meat is necessary to our society, fully confirming Melanie Joy's theory of carnism. Advocating for the reduction of salt and fats or better school dinners does not constitute a threat to Western cultural necessities; instead, meat reduction, vegetarianism, veganism, health alarms from the scientific community, and any other message underlining the dangers of meat would call into question many of our fundamental assumptions about Western society. It is also for this essential reason that many celebrity chefs have built their commercial interests on meat. The fact that meat is seen as being foundational to our society has also guaranteed its long-lasting commercial success, even though today this item of food has evidently lost its dominant role on the table.

Beyond this outcome, however, there is another result that links to the way that Westerners perceive ourselves in relation to the environment around us. Our society, and therefore also our TV shows and cultural theories, continually represents a split between humans and Nature, or between Nature and Culture. As argued by Simon During, "Nature is defined first against artifice and technology" while Culture "is productive human intervention in nature."[61] As I demonstrate below, it is around this split that we construct our role and position in relation to the other living beings.

Oliver's program depicts a "natural" nation, and the human willingness to belong to it. Representing the nation as a natural entity and not as a human construction has always been a strategy to persuade people to die for it and to reinforce nationalism.[62] In Blumenthal's show, participants don't want to recognize that meat is the flesh of a dead animal. The program, instead, stresses the link between meat and the living being, and represents Culture (cooking and eating meat) as more powerful than Nature (the dead animal). By doing so, the program shows that not only is there a split between Nature and Culture, but also that the two elements are in stark contrast to each other. In Ramsay's program, the natural, pure past is threatened by modernity. In Ray's show, finally, human naturalness and animality are extolled as models of beauty and sensuality.

In all of these examples, Nature and Culture are split, no matter which of the two is represented as more powerful. Considering Nature and Culture as two separate and in contrast entities is a comfortable excuse for selfish behavior. For example, if we consider that our field is Culture and that consequently Nature is detached from us, we can engage more easily in harmful behaviors, such as pollution, without any sense of guilt, as Nature is not part of us.

Regarding the way in which we perceive Nature and Culture, semiotic analysis challenges the view seeing the two concepts as separate and even in conflict. Morten Tønnesen and Kadri Tũũr point out that Nature is composed of signs and codes that represent, communicate, and signify.[63] Thus, Nature is a system of signs. But what about Culture? Is it part of Nature or are Nature and Culture two different systems?

As already explained in another study,[64] ecosemiotics and biosemiotics have developed interesting answers to these questions. Thomas Sebeok[65] considers Culture not in contrast to Nature, but as a part of it. Following on from him, semioticians have analyzed the relationships between Nature and Culture as twofold, reciprocal, and in a state of flux. Specifically, "ecosemiotics focuses on the engagement of culture and nature through signs,"[66] and clarifies the relationship between the two elements as a continuous exchange. It assumes that "thought semiotically manifests self environmentally."[67] Thus, ecosemiotics sees that Nature and Culture are not detached from each other, but, on the contrary, that "culture can be visualized as being produced by nature."[68] Therefore, it does not make sense "the untouchable dualism Nature–Culture. Nature allowed, Culture not."[69] In fact, for Dario Martinelli "it is when we divide the world in two that we are being superficial."[70] On Nature and Culture, he says that "it is unacceptable to treat them separately, because too many and too complex are the relations between the two. We *cannot* analyze any cultural phenomenon as completely untied from natural context."[71]

In conclusion, we may say that meat does not cause problems to both, the external environment and our bodies, but to the unique element that involves what is around us and what is inside us, in a more balanced, holistic view. If we continue to consider Nature and Culture as separate, meat will be necessary to this kind of society, as Joy argues, in order to underline this split and to support dominant views on the past, authenticity, sexuality, and so on. Only by rewriting our perceptions of Nature and Culture, and by considering them as a whole, will meat lose this social necessity. That day we will not need to associate meat with "authentic" nations, unconscious truths, mythical pasts, or supposed animal sexuality. Meat will only be an item of food, and choosing whether or not to eat it will only be a nutritional and ethical decision.

Notes

1. Melanie Joy, *Why We Love Dogs, Eat Pigs and Wear Cows: An Introduction to Carnism, the Belief System that Enable Us to Eat Some Animals and not Others* (San Francisco, CA: Conari Press, 2010).
2. Joseph R. Desjardins, *Environmental Ethics: Concepts, Policy, and Theory* (Mountain View, CA: Mayfield, 1999); Sjur Kasa, "Globalizing Unsustainable Food Consumption: Trade Policies, Producer Lobbies, Consumer Preferences, and Beef Consumption in Northern Asia," in *The Global Governance of Food*, eds. Sara R. Curran et al., (London, Routledge, 2013); Philip Lymbery and Isabel Oakeshott, *Farmageddon: The True Cost of Cheap Meat* (London, Bloomsbury, 2014).
3. The Guardian "A Bag of 'Fat, Chemicals – and Hepatitis': Why Britain Has Stopped Eating Sausages." *The Guardian*, 2015, accessed 2 November 2015, http://www.theguardian.com/lifeandstyle/short-cuts/2015/jul/21/why-britain-has-stopped-eating-sausages.
4. Eileen Hoffman, *Our Health, Our Lives: A Revolutionary Approach to Total Health Care for Women* (New York: Simon & Schuster, 1995), 362.
5. Douglas Kellner, *Media Culture: Cultural Studies, Identity and Politics Between the Modern and the Postmodern* (London and New York: Routledge, 1995).
6. Lindsay M. Banco, *Travel and Drugs in Twentieth-Century Literature* (New York: Routledge, 2010).
7. Claudia Springer, *James Dean Transfigured: The Many Faces of Rebel Iconography* (Austin, TX: University of Texas Press, 2007).
8. Deirdre M. Condit, "Tugging at Pregnant Consumers: Competing 'Smoke!' 'Don't Smoke!' Media Messages and their Messengers," in *Evaluating Women's Health Messages*, eds. Roxanne L. Perrott and Celeste M. Condit (London: Sage, 1996).
9. Ian Aitken, *The Concise Routledge Encyclopedia of the Documentary Film* (Abingdon: Routledge, 2006), 502.
10. Joy, *Why We Love*.
11. Jeff Miller and Jonathan Deutsch, *Food Studies: An Introduction to Research Methods* (London: Berg, 2009), 3.
12. Claude Lévi-Strauss, "The Culinary Triangle," *Partisan Review* 33 (1966).
13. Nick Fiddes, *Meat: A Natural Symbol* (London: Routledge, 1991).
14. Norbert Elias, *The Civilising Process: The History of Manners* (New York: Urizen, 1939).

15. Jack Goody, *Cooking, Cuisine and Class: A Study in Comparative Sociology* (New York: Pantheon Books, 1982).
16. Noëlie Vialles, *Animal to Edible* (Cambridge: Cambridge University Press, 1994).
17. Stephen Mennell, *All Manners of Food: Eating and Taste in England and France from the Middle Ages to the Present* (second edition) (Urbana, IL: University of Illinois Press, 1996).
18. Francesco Buscemi, "From Killing Cows to Culturing Meat," *The British Food Journal* 116(6) (2014).
19. Carolyn Marvin and David W. Ingle, *Blood Sacrifice and the Nation: Totem Rituals and the American Flag* (Cambridge: Cambridge University Press, 1999), 74.
20. Francesco Buscemi, "Jamie Oliver and the Gastrodiplomacy of Simulacra," *Public Diplomacy Magazine*, Special Issue Gastrodiplomacy, winter (2014).
21. Carol J. Adams, *The Sexual Politics of Meat (20th Anniversary Edition): A Feminist-Vegetarian Critical Theory* (New York: Continuum International Publishing, 2010).
22. Francesco Buscemi, "Edible Lies: How Nazi Propaganda Represented Meat to Defame the Jews," *Media, War and Conflict*, accepted (2016).
23. Pierre Bourdieu, *La Distinction: Critique Sociale du Jugement* (Paris: Led Editions du Minuit, 1979).
24. Peter Naccarato and Kathleen LeBesco, *Culinary Capital* (New York: Bloomsbury Academic, 2012).
25. Joy, *Why We Eat.*
26. Bob Ashley et al., *Food and Cultural Studies* (London: Routledge, 1994): Signe Rousseau, *Food Media: Celebrity Chefs and the Politics of Everyday Interference* (London: Berg, 2012).
27. Lindsay Stringfellow et al., "Conceptualizing Taste: Food, Culture and Celebrities," *Tourism Management* 37 (2013).
28. James Leggott and Tobia Hochscherf "From the Kitchen to 10 Downing Street: Jamie's School Dinner and the Politics of Reality Cooking," in *The Tube Has Spoken: Reality TV and History*, eds. Julie A. Taddeo and Ken Dvorak (Lexington, KY: University Press of Kentucky, 2010).
29. Peter Ruddick, "Dinner by Heston Blumenthal Congratulated for Low Levels of Hidden Salt," *Big Hospitality*, March 12, 2013, accessed January 14, 2016, http://www.bighospitality.co.uk/Trends-Reports/Dinner-by-Heston-Blumenthal-congratulated-for-low-levels-of-hidden-salt.
30. Gordon Ramsay, *Gordon Ramsay's Healthy Appetite* (Toronto: Key Porter Books, 2009).

31. Rousseau, *Food Media*.
32. Jamie Oliver, "Jamie's Great Britain," accessed January 14, 2016, http://www.channel4.com/programmes/jamies-great-britain/.
33. Micheal Billig, *Banal Nationalism* (London: Sage, 1995).
34. Heston Blumenthal, *Heston's Feasts*. London: Channel 4 TV, March 3, 2009—May 18, 2010.
35. Buscemi, "From Killing."
36. Adams, *The Sexual*.
37. Gordon Ramsay, "The F Word," accessed January 14, 2016, https://www.youtube.com/watch?v=SIX3JSqu64E.
38. Rachael Ray, "Rachael Ray Show," accessed January 14, 2016, https://www.youtube.com/watch?v=ZkUSpJEiLoo.
39. Buscemi, "Gastrodiplomacy."
40. You Tube, "Jamie's Great Britain – How To Butcher a Lamb," accessed January 14, 2016, https://www.youtube.com/watch?v=F3vjvrR9khE.
41. Channel 4, "Jamie's Great Britain – Episode 6," accessed January 14, 2016, http://www.channel4.com/programmes/jamies-great-britain/on-demand/50434-006.
42. Marvin and Ingle, *Blood Sacrifice*.
43. Buscemi, "Gastrodiplomacy."
44. Buscemi, "From Killing."
45. Elias, *The Civilising Process*.
46. Robert A. Schultz, *Technology* Versus *Ecology: Human Superiority and the Ongoing Conflict with Nature* (Hershey, PA: IGI Global, 2014), 120.
47. Emma Sturgess, "Don't Mourn the Sun Going Down on Jamie Oliver's Union Jack Empire," *The Guardian*, 29 January, 2014, accessed January 14, 2016, http://www.theguardian.com/lifeandstyle/wordofmouth/2014/jan/29/dont-mourn-jamie-oliver-union-jacks-empire.
48. Birgit Nordtug, "Subjectivity as an Unlimited Semiosis: Lacan and Peirce," *Studies in Philosophy and Education*, 23(2-3) (2004).
49. Joy, *Why We Eat*, 11.
50. Gay Adams, "Ramsay Reduced to Tears as Pigs Go under Knife," The Independent, August 8, 2006, accessed January 14, 2016, http://www.independent.co.uk/news/media/ramsay-reduced-to-tears-as-pigs-go-under-knife-411122.html.
51. John Frow, "Tourism and the Semiotic of Nostalgia," *October* 57 (1991): 130.

52. Lori Gruen and Robert C. Jones, "Veganism as an Aspiration," in *The Moral Complexities of Eating Meat*, eds. Ben Bramble and Bob Fischer (Oxford University Press, 2016).
53. Thomas Rogers, "How Food Television Is Changing America," *Saloon*, February 26, 2010, accessed January 14, 2016, http://www.salon.com/2010/02/26/food_network_krishnendu_ray/.
54. Amir Khan, "Beating Childhood Obesity Takes More than Cutting Calories, says Rachael Ray," Everyday Health, February 28, 2013, accessed 14 January 2016, http://www.everydayhealth.com/weight/beating-childhood-obesity-takes-more-than-cutting-calories-says-rachael-ray-9566.aspx.
55. Joy, *Why We Eat*.
56. Fiddes, *Meat*, 147.
57. Fiddes, *Meat*, 154.
58. Fiddes, *Meat*, 156.
59. Joy, *Why We Eat*, 96.
60. David Morley, "Broadcasting and the Construction of the National Family," in *The Television Studies Reader*, eds. Robert C. Allen and Annette Hill (London: Routledge, 2004); David Morley, "At Home with Television," *in Television After TV: Essays on a Medium in Transition*, eds. Lynn Spigel and Jan Olsson (Durham, NC: Duke University Press, 2004).
61. Simon During, *Cultural Studies: A Critical Introduction* (Abingdon: Routledge, 2005), 208.
62. Tim Edensor, *National Identity, Popular Culture and Everyday Life* (London: Berg, 2002); Kenneth R. Olwig "Natural Landscapes in the Representation of National Identity," in *The Ashgate Research Companion to Heritage and Identity*, eds. Brian Graham and Peter Howard (Adelshot: Ashgate, 2008).
63. Morten Tønnesen and Kadri Tũũr, "The Semiotics of Animal Representations: Introduction," in *The Semiotics of Animal Representations*, eds. K. Tũũr and M. Tønnesen (Amsterdam and New York: Rodopi, 2014).
64. Francesco Buscemi, "New Meat and the Media Conundrum with Nature and Culture, "*Lexia, Journal of Semiotics*, special Issue "Food and Cultural Identity," 19–20 (2015).
65. Thomas A. Sebeok, "Communication" in T.A. Sebeok, *A Sign is Just a Sign* (Bloomington: Indiana University Press, 1991).

66. Alfred K. Siewers, "Introduction: Song, Tree and Spring: Environmental Meaning and Environmental Humanities," *in Re-imagining Nature: Environmental Humanities and Ecosemiotics*, ed. Alfred K. Siewers (Lanham, MD: Bucknell University Press—Rowman and Littlefield, 2014), 5.
67. Siewers, "Introduction", 6.
68. Nandita Chaudhary, "Making Sense of the Bindi': Urban Indians' Appraisal of a Culturally Valued Symbol," in *Cultural Psychology of Human Values*, eds. Angela Uchoa Branco and Jan Valsiner (Charlotte, NC: Information Age Publishing, 2012), 114.
69. Dario Martinelli, *A Critical Companion to Zoosemiotics: People, Paths, Ideas* (Dordrecht: Springer, 2010), 35.
70. Martinelli, *A Critical Companion*, 58.
71. Martinelli, *A Critical Companion*, 58, original emphasis.

"The Worst Offense Here Is the Misrepresentation": *Thug Kitchen* and Contemporary Vegan Discourse

Alexis Priestley, Sarah K. Lingo, and Peter Royal

The language we use to conceive of and communicate about ourselves and our environments has material implications for the ways we navigate the world and interact with others, because, as Lakoff and Johnson write, "we act according to the way we conceive of things."[1] Given the increasing public attention to conscientious food consumption, it is imperative to examine the language surrounding, and consequently, the conceptual system structuring, cultural food habits, especially for those who, by nature of their "vegan" designation, are attentive to food practices. *Thug Kitchen* (*TK*), a popular vegan blog that frequently uses aggressive and racialized language to popularize its recipes, prompted controversy

Bryant Terry, "The Problem with 'Thug' Cuisine," *CNN*, October 10, 2014, accessed October 13, 2015, http://www.cnn.com/2014/10/10/living/thug-kitchen-controversy-eatocracy/index.htm.

A. Priestley (✉)
Rhetoric and Writing, Virginia Tech, Blacksburg, VA, USA
e-mail: priestal@vt.edu

S. K. Lingo • P. Royal
English, Virginia Tech, Blacksburg, VA, USA

J. Castricano, R.R. Simonsen (eds.), *Critical Perspectives on Veganism*,
The Palgrave Macmillan Animal Ethics Series,
DOI 10.1007/978-3-319-33419-6_16

when the food and cooking website *epicurious* revealed the bloggers to be a young white couple—Michelle Davis and Matt Holloway—in September 2014.[2] In the months since the *epicurious* article, various writers, bloggers, and other commentators have discussed the implications of *TK*'s caricatured use of African-American Vernacular English (AAVE) throughout its posts. In spite of this controversy, *TK* has become one of the most popular vegan blogs. If, as Lakoff and Johnson argue, the metaphors that govern the language we use to communicate—here we highlight veganism—are bound up in a broader conceptual system,[3] then, even though *TK* has attracted new audiences to vegan practices, the language it uses to promote veganism needs to be examined because the tropes imbricated in that language point to broader cultural issues concerning power, oppression, and patterns of consumption.

Since the *epicurious* post, the number of likes for *TK*'s Facebook page has grown to nearly 700,000 (674,829 at the time of writing), and its Twitter account has garnered over 60,000 followers. An endorsement from Gwyneth Paltrow, who has touted *TK* on *The Rachael Ray Show* and in other interviews, contributed greatly to the blog's early success. Paltrow first promoted *TK* in a 4 April 2013 post in *Goop*, a weekly e-newsletter that she curates. In the post, Paltrow provides an image from one of *TK*'s recipes that links to the blog followed by the words, "This might be my favorite thing ever."[4] This led to an appearance by Holloway and Davis on *The Rachael Ray Show* on 5 June 2015 to promote their cookbook, during which they briefly addressed criticism of their use of the word "thug."[5] Prior to this, *TK* had already achieved considerable critical success among several food publications and in prominent newspapers. In 2013, *TK* received the Best Food Blog Award from *Saveur*, an American gourmet, food, wine, and travel magazine,[6] as well as a nomination for a 2014 Veggie Award for Favorite Blog from *VegNews*, a prominent vegan magazine and website.

A few days before the *epicurious* post, the bloggers released a video trailer advertising the cookbook.[7] In the video, a number of staid, white suburbanites deliver profanity-laden lines in a deadpan tone. The trailer opens on a man walking through his lush backyard as he addresses the audience in a monologue that parodies pharmaceutical commercials. After explaining that he struggles to control his high cholesterol, the man says that he visited the doctor, who diagnosed him with "not giving a fuck about what I eat." He admits that he "should know better." A suburban mother, after reproaching herself for thinking that her "children got enough nutrition from that pre-packaged bullshit I bought at the store," asserts, "I don't play that shit anymore." She laments that she "was such a lazy fuckin' asshole" before. Finally, an elderly woman praises the cookbook for helping her "cut through the bullshit with language that I can understand." The monologues in the trailer all follow a similar formula: I was stupid/ignorant/irresponsible about food; I learned about the negative effects of my dietary habits; the *Thug Kitchen Cookbook* has rescued me from the bad food I was eating.

The cookbook trailer represents several features of the blog itself: liberal use of profanity, language that mimics AAVE, and the contrast between that language and the context in which it is used. Herein lies an oft-cited source of *TK*'s popularity: the unexpected and humorous combination of swear words and health food delivered by an abrasive narrator. For instance, Bethonie Butler of the Washington Post—representing a typical positive response to *TK*—writes of the blog's style, "It's as if Samuel L. Jackson went on a health kick and started a Tumblr."[8] This seemingly contradictory and aggressive combination resonates with a large audience, which notably includes many non-vegans. Humorously exaggerated shaming in *TK*'s recipes (and self-shaming in the case of its trailer) also contributes to its aggressive tone.

Conversely, negative online reactions to *TK* demonstrate that the language the blog employs is worthy of critique because of the harmful stereotypes it perpetuates. Online discussions of *TK* have often centered on its use of pseudo-AAVE and black stereotypes for the sake of humor. Commentators—including Dr. Amie "Breeze" Harper,[9] Laur M. Jackson,[10] Maya K. Francis,[11] and Akeya Dickson[12]—argue that *TK*'s language and persona constitute a form of "digital blackface," a term that refers to the mimicking and appropriation of black racial stereotypes in online fora; or, as blogger Laur M. Jackson puts it, "the odd and all-too-prevalent practice of white and non-Black people making anonymous claims to a Black identity through contemporary technological mediums such as social media."[13] She goes on to argue that digital blackface is both clumsy and potentially very harmful: "These attempts, while hilariously transparent, take advantage of the relative anonymity of the internet to perpetuate decontextualized stereotypes and project an image of Black people that fits the desire of anti-Black individuals."[14] Much of the discussion of *TK*'s relationship to digital blackface has centered on the use of the term "thug" and the distinction between its denotation and connotations. Akeya Dickson is a contributor to *The Root*, an online magazine launched by Henry Louis Gates, Jr. and Donald E. Graham; she writes of *TK*, "It's deceptive and feels a lot like the latest iteration of nouveau blackface."[15]

The blog omits the racial implications of the language it uses and avoids politically charged issues pertaining to veganism, such as animal cruelty, classism, speciesism, environmentalism, and the material and social conditions that limit access to a vegan diet—presumably in an effort to appeal to the widest possible audience. While these issues are conspicuously absent from *TK*'s rhetorical repertoire (unlike several other less popular vegan blogs), the language and imagery the bloggers employ reinforces unequal power dynamics of race, class, and gender. In our analysis, we demonstrate how the language of the blog performs this work.

Symbolic Language and Food

When symbolic language is associated with food preparation and consumption and spoken about in public spaces, some tropes gain popularity and become a part of the culture. Because food and foodways are a culturally embedded "system of communication,"[16] through which meaning is produced and circulated, they are not value-neutral. The meanings circulating in such cultural narratives are bound to different definitions of food, which are spatially and historically contingent. This means, for instance, that the food objects associated with veganism carry different meanings in different times and places and therefore require contextualized understanding. Such definitions are "regulated and distributed through normatively sanctioned institutional systems [...] in accordance with culturally defined categories."[17] If definitions of food and how it is produced and consumed are not just spatially and historically bound, but are also open to contestation, then they are "subject to political processes between different actors and interests."[18] The definitions that emerge as dominant become woven into the economic and cultural narratives of the societies to which they belong. For instance, there is a widely accepted metaphor about food that designates it as "fuel" for the body, casting the body as a machine.[19] The repertoire of figurative language that constructs the body as a machine allows different stakeholders to make specific claims about the nutritional value of particular food items and leads to the creation of new food products, like protein bars for the post-workout body, or the frozen breakfast sandwich for the on-the-go body. Consequently, eating habits are directly affected by the definitions of food that circulate within and become embedded in narratives that are reproduced in a particular society.

Food blogs are threaded through with those economic and cultural narratives, and food bloggers who have built a persuasive ethos circulate influential and contested narratives to large audiences. In our analysis of *TK*, we define a narrative of consumption as any description or anecdote attached to food. As Cynthia Enloe argues, narratives of consumption "mirror changes in global dynamics" and "help shape those dynamics."[20] These narratives of consumption are then woven into broader cultural, political, and economic narratives; therefore, as Paula Mathieu writes, "[it] is worth considering the roles language and persuasion play in defining habits of consumption."[21]

The political potential of food has long been examined in vegan scholarship; notably, Carol Adams argues that "ideas about meat, discussions about meat, are ideas about power, discussions about power [...] and not just power over animals, not just ideas about animals."[22] Though Adams specifically addresses the ethics of meat eating with respect to vegan practices, and *TK* avoids this conversation, we assert that *TK*, nevertheless, advances an argument about the relationship between consumption choice and power; this argument is present in the language and means of persuasion these bloggers utilize. In our analysis of this discourse, we draw on Eivind Jacobsen's work regarding the political potential of three main types of food tropes: nature, culture, and commodity.[23] Jacobsen argues that food tropes in Western public discourse largely fall within these three categories.[24] We use these categories as a baseline structure for our analysis of the rhetorical moves *TK*'s bloggers make in order to position their own vegan lifestyle choices within a broader conversation about veganism and healthy eating.

Food as Nature: *Thug Kitchen*'s "Spinach Cooler"

Food acts "as a vehicle for nutrients, [and] interacts with the human body [to help] provide the physiological requirements that sustain life."[25] Consequently, in its "Spinach Cooler" recipe,[26] *TK* presents a homeopathic view of food as nature, stating that this recipe can help fulfill vitamin requirements, fix undesirable skin issues, and prevent cancer. *TK* draws particular attention to the benefits of spinach, using phrases such as "[y]eah spinach makes you swoll as fuck, we know that." At the sentence level, *TK* presents spinach as coming "to the mother fucking rescue," to "repair" damaged skin and "fight" various illnesses. No longer simply a leafy green, spinach becomes an active ingredient in this homeopathic elixir. Within this narrative, food is presented as a natural resource, from which the blog's reader can mine essential nutrients that her "punk ass" doesn't get "enough" of. *TK* temporarily steps into the role of medical adviser, asking its readers, "did you know just one cup of spinach is over 300 % of your daily recommended Vitamin A?" Posing this nutritional fact as a question, *TK* suggests that its audience lacks access to nutritional knowledge and that *TK* serves as access to that knowledge. *TK* positions itself as an aggressive yet caring authority on and spokesperson for good nutrition and health, providing both information and medical advice ("But then again, smoking drastically increases your risk for lung cancer. So quit that shit").

One could easily assume the bloggers simply mean to infuse a potentially dry subject—the nutritive value of food—with innocuous humor. Part of the humor lies in the dissonance between the health terminology and the humor framed by the "thug" persona, because this combination is unexpected. One doesn't often get called a "punk ass" while receiving advice about Vitamin A intake.

The blog continues to adopt medicinal terminology as it explains, "Spinach has these plant-based compounds called 'flavonoids' that not only repair damaged skin but also fight multiple types of cancer." The blog struggles to strike a balance between the "thug" and the medical advisor as its language veers toward medical jargon and natural remedies, and its writers seem aware of this: the word "these," unnecessary in terms of clarity, makes the explanation potentially more conversational and thus less intimidating to a general audience. *TK* concludes its prescription by further distancing itself from the medical terminology of the previous paragraphs, invoking the health of the male speaker and reverting back to a "thug" tone: "Everybody knows I ain't even fucking playing when it comes to dick cancer, I gotta have my shit in tact [*sic*]."

Food as Commodity: *Thug Kitchen*'s "Peanut Tempeh Summer Roll"

When food is presented as a commodity in Western societies, as Jacobsen argues, it is "distributed through several middlemen, each with his or her own interests and institutional agendas."[27] *TK* acts as an intermediary in the "very first post that started [the blog]," "Peanut Tempeh Summer Rolls,"[28] which is laden with conflicting messages about food and commodity. The post begins with an anecdote: "My girl and I were cleaning out the fridge and whipped up ten of these motherfuckers." Immediately, the reader is given a glimpse into the writer's fridge, which, short of its usual fare, still features ingredients like tempeh, spring roll wrappers, rice vinegar, ginger, and fresh garlic. If readers are not yet feeling ashamed of their own almost empty fridges, likely unexceptional by comparison, *TK* promptly remedies this by asking, "What did you cook the last time you cleaned out your dirty ass fridge? A PICKLE AND KETCHUP SANDWICH? FUCK YOU." Here, the bloggers again juxtapose consumption habits with outrageous humor; ostensibly comparing their peanut tempeh summer rolls to inelegant alternatives, they criticize not only the reader's competence in the kitchen if she resorts to pickle and ketchup sandwiches but also her existing fridge contents. Undeniably, one's fridge contents are directly linked with consumption, and by literally saying, "Fuck you," *TK* judges its readers' fridge contents and therefore their economic status.

This criticism conflicts with the blog's frequent assertions that its recipes are budget-friendly. Here, and in many other posts, *TK* utilizes the cost-benefit ratio when justifying its ingredient choices, which is an essential component in the rhetorical repertoire of food-as-commodity.[29] The bloggers boast that the cost of their recipes is low and worth the resulting health benefits and taste. For example, consider *TK*'s "Smokey Eggplant Dip" recipe,[30] in which the bloggers point out that "[e]ggplant is abundant as fuck this time of year so you can buy them on the cheap." Even in the ingredients list of its peanut tempeh summer rolls, the bloggers, in describing what "large spring rolls wrappers/rice paper wrappers" are and where to find them, state, "[t]hey are cheap as fuck so don't stress." In addition, the bloggers emphasize the flexibility of utensil requirements; in this post, they instruct, "Pour half of the peanut sauce marinade into a shallow dish like a pie pan or whateverthefuck you have at your place," and in their ingredients list they write, "grated ginger (you can just cut this up all tiny if you don't have a grater)."

However, *TK* repeatedly departs from the cost-benefit argument it foregrounds in this post. The ingredient list, for instance, includes "1/3 cup peanut butter," with the stipulation "nothing full of sugar or a shit ton of salt," which might exclude many less-expensive peanut butters. Further, the recipe calls for additional vegetables and herbs (not included in the ingredients list), thus increasing the cost of the meal: "I used 2 medium carrots, 1 cucumber, 6 lettuce leaves, green onions, basil, cilantro, and avocado but use whatever you have hanging around." In a humorously nonchalant attempt to assuage the reader's potential concern about lacking these ingredients, the author clarifies, "I would make sure that you have some lettuce, something crunchy, and at least one herb but don't fucking stress about it." Nevertheless, the author assumes the reader has fresh vegetables and herbs left over when cleaning out her fridge. *TK* simultaneously accuses its reader of having a "dirty ass fridge" and expects the reader to have these perhaps prohibitively expensive ingredients on hand.

Food as Culture: *Thug Kitchen*'s "Roasted Brussels Sprouts"

As we have observed, *TK* frequently positions itself as an authority on food preparation by providing a means of accessing healthy, cost-effective, and appetizing meals. In its "Roasted Brussels Sprouts" recipe,[31] *TK* continues to act as this access point, but promises even more than improved nutrition and taste: elevated social standing. This recipe's accompanying image establishes a relationship between food consumption as a social activity and an individual's value: "SHAME THE SHIT OUT OF OTHER SIDE DISHES." The bloggers repeatedly cite popularity as a reason to serve this dish at gatherings. Invoking the Thanksgiving dinner table, the bloggers write, "GET THAT SOUPY GREEN SHIT OUTTA HERE," belittling the common green bean casserole. They continue, "C'mon, anybody showing up with that casserole from a can didn't even fucking try," thus condemning any dish that is easily made. In this way, social status is linked to the taste of a dish as well as the effort required to make it. The bloggers write, "Bring this bastard to Thanksgiving and nobody is going to ask you to do the goddamn dishes."

By equating a dish's aesthetic and nutritional quality to its maker's social standing, *TK* here recalls Jacobsen's discussion of food and culture. While referring to Bourdieu's concept of cultural capital, Jacobsen argues that "[t]aste and distinctions of taste [...] tend to be hierarchical-structured, reflecting power/impotence and dominance/marginality in society."[32] Furthermore, popular taste preference, as Jacobsen argues, "tends to be defined by the middle classes with their high levels of cultural capital. But it is also a means for the execution of dominance, whereby the marginality of the working class is made manifest."[33] *TK* recreates these structures by claiming that its dishes will save readers from embarrassment and also make them more popular within their own social circles. The recipe indicates, for example, that it "[s]erves 4–6 as a side, double that shit and bring it to Thanksgiving if you feel like being a popular motherfucker." *TK* promises not only popularity for its readers but also membership (albeit temporary) in Jacobsen's middle class, where readers acquire cultural capital by making culturally valued foods with culturally valued methods and materials. Thus, the passive, marginalized reader becomes, after digesting *TK*'s trendy directives, an active, centered taste-maker.

Discussion

We frame *TK*'s ethical argument within Jacobsen's three categories of food-related tropes because these are the three main areas within which food tropes have the most political potential. Within these grouped tropes, we can see the bloggers redefining what belongs at the center of good consumption practices: unrefined food that contains vital nutrients, is reasonably priced, and will help the reader gain credibility within her own social spheres. *TK* presents itself as an authority figure on the nature, commodity, and culture of vegan food and marginalizes products that do not fall within its definitions of veganism by shaming its readers into aligning with these proffered best consumption practices. In doing so, they are perpetuating oppressive and narrow-minded ideas about access and class. We mean here that *TK*'s very style appropriates a stereotypically lower-class dialect (AAVE) to promote a lifestyle that is often inaccessible to lower-class people. Although *TK* represents itself as breaking down barriers of access, it fails in this regard because its humor relies on the contrast between a lower-class persona and a middle- to upper-class lifestyle.

TK shames its readers into better food practices by marginalizing specific practices not only with regard to food and culture, as we have seen in the "Roasted Brussels Sprouts" post, but also with regard to nature and commodity. In terms of nature, consider its "Smokey Eggplant Dip" recipe, which orders readers to "STOP THE SHAMEFUL SNACKING," and then "[s]tow those prepackaged sad excuses for a snack and **GET FUCKING SERIOUS**." Using this language, *TK* makes a value judgment about its reader and the level of investment she has for her own eating habits and health. Alternatives to *TK*'s eggplant dip, those "prepackaged sad excuses," are imagined as "shameful snacking."

TK also directs inflammatory commentary at companies like McDonald's and Jamba Juice by saying "[o]nly [Jamba Juice] could make smoothies as unhealthy as McDonald's made oatmeal," and a "SEVEN DOLLAR

SMOOTHIE? FUCK YOU JAMBA JUICE."[34] By criticizing mainstream companies like McDonald's and Jamba Juice, which offer customers sustenance on-the-go, *TK* prompts its readership to substitute healthier foods. However, by telling their readers to not "BUY INTO FRITO LAY FUCKERY," they are asking their readers to buy into *TK's* own "fuckery," neglecting to consider that, for their readers, time might be a precious commodity. *TK* offers its recipes as healthy alternatives to Jamba Juice and McDonald's, constructing narratives of nutritive necessity by occupying a dual role of the knowledgeable and caring nutritionist and the judgmental critic.

Though *TK* claims, at least, to offer healthier, cheaper, and better-tasting alternatives to prepackaged and fast food, it fails to consider limitations of its readership's access to prohibitively expensive ingredients and specialty items such as Bragg's Liquid Aminos and nutritional yeast. However, this apparent lack of consideration is not something that *TK* necessarily needs to be held accountable for—the balance between health, cost, and taste is a frequent motif in conversation about food consumption. Yet *TK* creates a false dichotomy, wherein the reader is either an active participant in the eating practices *TK* espouses or she is against those practices. *TK* decontextualizes its arguments about food choice in a way that ignores the material conditions of the reader. By creating an all-or-nothing scenario in which these material conditions have no significance, *TK* delimits access in a way that ignores the very real effects of intersectionality.

In online responses to *TK*, the relationships among language, culture, material conditions, and embodied experience have been at the forefront. Many online commentators[35] argue that the *TK* bloggers and their allies defend the use of "thug" primarily because they enjoy a position of socio-economic and racial privilege that allows them to cherry-pick elements of black culture at their convenience without experiencing any of the disadvantages of actually being black. Several commentators[36] cite the deaths of Trayvon Martin, Michael Brown, and Oscar Grant as evidence of the cultural baggage associated with the word "thug." Jordan Sowunmi offers the case of Richard Sherman.[37] Sherman, a football player for the Seattle Seahawks, was labeled a "thug" following an emotional speech he gave after the Seahawks' 2015 Super Bowl win:

> Sherman played the game of a lifetime and served an integral role in sending the Seattle Seahawks into the Super Bowl, [*sic*] he gave an impassioned post-game speech in which he called out his opponents' perceived slights against him. He didn't even curse, but he was immediately labeled a thug, presumably because his skin color and dreadlocks fit the description of what people typically associate with that word.

Sowunmi also cites media treatment of Corey Pegues as an example of the racially charged connotations of the word "thug":

> More recently, I wrote about how the *New York Post* smeared storied New York City police officer, former drug dealer, and current community advocate Corey Pegues, describing him as a "thug cop" on the cover of their paper after he appeared on the Combat Jack Show and shared the story of how he transitioned from a victim of the trap to becoming an executive in the world's largest police force.[38]

These men, and many others like them, are targets of conscious or unconscious fear, particularly among white people, because they match a stereotype of young black men as violent, aggressive, and prone to criminal behavior—men whose very existence warrants preemptive action to prevent them from causing harm to others.[39] Sowunmi compares *TK*'s style to a kind of cultural and racial tourism: "One thing is clear: For the upwardly mobile white Angelinos behind *TK*, the word thug is ironic and funny, a bit of culturally exploratory fun. But for men like Sherman and Pegues, it's a putdown meant to demonize and dehumanize."[40] In addition, a wide array of writers[41] argue that "thug" is often a politically correct, coded, or euphemistic (albeit thinly veiled) way to call someone the *n*-word.

A common defense of *TK*'s use of "thug" cites the word's derivation from the name of an ancient Indian religious sect of robbers and assassins.[42] An example of this defense comes from Daniel Power, the CEO of Powerhouse Arena, a venue that hosted a release party for *Thug Kitchen: The Official Cookbook*.[43] When a group of protesters organized a boycott of the event to criticize the authors' offensive word choice, Power accused them of both misconstruing the intended effect of the word as it appears in the cookbook and wrongly insisting that the term "thug" "is a code word for the n-word."[44] Power suggests that the authors use the word to conjure a "boorish, bullying, and domineering"[45] character of indeterminate race as "a form of parody—not of the cultural origins of the contemporary use of the language and of the demeanor, but rather the extrapolation from their origins and application to a totally different worldview."[46] In other words, the authors' use of the word serves to satirize white culture rather than to exploit a black stereotype by inserting an aggressive, hyperbolic character into the "traditionally staid setting of cooking and cuisine publishing."[47] Certainly, the humor of *TK* arises in part from the contrast between its subject and style, but to suggest that this style does not evoke black stereotypes is to ignore the voices of many black persons who daily confront the very real and prominent racial connotations of "thug" and who have repeatedly and convincing argued that these connotations are undeniably central to *TK*'s style.[48]

In addition to the more salient problems with *TK*'s language choices, the blog's persona and humor misrepresent the material and social conditions in which veganism has been historically practiced, including its origins in American culture. In an article published shortly after the release of *Thug Kitchen: The Official Cookbook*, Bryant Terry—an author, food advocate, and CNN contributor—acknowledges "the important role of pop culture in changing people's attitudes, habits and politics around food": "'Start with the visceral, move to the cerebral and then the political' has been the mantra guiding most of my efforts."[49] However, Terry doubts whether *TK*'s humor is conducive to the cerebral and political development of vegan discourse in the public sphere. Terry agrees with many of the charges leveled against *TK* by other commentators, but he argues that "[t]he worst offense here is the misrepresentation."[50] He specifically targets the essential source of *TK*'s humor:

> The contrast drawn between the consciously progressive dishes shown and the imagined vulgar, ignorant thug only works if the thug is the kind of grimy person of color depicted in the news and in popular media as hustling drugs on a dystopian block, under the colorful glow of various burger stands, bulletproof take-out spots or bodega signs. "Those kind of people," the visual gag suggests, "intimidating you into…preparing arugula or tempeh? How absurd, how shocking, how hilarious!"[51]

The problem with this, Terry argues, is that, historically, African-American cuisine, particularly during the era of slavery, was almost exclusively plant-based. African slaves and their descendants adopted plant-based food practices out of necessity, which have since been appropriated by mainstream—and usually white—culture and become very popular: "Whether or not the hipsters and health nuts charmed by *Thug Kitchen* realize this, vegetarian, vegan and plant-strong culture in the black experience predates pernicious thug stereotypes. Said another way, the *Thug Kitchen*'s central comic conceit doesn't jibe with reality."[52]

The use of caricatured, decontextualized stereotypes to establish a brand and sell products seems at odds with the philosophical underpinnings of ethical veganism. The absence or omission of ethics in vegan discourse and products makes it more possible for well-intentioned people to fall into the trap represented by *TK*: using one form of oppression to combat another.

By using a racial stereotype, essentially drawn from clumsy, imitative AAVE memes, and effacing the more radical aspects of veganism such as the morality of exploiting animals, *TK* avoids any explicit internal contradiction between the use of that stereotype and its promotion of vegan food. However, if we consider the theoretical underpinnings of anti-racism and ethical veganism, then there is an obvious contradiction between exploiting a black stereotype to become more popular and promoting vegan food. *TK* disconnects the issues of animal consumption and racism from the history of oppression that relates them, focusing instead on the nutrition, taste, and prestige of veganism and the humor that mocking AAVE lends them.

Thug Kitchen's Rhetorical Choices as a Vegan Food Blog

In light of *TK*'s simultaneously oppressive and popular language, it will be useful now to examine their rhetorical strategies alongside those currently used by vegans in public and private spaces. As a trendy vegan food blog, *TK* adheres to and departs from traditional ways that vegans communicate with non-vegans and thus may at first appear to be a model for future vegan advocacy. We ask what, if anything, can vegans take from *TK*'s rhetoric, and what does its popularity mean to future vegan discourse?

The ways in which vegans can successfully promote veganism among non-vegans are frequently considered within vegan scholarship. Responding to concerns about potentially polarizing conversation between vegans and non-vegans, scholars have sought to define how non-vegans perceive vegan practice. Twine, drawing from Ahmed's "feminist killjoy," presents the "vegan killjoy," which "destabilize[s] an assumed shared sense of happiness" related to meat eating.[53] "In willfully speaking up," Twine writes, the "[vegan] killjoy may engender anxiety, discomfort, guilt, and risks exclusion for doing so."[54] Similarly, Greenebaum notes that "[v]egan bloggers and activists agree that omnivores (and some vegetarians) are comfortable talking about vegetarianism and veganism as long as the conversation focuses on diet, lifestyle, and personal choice."[55] Consequently, the successful promotion of veganism hinges on the exclusion of its more politically contentious components.

Scholars like Twine and Greenebaum have honed in on communication strategies that are employed to ameliorate existing tensions. Greenebaum identifies four more specific strategies that "'save face' and protect both parties from attack and alienation": "avoiding confrontation, waiting for an appropriate time, focusing on health benefits, and leading by example."[56] Twine similarly identifies strategies that "[build] bridges" between vegans and non-vegans, including "the mode of performing veganism in a *demonstrative* manner that draws omnivores or vegetarians into the sensual experience of vegan food."[57] Regarding this strategy, Twine suggests that vegans cooking for non-vegans is an especially effective way to "normalise"[58] vegan food: "This had the effect of bringing vegan eating into a familiar space and as a known, appreciated aesthetic experience once again expressed via relations of care."[59]

As our analysis shows, *TK* most frequently cites "diet, lifestyle, and personal choice"[60] and utilizes many "bridge-building" strategies[61] to persuade non-vegan readers to embrace a vegan lifestyle. Perhaps most notably, the blog excludes ethical veganism[62] from its discourse, focusing instead on how their recipes benefit individual readers such as by improving health and increasing popularity. As Greenebaum observes, "veganism [...] is mostly tolerated as long as it is being presented as a diet that does not include a moral agenda,"[63] and *TK*'s bloggers seem well-aware of this, even avoiding the word "vegan" in their posts. The term is featured most prominently in the FAQ,[64] where *TK* responds to the question, "Are all your recipes vegan?" (the fifth question of ten and just under "Do you respond to fan mail?"). *TK* answers, "You bet your sweet ass everything we do is vegan. Every recipe on our site is completely plant-based. Most ingredients we use can be easily substituted for omnivores." *TK* adopts the term "plant-based," which does not carry the same political charge as the term "vegan"; instead of conjuring up images of what will not be eaten (animals), the term calls attention to what *will* be eaten (plants). This rhetorical slippage is a means of countering potential suspicion from non-vegans who may have unfavorable associations with the word "vegan."

TK also naturalizes vegan food for non-vegan audiences in several ways. First, *TK* rhetorically distances itself from any ingredient that seems inaccessible either physically or socially—like the aforementioned Bragg's Liquid Aminos—by labeling it "old school hippie shit," and further by giving details about where this seemingly unfamiliar ingredient can be found within the familiar space of the grocery store: "near the vinegars or soy sauce."[65] Both of these are relatively innocuous, which potentially neutralizes any hesitation the audience may have about locating it. Second, *TK* calls attention to benefits of improved taste and includes aesthetically appealing photography in its recipes to emphasize the "sensual experience of vegan food" that Twine identifies as an effective strategy.[66]

Though *TK* "avoid[s] confrontation"[67] in the sense that it does not allude to violations of animal rights, it embraces confrontation through an aggressive approach to dietary change. As our analysis shows, *TK* frequently utilizes shame to motivate its readers to try its recipes and eat more plant-based dishes. This shaming strategy, combined with aggressive language and profanity, allows *TK* to rhetorically situate people who don't align with their food choices simultaneously as outsiders and as sabotaging their own bodies. However, *TK*'s aggression differs from that of the vegan killjoy; on the contrary, popular commentators, like Rachel Ray and Gwyneth Paltrow, have demonstrated that much of *TK*'s readership perceives its aggressive tactics as humorous and engaging. In addition to embracing confrontation, *TK* also refuses to "wait for an appropriate time"[68] to challenge meat eating. Instead, the blog consistently, aggressively forwards a vegan diet; according to *TK*, there is no appropriate time to eat meat, animal products, snacks, or fast food—including holidays, special occasions, and social gatherings (which are all featured prominently in their posts)—lest the audience be alienated from their own bodies or their social groups.

Conclusion

TK's popularity can be partially explained by their use of techniques already successfully employed by vegans, including exclusion of references to animal ethics, emphasis on how a vegan lifestyle benefits the individual, and visual evidence that vegan food can be aesthetically appealing, even to meat eaters. These techniques are clearly effective means of making veganism more inviting. However, *TK*'s popularity is also—perhaps even primarily—attributable to its use of AAVE for the sake of humor, which many have observed is at best offensive and at worst complicit in oppression. In either scenario, this language makes veganism more attractive to some at the expense of others.

TK represents what can happen in an era when veganism has become more mainstream by divorcing it from its radical ethical roots. Language can be a powerful tool of oppression, and *TK*'s racist language—which also reinforces harmful assumptions about class and food access—is particularly insidious. While it may appear innocuous to a large audience, casually using a word like "thug"—with its many offensive connotations—to promote a lifestyle that, at its essence, opposes the oppression and exploitation of bodies ultimately damages both the vegan and anti-racist movements. Perpetuating the word "thug" in popular culture as if it is not deeply embedded in racist discourse removes the black body from the language used to define, categorize, and oppress it. Consequently, the black body is constructed as, to borrow the term from Carol J. Adams, the "absent referent." Adams invokes the absent referent to describe the violence done to animal bodies for the sake of meat consumption:

"Through butchering, animals become absent referents. Animals in name and body are made absent *as animals* for meat to exist."[69] In other words, animals are more easily exploited when the living animal that precedes the butchering process is made linguistically and symbolically absent after that process. Take, for example, the cow that becomes beef or the pig that becomes pork. Similarly, it becomes apparent when juxtaposing Adams's definition with the absent referent of the thug that both meat consumption and the appropriation of AAVE by *TK*'s bloggers require a harmful disassociation of words from their consequences.

The above analysis, coupled with popular media response, demonstrates that *TK*'s brand of humor has alienated and offended audiences concerned with issues of race and oppression. However, vegans should recognize that this humor, while oppressive, has popularized veganism among non-vegan audiences. We are not by any means suggesting that vegan discourse should be humorless but that any humor used to promote veganism should align with its ethical principles. The promotion of veganism should not come at the expense of ethical compromises that contribute to oppression. If, as ethical vegans, we cannot ignore "the connections between the social constructions of whiteness, racialization, and racisms (as well as sexisms, nationalisms, etc.), and animal abuse,"[70] then as we move forward, we need to be more attentive to and reflexive about our communication practices in an effort to expunge oppressive language from popular vegan discourse.

Notes

1. George Lakoff and Mark Johnson, *Metaphors We Live By* (Chicago: University of Chicago Press, 1980), 3.
2. Matt Duckor, "Thug Life: A Behind-the-Scenes Look at the Masterminds Of Thug Kitchen," *Epicurious*, September 29, 2014, accessed October 13, 2015. http://www.epicurious.com/archive/blogs/editor/2014/09/thug-kitchen-author-real-names-revealed.html.
3. Lakoff and Johnson, *Metaphors We Live By*.
4. Gwyneth Paltrow, "Thug Kitchen," *Goop*, April 4, 2013, accessed October 13, 2015, http://goop.com/thug-kitchen/.

5. Michelle Davis and Matt Holloway, "A Healthy Pasta Dish from "Thug Kitchen' Bloggers," *The Rachael Ray Show*, by Rachael Ray, CBS, June 5, 2015, http://www.rachaelrayshow.com/food/18759_a_healthy_pasta_dish_from_thug_kitchen_bloggers/.
6. "Meet the 2013 BFBA Winners: Thug Kitchen," *Saveur*, May 28, 2013, accessed October 13, 2015, http://www.saveur.com/article/blog/Best-Food-Blog-Award-Winner-Thug-Kitchen.
7. "Thug Kitchen Cookbook Trailer (explicit)," YouTube video, 1:53, posted by "Thug Kitchen Team," September 25, 2014, http://www.youtube.com/watch?v=Ar7g_26QWu0.
8. Bethonie Butler, "Thug Kitchen, the Blog That Swears by Veganism. A Lot," *The Washington Post*, July 23, 2013, Food sec., accessed October 13, 2015, http://www.washingtonpost.com/lifestyle/food/thug-kitchen-the-blog-that-swears-by-veganism-a-lot/2013/07/22/40053460-eee9-11e2-9008-61e94a7ea20d_story.html.
9. Amie Breeze Harper, "On Ferguson, Thug Kitchen & Trayvon Martin: Intersections of [Post] Race-Consciousness, Food Justice and Hip-Hop Veganism," *The Sistah Vegan Project: A Critical Race Feminist's Journey Through the "Post-Racial" Ethical Foodscape…and Beyond*, November 4, 2014, accessed October 13, 2015, http://sistahvegan.com/2014/11/04/video-on-ferguson-thug-kitchen-trayvon-martin-intersections-of-post-race-consciousness-food-justice-and-hip-hop-veganism/.
10. Laur M. Jackson, "Memes and Misogynoir," *The Awl*, August 28, 2014, accessed October 13, 2015, http://www.theawl.com/2014/08/memes-and-misogynoir.
11. Maya Francis, "Thug Kitchen's Brand Of Technicolor Blackness," *Very Smart Brothas*, September 30, 2014, accessed October 13, 2015, http://verysmartbrothas.com/thug-kitchens-brand-of-technicolor-blackness/.
12. Akeya Dickson, "Thug Kitchen: A Recipe in Blackface," *The Root*, September 30, 2014, accessed October 13, 2015, http://www.theroot.com/articles/culture/2014/09/thug_kitchen_a_recipe_in_blackface.html.
13. Jackson, "Memes and Misogynoir."
14. Ibid.
15. Dickson, "Thug Kitchen: A Recipe in Blackface."
16. Roland Barthes, *Mythologies* (New York: Hill and Wang, 1972), 21.
17. Eivind Jacobsen, "The Rhetoric of Food: Food as Nature, Commodity and Culture," in *The Politics of Food*, ed. Marianne E. Lien and Brigitte Nerlich (New York: Berg, 2004), 61.

18. Ibid., 60.
19. Ibid., 59.
20. Cynthia Enloe, *Bananas, Beaches, and Bases: Making Feminist Sense of International Politics* (Berkeley: University of California Press, 1990), 197.
21. Paula Mathieu, "Economic Citizenship and the Rhetoric of Gourmet Coffee," *Rhetoric Review* 18 (1999): 114.
22. Carol J. Adams, "Why Feminist-Vegan Now?" *Feminism & Psychology* 20 (2010): 306.
23. Jacobsen, "The Rhetoric of Food," 63.
24. Ibid., 61.
25. Ibid.
26. "Spinach Cooler," http://www.thugkitchen.com/spinach_cooler.
27. Jacobsen, "The Rhetoric of Food," 61.
28. "Peanut Tempeh Summer Rolls," http://www.thugkitchen.com/peanut_tempeh_summer_rolls.
29. Jacobsen, "The Rhetoric of Food," 73.
30. "Smokey Eggplant Dip," http://www.thugkitchen.com/smokey_eggplant_dip.
31. "Roasted Brussels Sprouts," http://www.thugkitchen.com/roasted_brussels_sprouts_quinoa_cranberries.
32. Jacobsen, "The Rhetoric of Food," 70.
33. Ibid.
34. "Spinach Cooler."
35. See, as examples, Harper; Dickson; Sowunmi; and Twitty.
36. See, as examples, Harper; Jackson; Francis; and Dickson.
37. Jordan Sowunmi, "'Thug Kitchen' Is the Latest Iteration of Digital Blackface," *Vice*, October 3, 2014, accessed October 13, 2015, http://www.vice.com/read/thug-kitchen-is-the-latest-iteration-of-people-profiting-off-digital-blackface-909.
38. Ibid.
39. Harper, "On Ferguson, Thug Kitchen & Trayvon Martin."
40. Sowunmi, "'Thug Kitchen' Is the Latest Iteration of Digital Blackface."
41. See, as examples, Harper; Francis; Sowunmi; and Twitty.
42. "thug, n.". *OED Online*. December 2015. Oxford University Press. http://www.oed.com/view/Entry/201476?rskey=psYHU2&result=1&isAdvanced=false (accessed October 13, 2015).
43. Jordan Sargent, "Read a Publishing CEO's Condescending Defense of Thug Kitchen," *Gawker*, October 6, 2014, accessed October 15, 2015, http://gawker.com/bookstore-ceo-defends-thug-kitchens-use-of-thug-in-em-1642973425.

44. Ibid.
45. Ibid.
46. Ibid.
47. Ibid.
48. See, for example, Dickson; Francis; Harper; Jackson; Sowunmi; Terry; and Twitty.
49. Terry, "The Problem with 'Thug' Cuisine."
50. Ibid.
51. Ibid.
52. Ibid.
53. Richard Twine, "Vegan Killjoys at the Table—Contesting Happiness and Negotiating Relationships with Food Practices," *Societies* 4 (2014): 625.
54. Ibid.
55. Jessica B. Greenebaum, "Managing Impressions: 'Face-saving' Strategies of Vegetarians and Vegans," *Humanity & Society* 36 (2012): 310–11.
56. Ibid., 317.
57. Twine, "Vegan Killjoys at the Table," 636.
58. Ibid., 635.
59. Ibid., 636.
60. Greenebaum, "Managing Impressions," 311.
61. Twine, "Vegan Killjoys at the Table," 635.
62. That is, "the personal rejection of the commodity status of nonhuman animals, of the notion that animals have only external value, and of the notion that animals have less moral value than do humans" (Francione and Garner, *The Animal Rights Debate*, 62).
63. Greenebaum, "Managing Impressions," 310.
64. "FAQ," www.thugkitchen.com/faq.
65. "Roasted Chickpea & Broccoli Burrito," www.thugkitchen.com/roasted_chickpea_broccoli_burrito.
66. Twine, "Vegan Killjoys at the Table," 636.
67. Greenebaum, "Managing Impressions," 317.
68. Ibid.
69. Carol J. Adams, *The Sexual Politics of Meat: A Feminist-vegetarian Critical Theory* (New York: Continuum, 1990), 66.
70. Amie Breeze Harper, "Connections: Speciesism, Racism, and Whiteness as the Norm," in *Sister Species: Women, Animals, and Social Justice*, ed. Lisa Kemmerer (Champaign: University of Illinois Press, 2011), 76.

References

Adams, Carol J. 1990. *The sexual politics of meat: A feminist-vegetarian critical theory*. New York: Continuum.

Adams, Carol J. 2010. Why feminist-vegan now? *Feminism and Psychology* 20: 302–317.

Barthes, Roland. 1972. *Mythologies*. New York: Hill and Wang.

Butler, Bethonie. 2013. Thug Kitchen, the blog that swears by veganism. A lot. *The Washington Post*, July 23. Food sec. http://www.washingtonpost.com/lifestyle/food/thug-kitchen-the-blog-that-swears-by-veganism-a-lot/2013/07/22/40053460-eee9-11e2-9008-61e94a7ea20d_story.html. Accessed 13 Oct 2015.

Davis, Michelle, and Matt Holloway. 2015. A healthy pasta dish from 'Thug Kitchen' bloggers. *The Rachael Ray Show*. By Rachael Ray. CBS, June 5. http://www.rachaelrayshow.com/food/18759_a_healthy_pasta_dish_from_thug_kitchen_bloggers/.

Dickson, Akeya. 2014. Thug Kitchen: A recipe in blackface. *The Root*, September 30. http://www.theroot.com/articles/culture/2014/09/thug_kitchen_a_recipe_in_blackface.html. Accessed 13 Oct 2015.

Duckor, Matt. 2014. Thug life: A behind-the-scenes look at the masterminds of Thug Kitchen. *Epicurious*, September 29. http://www.epicurious.com/archive/blogs/editor/2014/09/thug-kitchen-author-real-names-revealed.html. Accessed 13 Oct 2015.

Enloe, Cynthia. 1990. *Bananas, beaches, and bases: Making feminist sense of international politics*. Berkeley: University of California Press.

Francione, Gary Lawrence, and Robert Gardner. 2010. *The animal rights debate: Abolition or regulation?* New York: Columbia University Press.

Francis, Maya. 2014. Thug Kitchen's brand of technicolor blackness. *Very Smart Brothas*, September 30. http://verysmartbrothas.com/thug-kitchens-brand-of-technicolor-blackness/. Accessed 13 Oct 2015.

Greenebaum, Jessica B. 2012. Managing impressions: 'Face-saving' strategies of vegetarians and vegans. *Humanity & Society* 36: 309–325.

Harper, Amie Breeze. 2011. Connections: Speciesism, racism, and whiteness as the norm. In *Sister species: Women, animals, and social justice*, ed. Lisa Kemmerer, 72–78. Champaign: University of Illinois Press.

Harper, Amie Breeze. 2014. On Ferguson, Thug Kitchen & Trayvon Martin: Intersections of [Post] race-consciousness, food justice and hip-hop veganism. *The Sistah Vegan Project: A critical race feminist's journey through the "post-racial" ethical foodscape … and beyond*. November 4. http://sistahvegan.

com/2014/11/04/video-on-ferguson-thug-kitchen-trayvon-martin-intersections-of-post-race-consciousness-food-justice-and-hip-hop-veganism/. Accessed 13 Oct 2015.

Jackson, Laur M. 2014. Memes and Misogynoir. *The Awl*, August 28. http://www.theawl.com/2014/08/memes-and-misogynoir. Accessed 13 Oct 2015.

Jacobsen, Eivind. 2004. The rhetoric of food: Food as nature, commodity and culture. In *The politics of food*, ed. Marianne E. Lien and Brigitte Nerlich, 59–78. New York: Berg.

Lakoff, George, and Mark Johnson. 1980. *Metaphors we live by*. Chicago: University of Chicago Press.

Mathieu, Paula. 1999. Economic citizenship and the rhetoric of gourmet coffee. *Rhetoric Review* 18: 112–127.

Meet the 2013 BFBA Winners: Thug Kitchen. *Saveur*, May 28, 2013. http://www.saveur.com/article/blog/Best-Food-Blog-Award-Winner-Thug-Kitchen. Accessed 13 Oct 2015.

Paltrow, Gwyneth. 2013. Thug Kitchen. *Goop*, April 4. http://goop.com/thug-kitchen/. Accessed 13 Oct 2015.

Sargent, Jordan. 2014. Read a publishing CEO's condescending defense of Thug Kitchen. *Gawker*, October 6. http://gawker.com/bookstore-ceo-defends-thug-kitchens-use-of-thug-in-em-1642973425. Accessed 15 Oct 2015.

Sowunmi, Jordan. 2014. 'Thug Kitchen' Is the latest iteration of digital black-face. *Vice*, October 3. http://www.vice.com/read/thug-kitchen-is-the-latest-iteration-of-people-profiting-off-digital-blackface-909. Accessed 13 Oct 2015.

Terry, Bryant. 2014. The problem with 'Thug' Cuisine. *CNN*, October 10. http://www.cnn.com/2014/10/10/living/thug-kitchen-controversy-eatocracy/index.htm. Accessed 13 Oct 2015.

Thug Kitchen. http://www.thugkitchen.com/. Accessed 13 Oct 2015.

Thug Kitchen Cookbook Trailer (explicit). YouTube video, 1:53. Posted by "Thug Kitchen Team," September 25, 2014. http://www.youtube.com/watch?v=Ar7g_26QWu0.

Twine, Richard. 2014. Vegan Killjoys at the table—Contesting happiness and negotiating relationships with food practices. *Societies* 4: 623–639.

Twitty, Michael. 2014. Thug Kitchen: It's not just about aping and appropriation, it's about privilege. *Afroculinaria*, October 22. http://afroculinaria.com/2014/10/22/thug-kitchen-its-not-just-about-aping-and-appropriation-its-about-privilege/. Accessed 13 Oct 2015.

Decolonizing Veganism: On Resisting Vegan Whiteness and Racism

Jennifer Polish

Are discourses and practices of veganism in the USA inevitably characterized by whiteness? From the infamous *Thug Kitchen* to People for the Ethical Treatment of Animals (PETA's) uncritical comparisons of slavery with factory farming, all the way to the practices—and often, the mere presence—of farmer's markets, community gardens, and Whole Foods stores in low-income neighborhoods of color that are enduring renewed colonization vis-à-vis gentrification, the unmarked whiteness of veganism is ubiquitous. Nonetheless, it is difficult, if not impossible, to critically reflect on veganism as both a politically charged foodways practice and a critical/ethical commentary on animality—how have Western cultures managed to fundamentally transform dead non/human animal bodies into "meat" rather than "corpses" or "carcasses"? —without attending to the racial implications of animality. As many activists and scholars of color (discussed below)

J. Polish (✉)
CUNY LaGuardia Community College
the CUNY Graduate Center, New York, NY, USA
e-mail: jpolish@gradcenter.cuny.edu

J. Castricano, R.R. Simonsen (eds.), *Critical Perspectives on Veganism*,
The Palgrave Macmillan Animal Ethics Series,
DOI 10.1007/978-3-319-33419-6_17

have already made clear, the very definition of "humanity" is irrevocably entrenched in Western conceptions of the value of life: to achieve (hu)Man status is fundamentally to achieve Westernized whiteness, and thus the very hierarchy of human versus non/human animal that veganism challenges is charged with the history of white supremacy from the outset. When we talk about racist language and physical violence as "dehumanizing," we are invoking the ways that structures of whiteness have and continue to position people of color (POC) as less than human: as "animal."

Yet, mainstream discourses persistently figure veganism as a racially unmarked (therefore, white) politic. It is essential, therefore, to interrogate veganism as an extension of whiteness and whiteness as a driving force of popularized veganism. The question becomes, Can the uncritical whiteness of mainstream veganism be challenged by examining the parallel structures of racist and anti-animal violence; the interlocked mechanisms of racist violence and meat-eating and dairy-producing; and the actions of vegans of color who are actively doing the work of making these connections explicit? This chapter is indebted to the work of scholars and activists of color, and will explore the ways that mainstream discourses and practices of veganism are inflected by whiteness. I will argue that it is absolutely vital to centralize POC models of veganism that already challenge this whiteness. This centralization will promote the integration of normalization (making veganism a much more common practice and praxis) and radicalization (emphasizing necessary connections between anti-racist and anti-speciesist work). By resisting unmarked whiteness, racially aware veganism can be a vehicle by which the politics of veganism can both become more normalized and strengthen its potential for radicalizing the politics of "meat."

Communities of color often push back against uncritical comparisons of POC with animals and against the role that the rising cultural popularity of white veganism plays in gentrification. Indeed, the relationship between corporatized veganism and gentrification (discussed below) further highlights the entrenchment of veganism and whiteness. How can POC ethically enter into a foodways practice that is popularly associated with violence against themselves? As a white, queer, vegan woman —also known as *that* stereotype that my girlfriend (understandably) swore she'd never date—my interest in unsettling the whiteness of veganism derives from my desire to enact a mode of veganism that prioritizes anti-racism as a fundamental cornerstone of any effort to deconstruct ideologies and practices of speciesism. This chapter will use the works of scholars of color like A. Breeze Harper and Sylvia Wynter to explore the ways that racism and speciesism are intertwined. The very concept of anthropocentrism is wrapped in the questions of what it is to *be human* and who is racially privileged enough to qualify as such. Further, the very same anthropocentrism that drives the physical consumption of animals encourages us to ask: why are (some of) the animals that are hegemonically deemed *human* thought to be worthier and more capable of life than the animals that are hegemonically deemed *non/human*? These questions—and therefore the very definitions of human and non/human, which so strongly inform the discourse of vegan activism—are fundamentally steeped in an ideology that enshrines whiteness as the defining characteristic of "the human." Since the very concept of humanity has its roots in racist ideology, so, too, does speciesism. As a result, genuine political, philosophical, and ethical commitments to dismantle the systemic violences of speciesism (these attempts often include veganism) must foreground these efforts in anti-racist rhetoric and praxis.

What *Is* Animal?

When pressed about their attempts to spread awareness about the devastating practices of factory farming and the sheer violence of reducing animal carcasses to mere "meat," (white) vegan activists and animal studies/post-humanist academics will often beg the question: "Do animals deserve equal consideration? Do their experiences *matter*?"[1] For many, these questions accompany graphic descriptions of the suffering of non/human animals who are forced to live their violently shortened lives in horrifying and torturous conditions on factory farms; in other words, the question "do non/human animal experiences matter?" is often utilized as a rhetorical tool to which the author of the question is providing the self-evident answer.[2] Amidst this kind of rhetorical strategy, one risks becoming positioned as a heartless speciesist who does not fundamentally "care" about the torture and murder of millions of creatures. While these strategies are obviously powerful—so many of us know someone who has become vegan (or at least vegetarian) because they randomly picked up a PETA pamphlet—they foreclose the possibility of being critical of veganism because of resistance to its sheer entrenchment in whiteness rather than resistance to violence against animals. The particular questions of whether animal lives *matter*—presented as they so often are as leading questions, with the expected answer of, "of course they do, this violence is terrible!"—is perhaps especially unsettling given that human beings still must ask (and receive negative answers to) the question, do Black lives *matter*? The phrasing of the movement—led by a coalition of Black women activists in the face of continued violence against Black communities in the USA—to make Black Lives Matter to the state offers an eerie and poignant counterpoint to the contrived question of whether animal experiences *matter*, especially given the historical "animalization" of racialized bodies.[3]

Because perhaps this is not the question. Perhaps the question—the question that pulls at the strings of both anti-racism and anti-speciesism—can be more accurately phrased as "what does it mean to be considered fully *human* (and therefore, considered *worthy* of 'mattering')? How can one *matter* without being considered (fully) human?" While scholars that engage in questions of humanism and animal studies, such as Cary Wolfe, sometimes position racism and speciesism as linked, the question of *how* they are linked is left inadequately explored. Wolfe, for example, uses the oppression of animals as a catch-all link to racism, classism, and sexism, stating:

> as long as it is institutionally taken for granted that it is all right to systematically exploit and kill nonhuman animals simply because of their species, then the humanist discourse of species will always be available for use by some humans against other humans as well, to countenance violence against the social other of whatever species—or gender, or race, or class, or sexual difference.[4]

While I certainly have no desire to claim that the linkages he gestures toward do not exist (surely they do), the linkages being made here are through analogizing at best. If "the humanist discourse of species will always *be available for use* by some humans against other humans as well," then speciesism is positioned as a distinct -ism that enables racism, not one which more intimately interacts with racism (and other institutional -isms) to promote itself. Scholars of color like Sylvia Wynter assert, more clearly, that even the question of *the human* itself is fundamentally shaped by racism, rather than simply existing as a logic that can permeate into inter-human violence (as Wolfe inadvertently suggests).

The kind of animal studies scholarship as represented by Wolfe suggests that there is some united realm of "the human" that is universally privileged above "the non/human animal." Scholars and vegans of color, however, have long been posing counterpoints to this neglect of the material history of the question, "what does it mean to be considered fully human?" Posing vital questions like this recognizes that the very category of *human* is not universal, but rather just as contentious as the category of *animal* which, as Jacques Derrida argues, is itself a violent universalization which makes monolithic the incomprehensibly vast range of animal experiences and modes of being. Applying the same logic to humanity by refusing to think of "humanity" as a monolithic category intimately intertwines the histories of speciesism and racism.

Using the insights and analyses of scholars of color Sylvia Winter, Frank Wilderson, and A. Breeze Harper as the primary points of departure, this chapter will argue that a pricipal problem in a lot of animal rights activism and scholarship resides in the assumption that humans are always already positioned above animals. To counteract this assumtion, this chapter will encourage a more nuanced understanding and praxis of ethical veganism, which inherently links ideologies and practices of racism with those of speciesism. In so doing, this work serves to amplify the work that has already been done in this area by scholars of color and as a call to (white) vegan scholars and activists to recognize the fundamental whiteness that taints any discussion of veganism that does not seek to expose and deconstruct the reliance of speciesism on the racisms involved in constructing "the human."

As a corollary to this, though whiteness currently taints much of vegan rhetoric and activism, it does not *need* to, and therefore the status quo of mainstream vegan whiteness *can* be decolonized and overturned. For example, this chapter will use Wynter, Wilderson, and Harper's insights to examine the gentrification of neighborhoods of color that is often led by corporate stores, "community" gardens, and advertising that assumes a white vegan audience; vegan recipe site and book *Thug Kitchen* (which performs digital and literary blackface while promoting veganism); and PETA's advertisement exhibits that uncritically use images of slavery to evoke emotional responses to violence agaisnt animals. While Wynter and Wilderson focus on the deconstruction of what it means to *be human,* Harper's writing and activism makes it clear that a full ethical and intellectual commitment to veganism is simultaneously a commitment to dismantling white supremacy.

In intertwining Wynter and Harper's writings, I want to take this moment to acknowledge my position of privilege: as a white vegan scholar, I run the risk of taking up space that could otherwise be occupied by scholars/vegans of color who have made similar arguments already better than I could. However, in the project of forcing whiteness to become evident where it seeks to make itself invisible, it is my responsibility to draw attention to whiteness generally. Specifically here, this entails highlighting the whiteness of mainstream veganism in the hopes that this can ameliorate some of its effects while I attempt to leverage my own privilege to expose rather than perpetuate the white supremacist agenda that my subject position cannot help but uphold.

Whiteness as Humanness

In his analysis of the fundamental ethico-political questions involved in examining what qualities and classificatory considerations constitute "the human," Frank Wilderson argues persuasively that "[w]hiteness [has been] the most impeccable embodiment of what it means to be Human."[5] Wilderson situates Blackness as being birthed amidst a "structural prohibition barring [it] from the conceptual framework of Human empathy."[6] In so doing, he asserts—like white vegan claims about animality—that POC[7] are systemically prevented from achieving recognition as possessing full humanity under the reign of white supremacy. This is because the fundamental basis of humanity, or "Man", cannot exist without the underlying, unmentionable whiteness that defines humanity itself.

Providing a foundation for Wilderson's understanding, Sylvia Wynter, largely in response to Franz Fanon's *Black Skin, White Masks*, articulates the ways that the term "human" is (and has always been) conflated with the Western ethnoclass "Man": in this way, "human beings *cannot* be defined in purely biogenetic terms" because even the very definition of the beings we consider *human* was created on the back of highly racialized conceptions of superiority.[8]

The exigency of this claim is enormous, as Wynter argues that "all our present struggles with respect to race, class, gender, sexual orientation, struggles over the environment, global warming, severe climate change, and sharply unequal distribution of the earth resources" are each "differing facets of the central ethnoclass Man vs. Human struggle."[9] This struggle, she asserts, can be articulated as the myriad ways that

> our present ethnoclass (i.e., Western bourgeois) conception of the human, Man, which overrepresents itself as if it were the human itself, and that of securing the well-being, and therefore the full cognitive and behavioral autonomy of the human species itself/ourselves.[10]

Her simultaneous usage of both "itself/ourselves" to describe self-(non) identity with humanness confronts her readers with the fundamental duality of her own subjectivity as a Jamaican woman, both as self-evidently human and as a person whose body bears the collective history upon which (white) humanity was created as a counterpoint. She further dramatizes this duality by separating these terms with a slash rather than an "and"; in so doing, Wynter both claims humanity for the very POC who have been treated as a foil to Man ("ourselves") *and* rejects it as something separate ("itself"). This gesture toward radical disidentification both acknowledges/rejects the violent material history of the concept of humanity and reasserts Wynter's own claim to humanity.

In similar gestures, Wynter redraws the lines of who gets to qualify for supposedly universal human identity in her exploration and fundamental restructuring of humanist and post-humanist thought. Wynter resituates (post-)humanism as fundamentally political and historical rather than philosophical, centralizing material histories of racist violence in her explorations of humanity, as opposed to the overwhelmingly Eurocentric constructions of Humanity found in the works of Giorgio Agamben such as *The Open: Man and Animal.* Wynter thereby encourages and performs the reorientation of the field(s) to prioritize the material histories of race that white male-dominated theories of humanity and post-humanity elide. Instead of assuming that Man is a universal category that uncritically includes all human animals—which Agamben does, for example, when he discusses the animalization of man and the humanization of animal as though humans of color are yet to be animalized—Wynter prioritizes the fundamental colonial project of the violent animalization of POC from which "humanity" draws its ostensibly "superior" identity.

In rewriting this animalization of racially marked human bodies (as well as the often overlapping bodies of women and people considered "mad") into the very definition of humanity, Wynter attends to the pitfalls of animalization in a highly anthropocentric way.[11] In so doing, she does draw uninterrogated differences between humans and non/human animals, which she refers to as "biological organisms", arguing that:

> each human Group-Subject knows the world, as do biological organisms, in relation to the securing of the conditions of the realization/actualization of their mode of being…genetically constituted in the case of biological organisms, rhetorico-symbolically in the case of humans.[12]

Tracing the evolution of (hu)Man-ity more as a series of socio-political reimaginings than as an essential and persistently progressive biological event, Wynter situates several distinct yet interrelated founding episodes in the conflation of Western Man as human, culminating with white European identity being "isomorphoic with Being human."[13] This Beingness relied in some epochs on the subjugation and obliteration of Indigenous Peoples in the "new world," and, in others, on the perpetuation of the colonial mission to "humanize" African peoples (vis-à-vis, ironically, dehumanizing processes of systemic violence).

These processes of violently dehumanizing POC largely in order to reify white European identity as *the* human identity are often, today, uncritically compared to the systematic violence inherent to the factory farming of non/human animals. There is tremendous potential here to interrogate seriously—without blithely analogizing and therefore inadvertently perpetuating the damaging associations of POC with negative assumptions about animality—the historical and material connections between the project of racism and the project of speciesism. There is potential here to generatively connect these projects and the traumatic experiences they create in order to disrupt the logic of human superiority over non/humans. A Wynter-esque approach to the history of humanism could be deployed to argue that promoting "humanity" as superior to non/humanity is, at its roots, promoting the universality of superior human whiteness. Dismantling one damaging assumption of superiority has the potential to pull at the threads that uphold the other. However, analogizing without questioning the role of whiteness in the production of both speciesism and racism covertly reinforces both systems. The next section will examine (un) successful attempts at calling attention to these intertwined oppressions.

#AllLionsMatter in the Face of #BlackLivesMatter

Given the overwhelming yet underacknowledged whiteness upon which the very premise of Humanity is based, it is perhaps unsurprising that vegan and related food justice movements are often also influenced by unmentioned white privilege. According to Rachel Slocum, "[a]lternative food networks articulate white ideals of health and nutrition, offer whitened dreams of farming and gardening that erase the past and present of race in agriculture…mobilize funding to direct programming toward non-white beneficiaries, and create inviting space for white people."[14]

These "whitened dreams" very often take the shape of gentrification, a tell-tale sign of which is often expensive, Whole Foods-type chain stores and/or white-dominated "community" gardens emphasizing "vegan options"; vegan-friendly operations like these often serve as the point-of-no-return for the mass displacement of POC through gentrification.[15] Some of these "community"-based urban farming collectives explicitly target neighborhoods of color for marketing and setup, such as the Ace of Spades urban farming collective, which built three gardens in South Seattle between 2010 and 2014.[16] According to Margaret Marietta Ramírez, the white spaces of these gardens and others like them—established and built by white volunteers in communities of color—"accumulate value and fuel gentrification, enabling landowners to profit as the black geographies of the CD [Central District], "a neighborhood east of downtown Seattle that is often considered to be the gateway to the 'south end' of Seattle", are displaced."[17] In contexts such as these, Slocum points out that "[r]acial inequalities are largely invisible."[18] These inequalities can easily be observed, however, in instances of organizations offering white savior-esque programming that often seek, like Ace of Spades, to "establish farms in 'neighborhoods that were lacking access to food and had high rates of diabetes and obesity": these organizational goals simultaneously pay lip service to and demean POC by assuming that white-dominated spaces are needed to "fix" POC communities.[19] Food justice collectives driven primarily by POC, however, often form powerful modes of resistance to both physical displacement and the disassociations of veganism and food justice concerns with communities of color.[20] Without genuine engagement with communities of color, primarily white food justice collectives that are often associated with veganism risk alienating POC and perpetuating violence against their communities through gentrification. Insofar as racism cannot be uncoupled from speciesism, gentrification—even when associated with the advancement of vegan collectives and stores—will therefore always be antithetical to full vegan praxis.

Despite the success of many of these POC food justice initiatives, uncritically white vegan mainstream forces consistently pit anti-racist agendas against anti-speciesist agendas, persistently challenging POC who are alarmed by predominantly white organizations comparing them to animals to "pick a side" to prove their anti-speciesist credentials. PETA's 2005 roving exhibit entitled "Are Animals the New Slaves?" compared Black people who had been lynched to slaughtered cows in a viscerally visual fashion. While the exhibit arguably did well to "disrupt settled notions of species difference [and] challenge the sacrosanct moral divide between humans and animals by asking people to recognize a connection between what are normally seen as separate and hierarchically ordered categories of beings," it carries a tremendous potential for damage.[21] Using images of Black people victimized by white supremacist violence as a selling point deploys dead Black bodies as objects of comparison to sell a particular message: this inadvertently subjects those victimized by lynching to a form of utililitarian commodification that PETA protests when deployed against animals. For this reason, scholars like Claire Jean Kim also acknowledge that projects like PETA's exhibit further "complicate[s] the project of building cross-group alliances in the context of fighting 'interlocking structures' of domination."[22]

In a similar inadvertently damaging fashion, in the midst of the #BlackLivesMatter movement's campaign to end state violence against POC, specifically Black Americans, white Twitter users, including many identifying themselves as vegan, rampaged on Twitter after the killing of Cecil the lion (murdered in July 2015 by an American hunter). The unironic 2015 #AllLionsMatter Twitter trend came in response to Haitian American author Roxanne Gay's Tweet that "I'm personally going to start wearing a lion costume when I leave my house so that if I get shot [by police], people will care."[23] In these and countless other microaggressive ways, the pitting of agendas of anti-racism and anti-speciesism against each other by white vegan mainstream forces is persistent. The knee-jerk reactions and almost automatic charges of POC speciesism that accompany white vegan responses to POC critiques of The racism implicit in the mainstream vegan language demonstrated by PETA, #AllLionsMatter, and defenses of recipe book and website *Thug Kitchen* (discussed below) all signify a broader tendency for white vegan rhetoric to dismiss anti-racist concerns as speciesist. These rhetorics alienate anti-specieism from anti-racism, even though—as the previous section demonstrates—racism and speciesism are inextricably linked if a critique of whiteness underlies their actualization.

White accusations of POC speciesism diminish and demean the experiences of POC who have historically been and daily continue to be animalized and treated as such. These accusations are informed by the fragility of whiteness's inability to be critiqued for its assumption that everyone experiences the privilege of being treated by the state and attendant systems as fully human. The "new" slaves implied in the PETA exhibit's question "Are Animals the New Slaves?" suggests strongly, of course, that slavery and its violent legacy is a thing of the past, that Black people living in the USA are no longer experiencing the violent, dehumanizing impact of slavery. Between mass incarceration and various other state mechanisms targeting Black people in the USA, uncritical comparison between POC and animals surely constitutes an artificially "colorblind" analogy that has been bleached several times over in fragile whiteness.[24] As Ain Drew argues, "PETA wasn't as concerned with helping Black folks overcome our health issues [caused by colonized non-vegan, processed diets and lack of health care] than they were about getting us to stop wearing mink coats or promoting dog-fighting."[25] Like white-run Ace of Spades taking over food-spaces in Seattle, PETA's condescension and lack of willingness to pay heed to foodways issues elide the ways that whiteness shapes and upholds speciesism. Speciesism is intimately supported by insititutions of whiteness, as evinced in the mass exploitation of immigrants of color who form the backbone of US food systems; destroyed Black health due to food-based colonization; and the relationship between white vegan-marketed outlets like Whole Foods and prison labor.[26] Making analogies that offend POC who have long been violently animalized threatens the ability of white vegan praxis to effectively guide more people and systems away from exploiting animals; paying less attention to sensational advertising and more attention to issues of immigration and colonization that more directly connect speciesism with racism is undoubtedly a more potentially fruitful and nonviolent way to address these interwined issues.

Automatic criticisms of POC reacting against comparisons with animals, however, dismisses the potential of truly exploring the interconnective tissue of speciesism and racism. Ignoring POC reactions against being compared to animals elides the pull—that burden of collective history and contemporary life, that affective instinct—that accompanies this comparison for people whose oppression has hinged (and continues to hinge) on white supremacy's relegating POC to a "genre" of humanity that is less complete than that of whiteness.[27] Similarly, connections between veganism and white privilege are evident given the vehemence of white rage toward Roxanne Gay when she drew critical attention to the white mainstream expressing more distress over Cecil's murder than the systematic state murders of POC across the country. Connections between mainstream veganism and white supremacy are made even clearer when dismissive rhetoric like the tweets slamming Gay's position accompanies actions that perpetuate the oppression of POC. For example, vegan-oriented stores and food collectives working their ways into neighborhoods of color enable and encourage displacement of POC from their neighborhoods via gentrification.[28] This demonstrates that, very often, POC accusations of mainstream vegan racism are rooted in analyses of mainstream vegan rhetoric about and treatment of POC rather than being rooted necessarily in speciesism. In these ways, the mainstream project of veganism is, precisely as Gay and others point out, one that accompanies and/or directly perpetuates colonization.

But veganism and anti-specieist rhetoric and practices need not, of course, be colonizing moves. In fact, the political potential generated by the work of vegans of color is powerful, and, Billy-Ray Belcourt suggests, even necessary to dismantling colonial violence. Given that "we cannot dismantle speciesism or re-imagine human-animal relations in the North American context without first or simultaneously dismantling settler colonialism and re-theorizing domesticated animal bodies as *colonial subjects* that must be centered in decolonial thought," Belcourt argues persuasively that racism and speciesism both render POC as non/human *and* deny animality its own subjectivities.[29] In a similar way, the choice that white vegan activists tend to give POC who object to white people comparing them to animals without respecting the violence accompanying those associations is revealed as artificial: it is not a "you are speciesist if you call out my racism" scenario, even though the #AllLionsMatter vegans and animal rights activists try to make it out to be such.

Tying together veganism and anti-racism—"because we don't have the luxury of being single-issue," according to the tag line of the Vegans of Color blog—vegan scholar and activist A. Breeze Harper makes explicit the link between the traumatic collective histories of POC in the USA and the ways that whiteness-inflected veganism reproduces the same racist violence that could otherwise be critically analyzed to support both anti-racist and anti-speciesist actions. In her discussion of the resistance by vegans of color to the infamous *Thug Kitchen*, a vegan recipe book with language written in verbal blackface by two white people, Harper argues:

> When reading about the controversy surrounding *Thug Kitchen* and how a group of vegans of color mobilized to shut the Bay Area reading down through protest, maybe we can understand how this protest wasn't some random anomaly; that it wasn't really about *Thug Kitchen* at all. These pro-

> tests are not single-issue and social phenomenon does not happen in a vacuum. [The outrage about] *Thug Kitchen* and vegans of color protest[ing it] is a microcosm that reflects the current racial climate in the USA. The book's support and "post-racial" comments by a significant number of mostly white people says a lot: it says "I don't have the trauma of racialized and state violence against my body that Black people do (and other racial minorities do). Why should I care about the word 'thug' and the racially violent history and recent events (i.e. Oscar Grant and Michael Brown) that trails behind it? As a matter of fact, I don't even have to realize that the term has been racialized and used against murder victims such as Michael Brown and Trayvon Martin to *justify* their deaths."[30]

The kind of "post-racialism" Harper dissects here mirrors the genres of post-humanism that elide the myriad ways that "human" is not, itself, a universally applied category to those bodies that are on the surface recognized as human, but are not treated as such in white supremacist society.

Harper's analysis of the response to *Thug Kitchen* makes explicit the verbal and potential physical violence that the white privilege of the authors perpetuates. Her incisive reading also, however, calls attention to a crucial factor that often is erased in both racist and anti-racist knee-jerk responses that are frequently elicited by popular vegan discourses: "this protest wasn't some random anomaly…it really wasn't about *Thug Kitchen* at all." Indeed, Harper's articulation of the authors' white privilege and relative safety from state violence that follows her bigger-picture comment highlights precisely the ways that white veganism often enacts upon the bodies of POC the same thoughtless devaluation that they accuse others of when they call for people to recognize that animal lives do, in fact, *matter*. Harper points out that white perspectives "don't have the trauma of racialized and state violence against my body" that POC do; is this not similar, then, to the rhetorical figurations of many vegans who charge that speciesists who see animal corpses as food don't realize that the term meat is used to cover up the fact that a living, feeling creature was murdered after living a life of torture? In calling for others to realize the violence against non/human animal bodies, many popularized white vegan accounts do not take into consideration the ways that they discount the violence they themselves perform on human bodies when gentrification colonizes POC neighborhoods with "vegan options" that price people out of their homes. This contemporary displacement reinforces the historical and contemporary traumas of forced movement associated with farms (which harken back to plantations, as Ramírez discusses).[31]

And yet this analogizing, too, is still not quite fulfilling Harper's call to recognize the bigger picture that all of this extends far, far beyond *Thug Kitchen*, "Are Animals the New Slaves", or even #AllLionsMatter. Colonial violence is not dead; slavery is not a thing of the past; humanity is not a uniform category (and nor is "animal"). Race is configured fundamentally as "the barring of nonwhite subjects from the category of the human as it is performed in the modern west."[32] The category of human is still contentious: if humanity is not a substantive category, then surely exposing this instability of the definition of "human" can only help the vegan cause. Ethical veganism must be interested in dismantling the anthropocentrism that is required to subjugate nonhumans in the context of factory farming, medical experiments, and other such testing and product-making. Since this anthropomorphism is wholly dependent not only upon the subjugation of nonhuman bodies, but also upon the colonization and subjugation of non-white, less-than-human human bodies, the protests that Harper engages with and the dialogues she later calls for are both necessary components of any vegan agenda committed to truly bringing down the reign of the human (which is coterminous with the reign of whiteness). To accomplish its goals, then, mainstream white veganism must dismantle its own whiteness.

"I Saw a Dead Body Today"

"Hands up, don't shoot!"

– *Black residents of Ferguson, Missouri, during protests following the police murder of Black teenager Michael Brown*

"Bring it, all you fucking animals! Bring it!"

– *White St. Louis police officer during the above-mentioned protest following the police murder of Black teenager Michael Brown*[33]

The common supermarket experience of walking past dead bodies—neatly packaged as food for human consumption, excluding the skin, muscles, fat, blood, and bones of the nonhuman animals—poses a brutal parallel with the way that the body of Black teenager Michael Brown was left on the street for hours by Ferguson police after he was murdered. As Roxanne Gay points out, though, perhaps if a murdered person of color had the accoutrements of a nonhuman animal that white American would more readily recognize as "innocent" and "deserving" of empathy and protection, outrage would be more universal.

In these gut-wrenching ways, the dictates and debates of humanism and post-humanism have quite literally life and death consequences daily. Both those considered nonhuman animals and those racialized as less-than-human by dominant US society are fundamentally placed at risk and are often forced to live lives in cages, in horrific conditions, in immediate, real terror. The origins of "humanity" ensures that the fates of anti-racist and anti-speciesist work must—if they are to be effective—rely on each other in order to dismantle the colonialism that prioritizes white supremacist agendas above any and all Other life. By drawing attention to and mobilizing action around issues such as colonization by gentrification, and by moving back to centralize the work of vegan activists and scholars of color, white vegans can work to highlight and thereby diminish the destructive impacts of white supremacy that currently define much of the underbelly of mainstream US veganism.

Notes

1. Sunaura Taylor, "Beasts of Burden: Disability Studies and Animal Rights" *Qui Parle: Critical Humanities and Social Sciences* 19, no. 2 (2011): 220.
2. Carol J. Adams, "The War on Compassion." In *The Feminist Care Tradition in Animal Ethics: A Reader*. Ed. Josephine Donovan (New York: Columbia University Press, 2007): 23.
3. Patrisse Cullers, Opal Tometi, and Alicia Garza, "A HerStory of the Black Lives Matter Movement" *BlackLivesMatter.com,* August 15 2015.
4. Cary, Wolfe, *Animal Rites: American Culture, The Discourse of Species, and Posthumanist Theory*. (University of Chicago Press, 2003): 8.
5. Frank B. Wilderson III, *Red, White & Black: Cinema and the Structure of US Antagonisms*. (Durham: Duke University Press, 2010): 25.

6. Wilderson, *Red, White & Black*, 12.
7. In this work, Wilderson focuses almost exclusively on Black Americans and Native Americans.
8. Sylvia Wynter, "On How We Mistook the Map for the Territory, and Reimprisoned Ourselves in Our Unbearable Wrongness of Being, of Desêtre: Black Studies Toward the Human Project." *A Companion to African-American Studies* (2006): 118.
9. Sylvia Wynter, "Unsettling the Coloniality of Being/Power/Truth/Freedom: Towards the Human, after Man, its Overrepresentation--An Argument." *CR: The New Centennial Review* 3, no. 3 (2003): 260–1.
10. Ibid.
11. Sylvia Wynter, "The Ceremony Must be Found: After Humanism." *Boundary* 2 (1984): 36.
12. Wynter, "The Ceremony", 23.
13. Wynter, "The Ceremony", 36.
14. Rachel Slocum, "Race in the Study of Food." *Progress in Human Geography* 35, no. 3 (2011): 314.
15. Margaret Marietta Ramírez, "The Elusive Inclusive: Black Food Geographies and Racialized Food Spaces." *Antipode* 47, no. 3 (2015): 757.
16. Ramírez 755.
17. Ramírez 753.
18. Slocum 314.
19. Ramírez 760.
20. Ramírez 750.
21. Claire Jean Kim, "Moral Extensionism or Racist Exploitation? The Use of Holocaust and Slavery Analogies in the Animal Liberation Movement." *New Political Science* 33, no. 3 (2011): 313.
22. Ibid.
23. Sarah Grey and Joe Cleffie, "Peter Singer's Race Problem." *Jacobin.* August 16 2015.
24. A. Breeze Harper, "Social Justice Beliefs and Addiction to Uncompassionate Consumption." In *Sistah Vegan: Black Female Vegans Speak on Food, Identity, Health, and Society.* Ed. A Breeze Harper. New York: Lantern Books (2010): 29.
25. Ain Drew, "Being a Sistah at PETA." In *Sistah Vegan: Black Female Vegans Speak on Food, Identity, Health, and Society.* Ed. A. Breeze Harper. New York: Lantern Books (2010): 63.

26. Rebecca J. Rosen, "How Dairy Milked by Prisoners Ends up on Whole Foods Shelves." *The Atlantic*. June 18 2014.
27. Wynter, "Territory", 117.
28. Ramírez 750.
29. Billy-Ray Belcourt, "Animal Bodies, Colonial Subjects:(Re) Locating Animality in Decolonial Thought." *Societies* 5, no. 1 (2014): 3.
30. A. Breeze Harper, "On Ferguson, *Thug Kitchen*, and Trayvon Martin: Intersections of [Post]Race-Consciousness, Food Justice, and Hip Hop Vegan Ethics." *The Sistah Vegan Project Blog*. October 14 2014.
31. Ramírez 751.
32. Alexander G. Weheliye, *Habeas Viscus: Racializing Assemblages, Biopolitics, and Black Feminist Theories of the Human*. (Durham: Duke University Press, 2014): 3.
33. Amanda Terkel, "Police Officer Caught On Video Calling Michael Brown Protesters 'F***ing Animals.'" The Huffington Post. August 12 2014.

References

Adams, Carol J. 2007. The war on compassion. In *The feminist care tradition in animal ethics: A reader*, ed. Josephine Donovan, 21–38. New York: Columbia University Press.

Agamben, Giorgio. 2004. *The open: Man and animal*. Trans. Kevin Attell. Stanford: Stanford University Press.

Belcourt, Billy-Ray. 2014. Animal bodies, colonial subjects:(Re) Locating animality in decolonial thought. *Societies* 5(1): 1–11.

Cullers, Patrisse, Opal Tometi, and Alicia Garza. 2015. A herStory of the black lives matter movement. *BlackLivesMatter.com*. 15 Aug 2015. Web.

Derrida, Jacques. 2008. *The animal that therefore I Am*, ed. Marie-Louise Mallet and Trans. David Wills. New York: Fordham University Press.

Drew, Ain. 2010. Being a Sistah at PETA. In *Sistah vegan: Black female vegans speak on food, identity, health, and society*, ed. A. Breeze Harper, 61–64. New York: Lantern Books.

Grey, Sarah, and Joe Cleffie. 2015. Peter Singer's race problem. *Jacobin*. 16 Aug 2015. Web.

Harper, A. Breeze. 2010. Social justice beliefs and addiction to uncompassionate consumption. In *Sistah vegan: Black female vegans speak on food, identity, health, and society*, ed. A. Breeze Harper, 20–41. New York: Lantern Books.

Harper, A. Breeze. 2014. On Ferguson, *Thug Kitchen*, and Trayvon Martin: Intersections of [Post]Race-consciousness, food justice, and hip hop vegan ethics. *The Sistah Vegan Project Blog.* 14 Oct 2014. Web.
Kim, Claire Jean. 2011. Moral extensionism or racist exploitation? The use of holocaust and slavery analogies in the animal liberation movement. *New Political Science* 33(3): 311–333.
Ramírez, Margaret Marietta. 2015. The elusive inclusive: Black food geographies and racialized food spaces. *Antipode* 47(3): 748–769.
Rosen, Rebecca J. 2014. How dairy milked by prisoners ends up on whole foods shelves. *The Atlantic.* 18 June 2014. Web.
Slocum, Rachel. 2011. Race in the study of food. *Progress in Human Geography* 35(3): 303–327.
Taylor, Sunaura. 2011. Beasts of burden: Disability studies and animal rights. *Qui Parle: Critical Humanities and Social Sciences* 19(2): 191–222.
Terkel, Amanda. 2014. Police officer caught on video calling Michael Brown Protesters 'F***ing animals.' *The Huffington Post*, Auguest 12. Web.
Weheliye, Alexander G. 2014. *Habeas viscus: Racializing assemblages, biopolitics, and Black feminist theories of the human.* Durham: Duke University Press.
Wilderson III, Frank B. 2010. *Red, white & black: Cinema and the structure of US antagonisms.* Durham: Duke University Press.
Wolfe, Cary. 2003. *Animal rites: American culture, the discourse of species, and posthumanist theory.* Chicago: University of Chicago Press.
Wynter, Sylvia. 1984. The ceremony must be found: After humanism. *Boundary* 2: 19–70.
Wynter, Sylvia. 2003. Unsettling the coloniality of being/power/truth/freedom: Towards the human, after man, its overrepresentation--An argument. *CR: The New Centennial Review* 3(3): 257–337.
Wynter, Sylvia. 2006. On how we mistook the map for the territory, and reimprisoned ourselves in our unbearable wrongness of being, of desêtre: Black studies toward the human project. *A Companion to African-American Studies*, 107–118.

Index

Note: Locators followed by 'n' refer to notes.

J. Castricano, R.R. Simonsen (eds.), *Critical Perspectives on Veganism*, The Palgrave Macmillan Animal Ethics Series, DOI 10.1007/978-3-319-33419-6

U

V

W